AF658375

Toxic Effects of Mercury

Shabnum Nabi

Toxic Effects of Mercury

Springer

Shabnum Nabi
Interdisciplinary Brain Research Centre (IBRC)
Jawaharlal Nehru Medical College
Aligarh Muslim University
Aligarh, UP, India

ISBN 978-81-322-1921-7 ISBN 978-81-322-1922-4 (eBook)
DOI 10.1007/978-81-322-1922-4
Springer New Delhi Heidelberg New York Dordrecht London

Library of Congress Control Number: 2014942337

© Springer India 2014
This work is subject to copyright. All rights are reserved by the Publisher, whether the whole or part of the material is concerned, specifically the rights of translation, reprinting, reuse of illustrations, recitation, broadcasting, reproduction on microfilms or in any other physical way, and transmission or information storage and retrieval, electronic adaptation, computer software, or by similar or dissimilar methodology now known or hereafter developed. Exempted from this legal reservation are brief excerpts in connection with reviews or scholarly analysis or material supplied specifically for the purpose of being entered and executed on a computer system, for exclusive use by the purchaser of the work. Duplication of this publication or parts thereof is permitted only under the provisions of the Copyright Law of the Publisher's location, in its current version, and permission for use must always be obtained from Springer. Permissions for use may be obtained through RightsLink at the Copyright Clearance Center. Violations are liable to prosecution under the respective Copyright Law.
The use of general descriptive names, registered names, trademarks, service marks, etc. in this publication does not imply, even in the absence of a specific statement, that such names are exempt from the relevant protective laws and regulations and therefore free for general use.
While the advice and information in this book are believed to be true and accurate at the date of publication, neither the authors nor the editors nor the publisher can accept any legal responsibility for any errors or omissions that may be made. The publisher makes no warranty, express or implied, with respect to the material contained herein.

Printed on acid-free paper

Springer is part of Springer Science+Business Media (www.springer.com)

The largest pond is as sensitive to atmospheric changes as the globule of mercury in its tube.

(Henry David Thoreau)

Dedicated to my parents and brothers for their endless affection and support

Foreword

It gives me immense pleasure and delight when I received a request from *Dr. Shabnum Nabi* to write a brief foreword to the book entitled *Toxic Effects of Mercury* written by her under the auspices of Springer publishing house. The work she has produced throughout these years in the form of books, book chapters and research papers is not only commendable but informative for one and all. The book deals with mercury toxicity and its ill hazards to our environment in general and humans in particular with a prime focus on oxidative stress in the degeneration of nervous and non-nervous tissues. The book further throws light on the severity of mercury through consumption of contaminated fish and other food products. It is worth to mention over here that methylmercury became a prime environmental health issue during 1950s. In spite of decades of research, the toxic mechanisms of methylmercury still remain an enigma. This book, therefore, has a wide scope, utility and potential as mercury is being used in agriculture, forestry and many potential industries, especially cosmetic industry.

In short, it is surely a work to treasure and valuable source of information for medical practitioners, toxicologists, clinical biochemists, researchers, scientists, and general public in particular.

(Dr. Tariq Ahmad)
Sr. Assistant Professor
Entomology Research Division
P. G. Department of Zoology
University of Kashmir, Srinagar-190006 (J & K), India

Preface

> The words of Mercury are harsh after the songs of Apollo.
>
> (William Shakespeare)

Mercury is among the best known most puzzling metals in the environment. It is a unique element that, unlike many metals, has no biological function. It is liquid at room temperature and is 13.6 times heavier than water. It occurs in all media in several forms, both organic and inorganic. It is relatively uncommon in the Earth's crust, from which it is liberated by natural processes such as erosion and volcanism as well as by mining. All mercury compounds are toxic to humans and animals, but the organic forms, particularly methylmercury and dimethylmercury, have the highest toxicity. Methylmercury is the form found most widely in nature, and it bioaccumulates in the food chain. It is the form to which most human exposure occurs. Mercury has potent compromising effects upon the immune system and is also associated with chronic overgrowth of Candida, anemia, forgetfulness, tremors, depression, drowsiness, insomnia, headache, loss of energy etc. Chronically its toxicity causes low body temperature, bleeding gums, loosening of teeth and other mouth sores, sore throat, joint pain, high blood pressure, nutritional disturbances, urinary disturbances etc., to list just a few of the 42 conditions identified. Mercury poisoning is involved in five categories of pathology: (1) neurological diseases, (2) cardiovascular diseases, (3) collagen diseases, (4) immunological diseases, and (5) allergies.

Many of us vividly recall the devastating effects of point source of mercury (Hg) pollution in Minimata Bay, Japan, documented by poignant photos of mothers caring for their children afflicted with crippling deformities. From 1952 to 1968, Hg waste from the production of acetaldehyde and vinyl chloride was discharged (200–600 t) into Minimata Bay. This resulted in Hg bioaccumulation by marine fish and shellfish, biomagnification in the marine food web, and subsequent ingestion by humans. Although this was perhaps the most severe case in history, the processes determining the fate of Hg in the environment are common to all aquatic ecosystems.

The Minimata catastrophe also revealed that the neurotoxicological effects on fetal development from pregnant women ingesting Hg-contaminated fish could be especially severe. This finding, as well as research over subsequent decades on the process of Hg bioaccumulation and biomagnification in aquatic and marine ecosystems, has led to today's warnings by many public health agencies about pregnant women limiting their consumption of fish like tuna, swordfish, shark, and other species that feed on the top of food chains.

Yet, as with most environmentally mediated health risks involving either chemical or biological agents, the ecological dimensions and health effects of Hg are far more subtle than either the tragic Minimata case or the health warnings suggest. This was illustrated more recently by pioneering transdisciplinary and participatory research carried out in Brazil to address chronic, low-level mercury exposure. These efforts demonstrated the efficacy of an "ecosystem approach" in revealing how deforestation and farming practices in the Amazon led to subclinical but nonetheless debilitating neurological manifestations. Few environmental health issues demand a transdisciplinary research agenda that spans such a wide range of disciplines – from biogeochemistry and ecosystem ecology to epidemiology and participatory action research – as environmental exposure to methylmercury (MeHg). Also, among the classic environmental health issues involving human and wildlife exposures to persistent and global contaminants (e.g., lead and organochlorides), MeHg remains the least resolved, in terms of both science and policy.

In 1997, the U.S. Environmental Protection Agency (EPA) issued two reports to the U.S. congress on mercury (Hg) and its effects on public health. The first of these reports, the mercury study report to congress, assessed the source and amount of mercury emissions in the United States, the detrimental effects of Hg on humans and wildlife and the feasibility of control technologies. The second report, the utility hazardous air pollutant report to congress, looked specifically at emissions from utility companies and cited Hg as a major contaminant, especially in emissions from coal-fired power plants. Once in the environment, Hg can be converted to MeHg, which bioaccumulates in the food chain. Such bioaccumulation can lead to high concentrations of MeHg in predatory fish. Because of concerns about MeHg exposure levels in the United States from the consumption of contaminated fish, particularly among sensitive populations, questions have arisen among federal agencies over what is an acceptable level of exposure to MeHg. Because of gaps in the scientific data regarding Hg toxicity, particularly MeHg, the potentially widespread implications from human health, and the high financial costs and feasibility problems associated with further regulating Hg emissions, Congress directed EPA in the House Appropriations Report for EPA'S Fiscal 1999 funding to contract with the national research council (NRC) to prepare recommendations on the appropriate reference dose for Hg exposure.

International conference on Mercury as a global pollutant (ICMGP) provides a needed global forum for the exchange of innovative ideas, and it provides an opportunity to communicate research results to public policy makers, industry experts and public representatives in order to promote the direct use of scientific and technical data to environmental protection and control. Intense interest in mercury is evident from the recommendations adopted by the European Union Mercury Strategy in June 2005; also, a Global Mercury Assessment was approved by the governing council of the United Nations environmental program in early 2002. The 7th ICMGP was held in Ljubljana Slovenia, the location of the second largest mercury mine in the world – the Idrija mercury mine. Although the mercury mine was the basis of prosperity of the town of Idrija for five centuries, it also caused extensive contamination of the town and its surroundings, leading to high levels of mercury exposure of the miners and the other inhabitants. The mercury mine was closed in the mid-1980s to avoid further releases of mercury to the environment. Mercury research in Slovenia was initiated by the Jozef Stefan institute in the early 1960s as a result of environmental and health concerns related to the operation of the mine. Initially, the main goal of the institute was to address the health status of the miners. Studies of the effects of mercury on the environment soon followed, and results have been published widely in the scientific literature.

In the summer of 2006, the "Eighth International Conference on Mercury as a Global Pollutant" was held in Madison, Wisconsin, to which over 1,000 scientists from around the world came to share their understanding of the environmental fate of this pervasive contaminant. The global nature of Hg is due to a number of factors: (1) it is generated largely from coal-fired power plants which are increasing worldwide; (2) it is atmospherically transported from these sources to eventual deposition on all land and water surfaces; and (3) it is transferred to humans through fish consumption which knows no international or socioeconomic boundaries.

Mercury in combination with allergens has a tendency to rupture white blood cells, precipitating allergic reactivity. The connection is made between dental mercury amalgams and chronic fatigue/immune depression syndrome. Mercury crosses the placental barrier to contaminate a developing fetus. Mercury exposure is not something to ignore.

> The quicksilver of creativity will not be solidified by legal pronouncement; it will necessarily flow into new and sometimes frightening fields (Mathew Tobriner)

Aligarh, India

Shabnum Nabi

Acknowledgements

> Real life isn't always going to be perfect or go our way, but the recurring acknowledgement of what is working in our lives can help us not only to survive but surmount our difficulties.
>
> (Sarah Ban Breatrnach)

All thanks are due to *Allah* – The creator, cherisher and sustainer of the universe. He created man and "taught him that which he knew not" (Al-Quran). He is most Gracious and most Merciful to all his creations. He endowed me with the requisite knowledge and ability to embark upon this piece of work and bestowed upon me the courage, patience and strength to carry it to its completion. It is indeed a great pleasure to acknowledge my humble sense of gratitude to *Springer* for encouraging me to pen down this work in the form of a book. Their encouragement to young budding authors and researchers for writing manuscripts, monographs and books without any financial implications to the authors is greatly appreciated. I thank my Lord for giving me such polite *Parents* whose adore, forfeits and sustained endeavors enabled me to acquire knowledge. Whatever I am today is the result of their prayers, love, care and sincere efforts. To my extended family, I hardly know from where to begin but here it goes. I sincerely acknowledge the generous help rendered by my elder biological brothers *Mr. Mushtaq Ahmad*, *Mr. Parvez Ahmad* and *Er. Mohammad Lateef*. Words seem to be inadequate for the immense appreciation and gratitude to my brother *Dr. Tariq* who encouraged, helped, tolerated and supported me in all stages of my endeavour and without his active cooperation this book would not have seen the light of day.

I strongly feel that no word of formal acknowledgement can possibly express the help rendered by my supervisors *Dr. Anjum Ara* (Reader, Department of Zoology, AMU, Aligarh) and *Dr. Shamim J. Rizvi* (Director, Interdisciplinary Brain Research Centre, J.N. Medical College, AMU, Aligarh). The support provided by them certainly deserves special mention. I thankfully acknowledge my debt to *Prof. Irfan Ahmad* (Chairman, Department of Zoology, AMU, Aligarh) and *Dr. G.G.H.A. Shadab* (Sr. Assistant Professor, Department of Zoology, AMU, Aligarh) for their continuous encouragement and valuable advice during the preparation of the manuscript.

I recognize the contributions of many of my *learned friends and colleagues* who kindly shared their special knowledge with me and volunteered their intellectual capabilities in shaping this book. Surely, there are too many to name, but finally I would like to dedicate this book to all those who indirectly provided indispensable support and continuous backup, showered with lots of affection and prayers. Distinctive thanks to them for immensely boosting my morality.

> Knowledge is the end based on acknowledgement.
>
> (Ludwig Wittgenstein)

Shabnum Nabi

Contents

Part II Review of Literature

Part III Toxicity Evaluation Methodology

About the Author

Dr. Shabnum has obtained her Ph.D. degree in Zoology (Cytogenetics) in 2012. The prime focus of her research has been neurotoxicity, genotoxicity and neurobehavioral toxicity. She is the author of several papers published in reputed and peer-reviewed journals. Besides having National and International conferences to her credit, she is the member of Zoological Society of India. She has participated as a young scientist in a congregation of Nobel Prize Winners at IIIT Allahabad. She has also availed a Research Fellowship (JRF) from University Grants Commission (UGC), New Delhi.

Part I

History of Mercury Toxicology

1 Occurrence and Properties

Sudden resolutions, like the sudden rise of mercury in a barometer, indicate little else than the variability of the weather.

(David Hare)

Mercury Element (The Liquid Metal)

Mercury (pronounced **/markjuri/** *MER-kya-ree*), also called **quicksilver** (*kwiksilver*) or **hydrargyrum** (*hye-DRAR-ji-rem*), is a chemical element with the symbol **Hg** (Latinized Greek *hydrargyrum*, meaning watery or liquid silver) and atomic number 80. A heavy, silvery d-block metal, mercury is one of the six chemical elements that are liquid at or near room temperature and pressure (Senese 2007; Norrby 1991), the others being cesium, francium, gallium, bromine, and rubidium. Mercury is the only metal that is liquid at standard conditions for temperature and pressure. With a melting point of −38.83 °C and boiling point of 356.73 °C, mercury has one of the widest ranges of its liquid state of any metal (Fig. 1.1).

Mercury occurs in deposits throughout the world mostly as cinnabar (mercuric sulfide), which is the source of the red pigment vermilion, and is mostly obtained by reduction from cinnabar. Cinnabar is highly toxic by ingestion or inhalation of the dust. Mercury poisoning can also result from exposure to soluble forms of mercury (such as mercuric chloride or methylmercury), inhalation of mercury vapor, or eating fish contaminated with mercury.

Mercury is used in thermometers, barometers, manometers, sphygmomanometers, float valves, and other scientific apparatus, though concerns about the element's toxicity have led to mercury thermometers and sphygmomanometers being largely phased out in clinical environments in favor of alcohol-filled, digital, or thermistor-based instruments. It remains in use in a number of other ways in scientific research applications and in amalgam material for dental restoration. It is used in lighting; electricity passed through mercury vapor in a phosphor tube produces short-wave ultraviolet light which then causes the phosphor to fluoresce, making visible light.

Properties

General

Name: Mercury
Symbol: Hg
Atomic number: 80
Atomic mass: 200.59 amu
Melting point: −38.83 °C (234.32 K, −37.89 °F)
Boiling point: 356.73 °C (629.88 K, 674.11 °F)
Number of protons/electrons: 80
Number of neutrons: 121

S. Nabi, *Toxic Effects of Mercury*, DOI 10.1007/978-81-322-1922-4_1,
© Springer India 2014

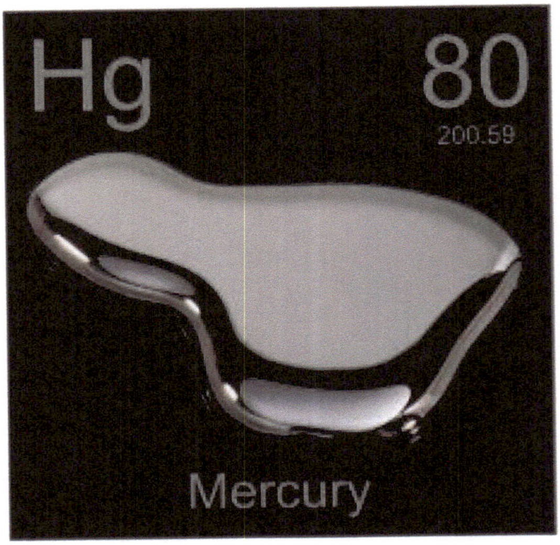

Fig. 1.1 Mercury poster sample (Source: English Wikipedia)

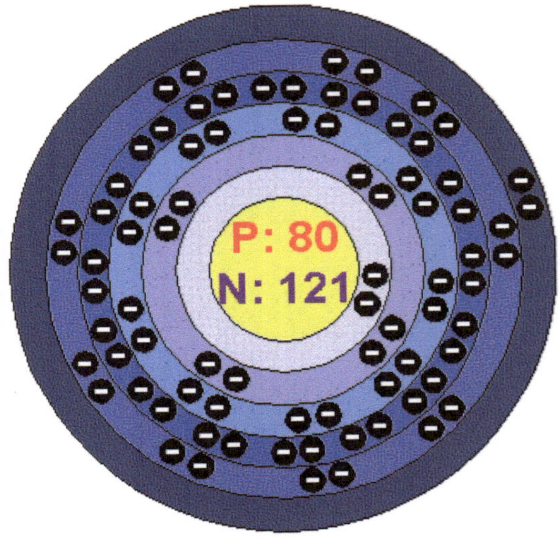

Fig. 1.2 Atomic structure of mercury; number of energy levels: 6, first energy level: 2, second energy level: 8, third energy level: 18, fourth energy level: 32, fifth energy level: 18, sixth energy level: 2 (Source: English Wikipedia)

Classification: transition metal
Crystal structure: rhombohedral
Density at **293 K**: 13.456 g/cm^{-3}
Color: silver

Physical

Mercury is a heavy, silvery-white metal. As compared to other metals, it is a poor conductor of heat, but a fair conductor of electricity (Hammond 2000). Mercury has an exceptionally low melting temperature for a d-block metal. A complete explanation of this fact requires a deep excurse into quantum physics, but it can be summarized as follows: mercury has a unique electronic configuration where electrons fill up all the available 1s, 2s, 2p, 3s, 3p, 3d, 4s, 4p, 4d, 4f, 5s, 5p, 5d, and 6s subshells. As such configuration strongly resists removal of an electron, mercury behaves similarly to noble gas elements, which form weakly bonded and thus easily melting solids. The stability of the 6s shell is due to the presence of a filled 4f shell. An f shell poorly screens the nuclear charge that increases the attractive Coulomb interaction of the 6s shell and the nucleus. The absence of a filled inner *f* shell is the reason for the much higher melting temperature of cadmium. Metals such as gold have atoms with one less 6s electron than mercury. Those electrons are more easily removed and are shared between the gold atoms forming relatively strong metallic bonds (Norrby 1991; Senese 2009) (Fig. 1.2).

Reactivity and Compounds

Mercury dissolves to form amalgams with gold, zinc, and many other metals. Because iron is an exception, iron flasks have been traditionally used to trade mercury. Other metals that do not form amalgams with mercury include tantalum, tungsten, and platinum. When heated, mercury also reacts with oxygen in air to form mercury oxide; this then can be decomposed by further heating to higher temperatures (Patnaik and Pradyot 2003).

Since it is below hydrogen in the reactivity series of metals, mercury does not react with most acids, such as dilute sulfuric acid, though oxidizing acids such as concentrated sulfuric acid and nitric acid or aqua regia dissolve it to give sulfate, nitrate, and chloride salts. Like silver, mercury reacts with atmospheric hydrogen sulfide. Mercury even reacts with solid sulfur flakes, which are used in mercury spill kits to absorb mercury vapors (spill kits also use activated charcoal and powdered zinc) (Patnaik and Pradyot 2003).

Higher oxidation states of mercury were confirmed in September 2007, with the synthesis of mercury(IV) fluoride (HgF_4) using matrix isolation techniques (Wang et al. 2007).

Laboratory tests have found that an electrical discharge causes the noble gases to combine with mercury vapor. These compounds are held together with van der Waals forces and result in Hg·Ne, Hg·Ar, Hg·Kr, and Hg·Xe. Organic mercury compounds are also important. Methylmercury is a dangerous compound that is widely found as a pollutant in water bodies and streams (National Research Council 2000).

Mercury and Aluminum

Aluminum in air is ordinarily protected by a molecule thin layer of its own oxide, which is not porous to oxygen. Mercury coming into contact with this oxide does no harm. However, if any elemental aluminum is exposed (even by a recent scratch), the mercury may combine with it and potentially damage a large part of the aluminum before it finally ends (Vargel et al. 2004; Gray 2004).

For this reason, restrictions are placed on the use and handling of mercury in proximity with aluminum. In particular, mercury is not allowed aboard aircraft under most circumstances because of the risk of it forming an amalgam with exposed aluminum parts in the aircraft (Vargel et al. 2004).

Isotopes

Isotope	Half-life
Hg-194	520.0 years
Hg-196	Stable
Hg-197	2.7 days
Hg-197 m	23.8 h
Hg-198	Stable
Hg-199	Stable
Hg-200	Stable
Hg-201	Stable
Hg-202	Stable

Isotope	Half-life
Hg-203	46.6 days
Hg-204	Stable
Hg-206	8.2 min

History

Hg is the modern chemical symbol for mercury. It comes from *hydrargyrum*, a Latinized form of the Greek word Ύδραργυρος (*hydrargyros*), which is a compound word meaning "water" and "silver" – since it is liquid, like water, and yet has a silvery metallic sheen. The element was named after the Roman god Mercury, known for speed and mobility. It is associated with the planet Mercury; the astrological symbol for the planet is also one of the alchemical symbols for the metal. Mercury is the only metal for which the alchemical planetary name became the common name (Stillman 2003).

The mines in Almadén (Spain), Monte Amiata (Italy), and Idrija (now Slovenia) dominated the mercury production from the opening of the mine in Almadén 2500 years ago until new deposits were found at the end of the nineteenth century (Eisler 2006).

Occurrence

Mercury is an extremely rare element in the Earth's crust, having an average crustal abundance by mass of only 0.08 ppm (Ehrlich and Newman 2008). However, because it does not blend geochemically with those elements that constitute the majority of the crustal mass, mercury ores can be extraordinarily concentrated considering the element's abundance in ordinary rock. The richest mercury ores contain up to 2.5 % mercury by mass, and even the leanest concentrated deposits are at least 0.1 % mercury (12,000 times average crustal abundance). It is found either as a native metal (rare) or in cinnabar, corderoite, livingstonite, and other minerals, with cinnabar (HgS) being the most common ore.

Mercury ores usually occur in very young orogenic belts where rocks of high density are forced to the crust of the Earth, often in hot springs or other volcanic regions (USGS 2009).

Mercury is extracted by heating cinnabar in a current of air and condensing the vapor. The equation for this extraction is

$$HgS + O_2 \rightarrow Hg + SO_2$$

In 2005, China was the top producer of mercury with almost two-thirds of global share followed by Kyrgyzstan (British Geological Survey, NERC 2001-05). Several other countries are believed to have unrecorded production of mercury from copper electrowinning processes and by recovery from effluents.

Because of the high toxicity of mercury, both the mining of cinnabar and refining for mercury are hazardous and historic causes of mercury poisoning. In China, prison labor was used by a private mining company as recently as the 1950s to create new cinnabar mercury mines. Thousands of prisoners were used by the Luo Xi mining company to establish new tunnels (Sheridan 2009). In addition, worker health in functioning mines is at high risk.

The European Union directive calling for compact fluorescent bulbs to be made mandatory by 2012 has encouraged China to reopen deadly cinnabar mines to obtain the mercury required for CFL bulb manufacture. As a result, new generations of Chinese, their livestock, and their crops are being poisoned, particularly in the southern cities of Foshan and Guangzhou and in the Guizhou province in the southwest (Sheridan 2009).

Abandoned mercury mine processing sites often contain very hazardous waste piles of roasted cinnabar calcines. Water runoff from such sites is a recognized source of ecological damage. Former mercury mines may be suited for constructive reuse. For example, in 1976, Santa Clara County, California, purchased the historic Almaden Quicksilver Mine and created a county park on the site after conducting extensive safety and environmental analysis of the property (Boulland 2006).

Releases in the Environment

Preindustrial deposition rates of mercury from the atmosphere may be in the range of 4 ng/(1 L of ice deposit). Although that can be considered a natural level of exposure, regional or global sources have significant effects. Volcanic eruptions can increase the atmospheric source by four to six times (Brooks 2007).

Natural sources such as volcanoes are responsible for approximately half of atmospheric mercury emissions. The human-generated half can be divided into the following estimated percentages: (Pacyna et al. 2006; Brooks 2007; Solnit 2006).

- Sixty-five percent from stationary combustion, of which coal-fired power plants are the largest aggregate source (40 % of US mercury emissions in 1999). This includes power plants fueled with gas where the mercury has not been removed. Emissions from coal combustion are between one and two orders of magnitude higher than emissions from oil combustion, depending on the country (Pacyna et al. 2006).
- Eleven percent from gold production. The three largest point sources for mercury emissions in the United States are the three largest gold mines. Hydrogeochemical release of mercury from gold mine tailings has been accounted as a significant source of atmospheric mercury in eastern Canada (Maprani et al. 2005).
- 6.8 % from nonferrous metal production, typically smelters.
- 6.4 % from cement production.
- 3.0 % from waste disposal, including municipal and hazardous waste, crematoria, and sewage sludge incineration. This is a significant underestimate due to limited information and is likely to be off by a factor of two to five.
- 3.0 % from caustic soda production.
- 1.4 % from pig iron and steel production.
- 1.1 % from mercury production, mainly for batteries.
- 2.0 % from other sources.

The above percentages are estimates of the global human-caused mercury emissions in the year 2000, excluding biomass burning, an important source in some regions (Pacyna et al. 2006).

Mercury also enters into the environment through the disposal (e.g., land filling, incineration) of certain products. Products containing mercury include auto parts, batteries, fluorescent bulbs, medical products, thermometers, and thermostats (Environmental Protection Agency 2007). Due to health concerns, toxic use reduction efforts are cutting back or eliminating mercury in such products. For example, most thermometers now use pigmented alcohol instead of mercury. Mercury thermometers are still used in the medical field because they are more accurate than alcohol thermometers, though both are being replaced by electronic thermometers. Mercury thermometers are still widely used for certain scientific applications because of their greater accuracy and working range.

One of the worst industrial disasters in history was caused by the dumping of mercury compounds into Minamata Bay, Japan. The Chisso Corporation, a fertilizer and later petrochemical company, was found responsible for polluting the bay from 1932 to 1968. It is estimated that over 3,000 people suffered various deformities, severe mercury poisoning symptoms, or death from what became known as Minamata disease.

References

Boulland M (2006) New Almaden Arcadia Publishing, p 8

British Geological Survey, NERC 2001-05. World Mineral Production, London

Brooks WE (2007) Mercury. U.S. Geological Survey (USGS)

Ehrlich HL, Newman DK (2008) Geomicrobiology. CRC Press, Boca Raton, p 265

Eisler R (2006) Mercury hazards to living organisms. CRC/Taylor and Francis, Science. p 312

Environmental Protection Agency (EPA) (2007) Mercury: spills, disposal and site cleanup

Gray T (2004) The amazing rusting aluminum. Popular Science. Bonnier Corporation, USA

Hammond CR (2000) The elements. In: Handbook of chemistry and physics, 81st edn. Taylor and Francis, Boca Raton, p 19

Maprani AC, Al TA, Macquarrie KT, Dalziel JA, Shaw SA, Yeats PA (2005) Determination of mercury evasion in a contaminated headwater stream. Environ Sci Technol 39:1679–1687

National Research Council (2000) Toxicological effects of methylmercury. National Academy Press, Washington, DC, Committee on the Toxicological Effects of Methylmercury

Norrby LJ (1991) Why is mercury liquid? Or, why do relativistic effects not get into chemistry textbooks? J Chem Educ 68:110

Pacyna EG, Pacyna JM, Steenhuisen F, Wilson S (2006) Global anthropogenic mercury emission inventory for 2000. Atmos Environ 40:4048

Patnaik, Pradyot (2003) Handbook of inorganic chemical compounds. McGraw-Hill, New York, pp 560–576

Senese F (2007) Why is mercury a liquid at STP? General chemistry Online at Frostburg State University

Senese F (2009) Why is mercury a liquid at STP? General chemistry Online at Frostburg State University

Sheridan M (2009) Green light bulbs poison workers: hundreds of factory staff is being made ill by mercury used in bulbs destined for the West. The Sunday Times, London

Solnit R (2006) Winged mercury and the golden calf. Orion Magazine

Stillman JM (2003) Story of alchemy and early chemistry. Kissinger Publishing, Whitefish, pp 7–9

USGS (2009) Mineral commodity summaries. U.S. Department of the Interior, United States Geological Survey

Vargel C, Jacques M, Schmidt MP (2004) Corrosion of aluminum. Elsevier, United States, p 158

Wang X, Andrews L, Riedel S, Kaupp M (2007) Mercury is a transition metal: the first experimental evidence for HgF4. Angew Chem Int Ed (Weinheim: Wiley) 46:8371–8375

2 Estimated Applications

Mercury is used primarily for the manufacture of industrial chemicals or for electrical and electronic applications. It is used in some thermometers, especially ones which are used to measure high temperatures. A still increasing amount is used as gaseous mercury in fluorescent lamps, while most of the other applications are slowly phased out due to health and safety regulations, and is in some applications replaced with less toxic but considerably more expensive Galinstan alloy (Fig. 2.1).

Present Use

Medicine

Mercury and its compounds have been used in medicine, although they are much less common today than they once were, now that the toxic effects of mercury and its compounds are more widely understood. The element mercury is an ingredient in dental amalgams. Thiomersal (called *thimerosal* in the United States) is an organic compound used as a preservative in vaccines, though this use is in decline (FDA 2004). Another mercury compound, merbromin (Mercurochrome) is a topical antiseptic used for minor cuts and scrapes and is still in use in some countries (Fig. 2.2).

Mercury(I) chloride (also known as calomel or mercurous chloride) has traditionally been used as a diuretic, topical disinfectant, and laxative. Mercury(II) chloride (also known as mercuric chloride or corrosive sublimate) was once used to treat syphilis (along with other mercury compounds), although it is so toxic that sometimes the symptoms of its toxicity were confused with those of the syphilis it was believed to treat (Pimple et al. 2002). It was also used as a disinfectant. Blue mass, a pill or syrup in which mercury is the main ingredient, was prescribed throughout the 1800s for numerous conditions including constipation, depression, childbearing, and toothaches (Mayell 2007). In the early twentieth century, mercury was administered to children yearly as a laxative and dewormer, and it was used in teething powders for infants. The mercury-containing organohalide merbromin (sometimes sold as Mercurochrome) is still widely used but has been banned in some countries such as the United States (Cecil 2004).

Since the 1930s some vaccines have contained the preservative thiomersal, which is metabolized or degraded to ethyl mercury. Although it was widely speculated that this mercury-based preservative can cause or trigger autism in children, scientific studies showed no evidence supporting any such link (Parker et al. 2004). Nevertheless, thiomersal has been removed from or reduced to trace amounts in all US vaccines recommended for children 6 years of age and under, with the exception of inactivated influenza vaccine (FDA 2007).

Mercury in the form of one of its common ores, cinnabar, remains an important component of Chinese, Tibetan, and Ayurvedic medicine. As problems may arise when these medicines are exported to countries that prohibit the use of

S. Nabi, *Toxic Effects of Mercury*, DOI 10.1007/978-81-322-1922-4_2,
© Springer India 2014

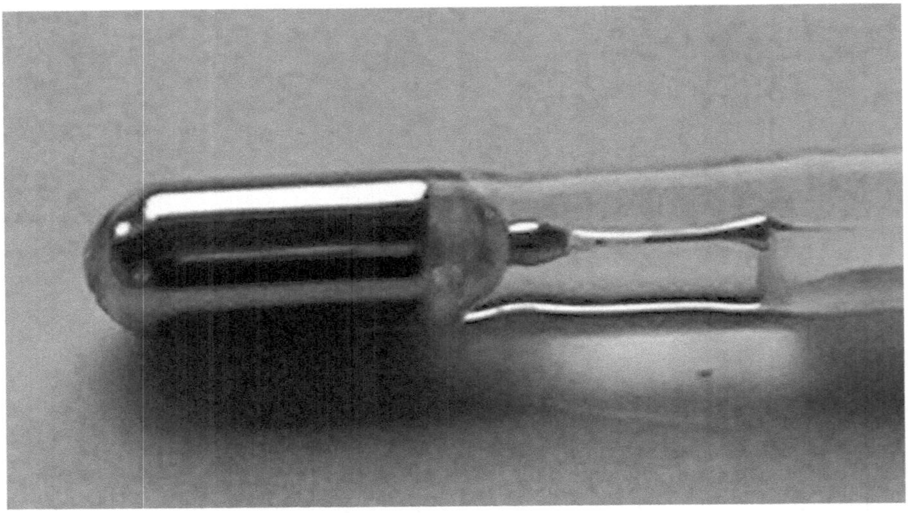

Fig. 2.1 The bulb of mercury in glass thermometer (Source: English Wikipedia)

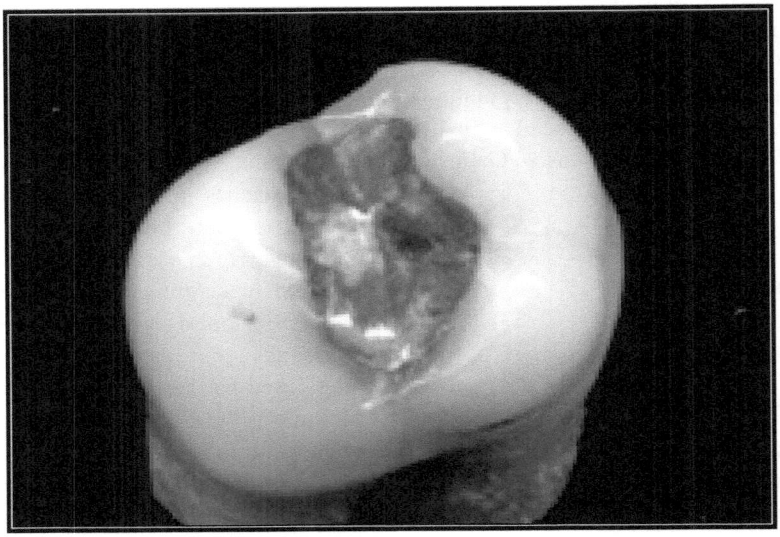

Fig. 2.2 Amalgam filling (Source: English Wikipedia)

mercury in medicines, in recent times, less toxic substitutes have been devised.

Today, the use of mercury in medicine has greatly declined in all respects, especially in developed countries. Thermometers and sphygmomanometers containing mercury were invented in the early eighteenth and late nineteenth centuries, respectively. In the early twenty-first century, their use is declining and has been banned in some countries, states, and medical institutions. In 2002, the U.S. Senate passed legislation to phase out the sale of nonprescription mercury thermometers. In 2003, Washington and Maine became the first states to ban mercury blood pressure devices (Health Care without Harm 2003). Mercury compounds are found in some over-the-counter drugs, including topical antiseptics, stimulant laxatives, diaper-rash ointment, eyedrops, and nasal sprays. The FDA has "inadequate data to establish general recognition of the safety and effectiveness" of the mercury ingredients in these products (FDA 2007). Mercury is still used in some diuretics, although substitutes now exist for most therapeutic uses.

Production of Chlorine and Caustic Soda

Chlorine is produced from sodium chloride (common salt, NaCl) using electrolysis to separate the metallic sodium from the chlorine gas. Usually the salt is dissolved in water to produce brine. By-products of any such chloralkali process are hydrogen (H_2) and sodium hydroxide (NaOH), which is commonly called caustic soda or lye. By far the largest use of mercury (The CRB Commodity Yearbook 2000; Leopold 2002) in the late twentieth century was in the mercury cell process (also called the Castner–Kellner process) where metallic sodium is formed as an amalgam at a cathode made from mercury; this sodium is then reacted with water to produce sodium hydroxide (Chlorine Online Diagram of mercury cell process 2006). Many of the industrial mercury releases of the twentieth century came from this process, although modern plants claimed to be safe in this regard (Leopold 2002). After about 1985, all new chloralkali production facilities that were built in the United States used either membrane cell or diaphragm cell technologies to produce chlorine.

Laboratory Uses

Some medical thermometers, especially those for high temperatures, are filled with mercury; however, they are gradually disappearing. In the United States, nonprescription sale of mercury fever thermometers has been banned since 2003 (Mercury Reduction Act of 2003). Mercury is also found in liquid mirror telescopes.

Some transit telescopes use a basin of mercury to form a flat and absolutely horizontal mirror, useful in determining an absolute vertical or perpendicular reference. Concave horizontal parabolic mirrors may be formed by rotating liquid mercury on a disk, the parabolic form of the liquid thus formed reflecting and focusing incident light. Such telescopes are cheaper than conventional large mirror telescopes by up to a factor of 100, but the mirror cannot be tilted and always points straight up (Govert Schilling 2003; Gibson 1991).

Liquid mercury is a part of popular secondary reference electrode (called the calomel electrode) in electrochemistry as an alternative to the standard hydrogen electrode. The calomel electrode is used to work out the electrode potential of half cells (Brans and Hay 1995). Last, but not least, the triple point of mercury, −38.8344 °C, is a fixed point used as a temperature standard for the International Temperature Scale (ITS-90) (Hammond 2005).

Niche Uses

Skin tanner containing a low-pressure mercury vapor lamp and two infrared lamps, which act both as light source and electrical ballast Gaseous mercury, is used in mercury vapor lamps and some "neon sign"–type advertising signs and fluorescent lamps. Those low-pressure lamps emit very spectrally narrow lines, which are traditionally used in optical spectroscopy for calibration of spectral position. Commercial calibration lamps are sold for this purpose; however, simply reflecting some of the fluorescent-lamp ceiling light into a spectrometer is a common calibration practice (Hopkinson et al. 2004). Gaseous mercury is also found in some electron tubes, including ignitrons, thyratrons, and mercury arc rectifiers (Howatson 1965). It is also used in specialist medical care lamps for skin tanning and disinfection (Milo and Casto 1990). Gaseous mercury is added to cold cathode argon-filled lamps to increase the ionization and electrical conductivity. An argon-filled lamp without mercury will have dull spots and will fail to light correctly. Lighting containing mercury can be bombarded/oven pumped only once. When added to neon-filled tubes, the light produced will be inconsistent red/blue spots until the initial burning-in process is completed; eventually it will light a consistent dull off-blue color (Shionoya 1999) (Fig. 2.3).

Fig. 2.3 Skin tanner containing low-pressure mercury vapor lamp and two infrared lamps, which act both as light source and electrical ballast (Source: English Wikipedia)

Cosmetics

Mercury, as thiomersal, is widely used in the manufacture of mascara. In 2008, Minnesota became the first state in the United States to ban intentionally added mercury in cosmetics, giving it a tougher standard than the federal government (CIDPUSA Foundation 2008).

Gold and Silver Mining

Historically, mercury was used extensively in hydraulic gold mining in order to help the gold to sink through the flowing water–gravel mixture. Thin mercury particles may form mercury–gold amalgam and therefore increase the gold recovery rates (Hammond 2000). Large-scale use of mercury stopped in the 1960s. However, mercury is still used in small-scale, often clandestine, gold prospection. It is estimated that 45,000 metric tons of mercury used in California for placer mining has not been recovered (Alpers et al. 2005). Mercury was also used in silver mining (Mercury amalgamation 2009).

Other Present Uses

Gaseous mercury is used in mercury vapor lamps and some "neon sign"–type advertising signs and fluorescent lamps. Those low-pressure lamps emit very spectrally narrow lines, which are traditionally used in optical spectroscopy for calibration of spectral position. Commercial calibration lamps are sold for this purpose; however, simply reflecting some of the fluorescent-lamp ceiling light into a spectrometer is a common calibration practice (Hopkinson et al. 2004). Gaseous mercury is also found in some electron tubes, including ignitrons, thyratrons, and mercury arc rectifiers (Howatson 1965). It is also used in specialty medical care lamps for skin tanning and disinfection (Milo and Casto 1990). Gaseous mercury is added to cold cathode argon-filled lamps to increase the ionization and electrical conductivity.

Some medical thermometers, especially those for high temperatures, are filled with mercury. The triple point of mercury, −38.83 °C, is a fixed point used as a temperature standard for the International Temperature Scale (ITS-90) (Hammond 2000).

Fig. 2.4 Mercury manometer to measure pressure (Source: English Wikipedia)

Proposed Uses

Liquid mercury has been proposed as a working fluid for a heat pipe type of cooling device for spacecraft heat rejection systems or radiation panels. A new type of atomic clock, using mercury instead of cesium, has been demonstrated. Accuracy is expected to be within 1 s in 100 Ma (BBC 2001; NIST 2001).

Historic Uses

Mercury was used for preserving wood, developing daguerreotypes, silvering mirrors, antifouling paints (discontinued in 1990), herbicides (discontinued in 1995), handheld maze games, cleaning, and road leveling devices in cars. Mercury compounds have been used in antiseptics, laxatives, antidepressants, and antisyphilitics. It was also allegedly used by allied spies to sabotage German planes. A mercury paste was applied to bare aluminum, causing the metal to rapidly corrode. This would cause structural failures (Gray 2004) (Fig. 2.4).

References

Alpers CN, Hunerlach MP, May JT, Hothem RL (2005) Mercury contamination from historical gold mining in California. U.S. Geological Survey (USGS)

BBC (2001) World's most precise clock developed

Brans YW, Hay WW (1995) Physiological monitoring and instrument diagnosis in perinatal and neonatal medicine. CUP Archive, p 175. ISBN 0-521-41951-4

Cecil A (2004) What happened to mercurochrome? The Straight Dope

Chlorine Online Diagram of mercury cell process. Euro Chlor. Archived from the original on 2006-09-02

CIDPUSA Foundation (2008) Mercury in your eye, dangerous cosmetics

FDA (2004) Mercury (Thimerosal) in vaccines. Medical News Today

FDA (2007) Thimerosal in vaccines. Center for Biologics Evaluation and Research, U.S. Food and Drug Administration

Gibson BK (1991) Liquid mirror telescopes: history. J R Astron Soc Can 85:158

Govert Schilling. (2003) Liquid-mirror telescope set to give stargazing a new spin. Archived from the original on 2003-08-18

Gray T (2004) The amazing rusting aluminum. Popular Science

Hammond CR (2000) The elements. In: Handbook of chemistry and physics, 81st edn. Taylor and Francis, Boca Raton, p 19

Hammond CR (2005) The elements. In: Lide DR (ed) CRC handbook of chemistry and physics, 86th edn. CRC Press, Boca Raton. ISBN 0-8493-0486-5

Health Care without Harm (2003) Two states pass first-time bans on mercury blood pressure devices

Hopkinson GR, Goodman TM, Prince SR (2004) A guide to the use and calibration of detector array equipment. SPIE Press, Bellingham, p 125. ISBN 0-8194-5532-6

Howatson AH (1965) "8". An introduction to gas discharges. Pergamon Press, Oxford. ISBN 0-08-020575-5

Leopold BR (2002) Chapter 3: Manufacturing processes involving mercury. Use and release of mercury in the United States. National Risk Management Research Laboratory, Office of Research and Development, U.S. Environmental Protection Agency, Cincinnati, Archived from the original on June 21, 2007

Mayell H (2007) Did Mercury in "Little Blue Pills" make Abraham Lincoln erratic? National Geographic News

Mercury amalgamation 2009. Corrosion doctors (corrosion information hub: the corrosion doctor's website)

Mercury Reduction Act of 2003. United States. Congress. Senate. Committee on Environment and Public Works

Milo GE, Casto BC (1990) Transformation of human diploid fibroblasts. CRC Press, Boca Raton, p 104. ISBN 0-8493-4956-7

NIST (2001) New atomic clock could be 1,000 times better than today's best, NIST update

Parker SK, Schwartz B, Todd J, Pickering LK (2004) Thimerosal-containing vaccines and autistic spectrum disorder: a critical review of published original data. Pediatrics 114:793–804

Pimple KD, Pedroni JA, Berdon V (2002) Syphilis in history. Poynter Center for the Study of Ethics and American Institutions at Indiana University-Bloomington

Shionoya S (1999) Phosphor handbook. CRC Press, Boca Raton, p 363. ISBN 0-8493-7560-6

The CRB Commodity Yearbook (annual) (2000) p 173. ISSN 1076-2906

Distribution, Metabolism, and Excretion

3

The general population is primarily exposed to methylmercury through the diet. However, air and water, depending upon the level of contamination, can contribute significantly to the daily intake of total mercury. In most foodstuffs, mercury is largely in the inorganic form and below the limit of detection (20 μg mercury/kg fresh weight). However, fish and fish products are the dominant source of methylmercury in the diet, and levels greater than 1,200 μg/kg have been found in the edible portions of shark, swordfish, and Mediterranean tuna. Similar levels have been found in pike, walleye, and bass taken from polluted freshwaters.

It has been estimated that humans have a daily intake of about 2.4 μg methylmercury from all sources and a daily uptake of approximately 2.3 μg. The total daily intake of *all* forms of mercury from all sources has been estimated to be 6.7 μg, with an added burden of 3.8–21 μg of mercury vapor from dental amalgams, if present. The level of mercury in fish, even for humans consuming only small amounts (10–20 g of fish/day), can markedly affect the intake of methylmercury. The consumption of 200 g of fish containing 500 μg mercury/kg will result in the intake of 100 μg mercury (predominantly methylmercury). This amount is one-half of the recommended provisional tolerable weekly intakes (WHO 1989).

Biochemical Aspects

Absorption

Data are available showing that methylmercury is readily absorbed from the gastrointestinal tract of humans and animals. Based on retention and excretion data from humans, Aberg et al. (1969) reported that 95 % of a single oral dose of methylmercuric nitrate was absorbed. Efficient absorption of methylmercury was also demonstrated in another study using human volunteers receiving an oral dose of protein-bound methylmercury (Miettinen et al. 1973). Up to 80 % of volatile methylmercury compounds such as methylmercury chloride vapor may be absorbed upon inhalation (Berlin 1983). Dermal absorption of methylmercury is known to occur both in humans and animals but quantitative data are lacking.

Distribution

Methylmercury is transported in red blood cells with a small fraction being bound to plasma proteins (Berlin 1983). The compound readily penetrates membranes resulting in widespread distribution in the body; however, higher concentrations (up to 10 % of total dose) accumulate in the central nervous system (CNS). In the CNS,

S. Nabi, *Toxic Effects of Mercury*, DOI 10.1007/978-81-322-1922-4_3,
© Springer India 2014

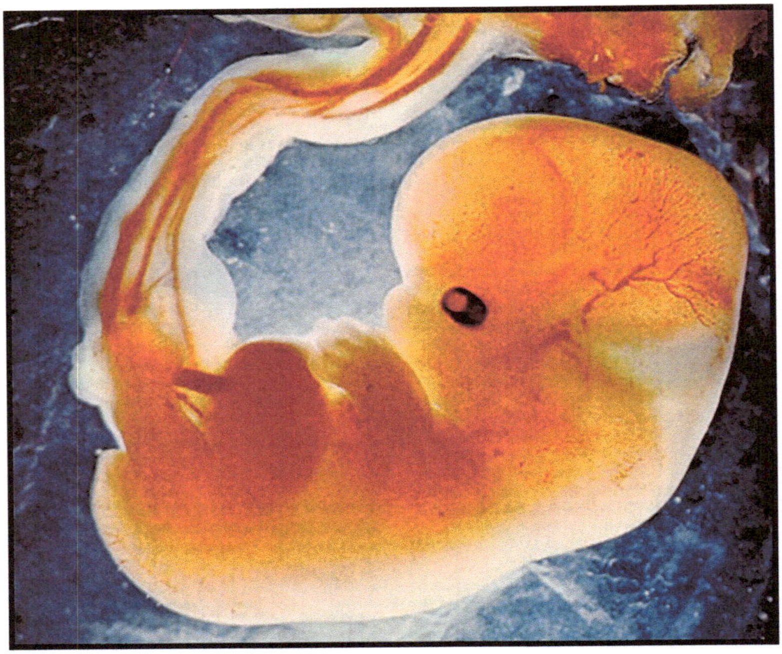

Fig. 3.1 Fetal umbilical cord (Source: English Wikipedia)

methylmercury remains in the organic form but in other tissues is converted and stored as inorganic mercury with the highest concentrations generally occurring in the liver and kidney. Methylmercury readily traverses the placenta and results in higher levels of the compound in fetal relative to maternal blood (ATSDR 1989). Incorporation of methylmercury in hair during hair formation in the follicle results in concentrations that are up to 250 times greater than that in other tissues. A report by Dutczak et al. (1991) provided data from guinea pigs, hamsters, and a macaque monkey indicating extensive absorption of methylmercury by the gall bladder and subsequent biliary–hepatic cycling of the compound, which may contribute to the long biological half-life of methylmercury.

Methylmercury enters the body through different routes of exposure and is able to easily cross membrane barriers including the blood–brain barrier because organic mercury is transported into tissues by a methylmercury–cysteine complex. The methylmercuric cation has high affinity for this particular amino acid because it contains sulfhydryl groups. The metabolism of organic mercury occurs at a slow rate, which allows time for the methylmercury to enter the membrane barriers. So as the methylmercury complexes enter the membrane barriers, they begin to metabolize into inorganic mercury. This results in a great degree of ionic mercury accumulation in tissues including brain tissues. The half-life of methylmercury in the blood and brain is very different due to this accumulation in brain tissue. Blood half-life for humans is 49–164 days and the half-life for brain may be significantly longer. In addition to the accumulation of mercury in the brain and other body tissues, mercury has also been found to accumulate in fetal umbilical cords (Fig. 3.1).

Metabolism

Methylmercury may be metabolized to inorganic mercury by the liver and kidneys, with the inorganic form then entering an oxidation–reduction cycle in the red blood cells, lungs, and liver resulting in formation of the divalent cation (Hg^{++}) (ATSDR 1989). Methylmercury remaining in the gastrointestinal tract is converted to inorganic mercury by the intestinal flora (Nakamura et al. 1977; Rowland et al. 1980). Available data suggest that metabolism of methylmercury is similar in animals and humans (ATSDR 1989). Methylmercury in the human diet is almost completely absorbed into the bloodstream and distributed to all tissues within about 4 days. However, maximum levels in the brain are only reached after 5–6 days. In humans, blood to hair ratios is about 1:250, with appreciable individual variation. Similarly, large individual differences are seen in cord to maternal blood mercury ratios, the levels generally being higher in cord blood. Species differences exist in the distribution of methylmercury between red blood cells and plasma (about 20:1 in humans, monkeys, and guinea pigs, 7:1 in mice, and >100:1 in rats).

Methylmercury is converted to inorganic mercury in experimental animals and humans. The duration of the exposure and the interval after its cessation determine the fraction of total mercury present in tissues in the Hg^{++} form. In humans, after high oral intakes of methylmercury for 2 months, the following values were reported. (Percentage of total mercury in tissues as inorganic mercury): whole blood, 7 %; plasma, 22 %; breast milk, 39 %; urine, 73 %; liver, 16–40 %.

In the case of continuous exposure, a single-compartment model with a 70-day half time predicts that the whole-body steady state (where intake equals excretion) will be attained within approximately 1 year and that the maximum amount accumulated will be 100 times the average daily intake. The validity of the single-compartment model is supported by the reasonable agreement between predicted and observed blood concentrations of methylmercury in single-dose tracer studies, single-dose fish intake experiments, and studies involving the extended controlled intake of methylmercury from fish. It is also supported by results from the longitudinal hair analysis of individuals with very high intakes of methylmercury.

Mean reference values for total mercury in commonly used indicator media are whole blood, 8 μg/l; hair, 2 μg/g; urine, 4 μg/l; and placenta, 10 μg/kg wet weight. Long-term fish consumption is the major determinant of methylmercury and, usually, total mercury levels in blood. For example, in communities in which there is a long-term daily consumption of 200 μg mercury/day from fish, blood mercury levels are approximately 200 μg/l and corresponding hair levels about 250 times higher (50 μg/g hair).

Excretion

Methylmercury is excreted primarily in the feces as inorganic mercury (Norseth and Clarkson 1971). This is the result of biliary excretion of the compound and subsequent conversion to the inorganic form by intestinal flora. Some of the methylmercury excreted in the bile may also be reabsorbed, thereby creating enterohepatic circulation of the organic form. Less than 1 % of the body burden of methylmercury is excreted daily, resulting in a biological half-life of approximately 70 days (Berlin 1983). Over a 4-day period, a human volunteer excreted only about 6 % of the ingested dose of radiolabeled, protein-bound methylmercury, the biological half-life ultimately being 76 days (Miettinen et al. 1973). Methylmercury is also secreted in breast milk with concentrations being about 5 % of that in the blood. Removal of inorganic mercury via exhalation, saliva, and sweat results from the metabolism of the organic form (ATSDR 1989).

The rate of excretion of mercury in both laboratory animals and humans is directly proportional to the simultaneous body burden and can be described by a single-compartment model with a biological half time, in fish-eating humans, of 39–70 days (average approximately 50 days). Lactating females have significantly shorter half times for mercury excretion than non-lactating ones. Mercury half times in hair closely follow those in blood but with wider variation (35–100 days, average 65 days). Suckling mice are incapable of excreting methylmercury, but they abruptly change to the adult rate of excretion at the end of the suckling period.

Effect on Enzymes and Other Biochemical Parameters

Cleavage of the Carbon–Mercury Bond

Of the organic mercury chemicals (methyl, dimethyl, phenyl, mercuric acetate), methylmercury is considered to be chemically the most stable. Bacterial demethylation of methylmercury can take place in the intestinal tract and involves the generation of both enzymatic and reactive oxygen species, resulting in decreased absorption of methylmercury (National Research Council 2000).

The reaction by which methylmercury is demethylated was investigated using several reactive oxygen modulators in a model using rat liver slices in vitro. The rate of conversion of methylmercury to inorganic mercury in vitro is similar to experiments in rats in vivo, indicating that the model is suitable for reproducing the biotransformation of methylmercury. Results with various reagents suggest that the demethylation of methylmercury may be aided by superoxide anions produced by the electron transfer system in the inner mitochondrial membrane (Yasutake and Hirayama 2001).

Complexes with Thiol Radicals

Methylmercury forms complexes with thiol ligands, including glutathione and L-cysteine, facilitating its extracellular transport and ability to bind to intracellular proteins, targeting sulfhydryl enzymes. Complexing with L-cysteine increases the concentration of methylmercury in the brain, liver, and kidney, while a decrease in renal *gamma*-glutamyl transpeptidase activity results in a decrease in renal methylmercury and increased urinary excretion.

References

Aberg B, Ekman L, Falk R, Greitz U, Persson G, Snihs JO (1969) Metabolism of methylmercury (^{203}Hg) compounds in man. Arch Environ Health 19:478–484

ATSDR (1989) Toxicological profile for mercury. Agency for Toxic Substances and Disease Registry. U.S. Department of Health and Human Services. Public Health Service, Atlanta, p 169

Berlin I (1983) Organic compounds of mercury. In: Parmeggiani L (ed) Encyclopedia of occupational health and safety, 3rd edn. International Labour Organization, Geneva, pp 1336–1338

Dutczak W, Clarkson TW, Ballatori N (1991) Biliary-hepatic recycling of a xenobiotic: gallbladder absorption of methyl mercury. Am J Physiol 260:G873–G880

Miettinen M, Turpeinen O, Karvonen MJ, Elosuo R, Paavilainen E (1973) Effect of diet on coronary-heart disease mortality. Lancet 2:1266–1267

Nakamura I, Hosokawa K, Tamura H, Miura T (1977) Reduced mercury excretion with feces in germfree mice after oral administration of methylmercury chloride. Bull Environ Contam Toxicol 17:528–533

National Research Council (2000) Toxicological effects of methylmercury. National Academy Press, Washington, DC, Committee on the Toxicological Effects of Methylmercury

Norseth T, Clarkson TW (1971) Intestinal transport of 203Hg-labeled methylmercury chloride; role of biotransformation in rats. Arch Environ Health 22:258

Rowland I, Davies M, Evans J (1980) Tissue content of mercury in rats given methylmercury chloride orally: influence of intestinal flora. Arch Environ Health 35:155

World Health Organization (1989) Environmental health criteria 86: environmental aspects of mercury. International programme of chemical safety. World Health Organization, Geneva, p 115

Yasutake A, Hirayama K (2001) Evaluation of methylmercury biotransformation using rat liver slices. Arch Toxicol 75:400–406

Toxic Effects and Remedy

4

Toxicological Studies

Acute Toxicity

The LD_{50} of methylmercury in rodents treated orally is usually 10–40 mg/kg BW. Methylmercury is also considered to be corrosive at high doses (WHO 2000). Alterations in various serum and urine biochemical parameters within 72 h after dosing were observed in male rats given mercury as a single oral dose (total mercury, 3.96 mg/kg body weight; methylmercury, 0.047 mg/kg body weight) in boiled whale blubber. The activity of serum lactate dehydrogenase was increased twofold, and there were significant increases in urinary volume and activity of N-acetyl-beta-D-glucosaminidase and albumin compared with the control group, which received total mercury in a single dose of 0.4 μg/kg body weight (Endo et al. 2003).

Long-Term Studies of Toxicity and Carcinogenicity

Long-term exposure to methylmercury has induced renal tumors in mice, but only at doses at which significant nephropathy was also evident. No significant increase in any tumor type, including renal tumors, has been reported in rats (WHO 2000).

Adverse Effects on Cellular and Nutritional Alterations

Mercury has the ability to cause changes at the cellular level, which has been seen in platelets and erythrocytes. These cells have been used as surrogate markers for mercury damage of neurological tissue. The addition of methylmercury to whole blood can cause dramatic dissolution of microtubules in platelets and red blood cells – an effect more pronounced in erythrocytes than platelets – which is consistent with the known sequestration of methylmercury in erythrocytes (Durham et al. 1997). This effect on microtubules has also been found in the brain (Falconer et al. 1994) and results in disruption of the cell cycle. This disruption can cause apoptosis (programmed cell death) in both neuronal and non-neuronal cells (Miura et al. 1999).

Mercury causes apoptosis in monocytes and decreases phagocytic activity (InSug et al. 1997). In one study, the percentage of cells undergoing apoptosis was dependent on the mercury content of the medium, regardless of the form of mercury. Methylmercury chloride exposure caused a decrease in the mitochondrial transmembrane potential within 1 h of exposure, leading to altered mitochondrial function. Methylmercury can also cause increased lymphocyte apoptosis. This mechanism includes a depletion of glutathione (GSH) content, which predisposes the cell to oxidative

S. Nabi, *Toxic Effects of Mercury*, DOI 10.1007/978-81-322-1922-4_4,
© Springer India 2014

damage, while activating death-signaling pathways (Shenker et al. 1998). On examination of synovial tissue, it was found that mercury (as well as cadmium and lead) caused a decrease in DNA content and an increase in collagenase-resistant protein formation (Goldberg et al. 1983), leading to increased risk for reduced joint function and decreased ability to repair joint damage.

Mercury is bound by selenium in the body, which can actually counteract mercuric chloride and methylmercury toxicity (Ganther 1978; Ganther et al. 1972). This appears to result in a reduced amount of available selenium, which compounds the oxidative burden on the body. Mercury decreases GSH levels in the body (De Souza Queiroz et al. 1998), which occurs by several mechanisms. Mercury binds irreversibly to GSH, causing the loss of up to two GSH molecules per molecule of mercury. The GSH–Hg–GSH complex is excreted via the bile into the feces. Part of the irreversible loss of GSH is due to the inhibition of GSH reductase by mercury (Zalups and Lash 1996), which is used to "recycle" oxidized GSH and return GSH to the pool of available antioxidants. At the same time, mercury also inhibits GSH synthetase, so a lesser amount of new GSH is created. Since mercury promotes formation of hydrogen peroxide, lipid peroxides, and hydroxyl radicals, it is evident that mercury sets up a scenario for a serious imbalance in the oxidative/antioxidant ratio of the body (Miller et al. 1991). Mercury's heavy oxidative toll on the body has been postulated to be a cause of increased rates of fatal myocardial infarctions and other forms of cardiovascular disease (Salonen et al. 1995). These interactions clearly show an increased need for selenium, glutathione, and vitamin E, which have been shown to reduce methylmercury toxicity (Goldberg et al. 1983; Welsh and Soares 1976).

Toxicity to Other Organs and Systems in the Body

Mercury-Induced Neurotoxicity

Mercury in both organic and inorganic forms is neurotoxic. Methylmercury accumulates in the brain and becomes associated with mitochondria, endoplasmic reticulum, Golgi complex, nuclear envelopes, and lysosomes. In nerve fibers, methylmercury is localized primarily in myelin sheaths (where it leads to demyelination) and in mitochondria (Chang 1997). Pathological examination of patients with methylmercury poisoning indicates the cerebellar cortex is prominently affected, with granule cells being more susceptible than Purkinje cells. Typically, glial cells are spared direct damage, although reactive gliosis may occur. Toxicity from mercury probably does not result from action on a single target. Instead, because of its highly reactive nature, a complex series of many unrelated (and some interrelated) effects may occur more or less simultaneously, initiating a sequence of additional events that ultimately lead to cell death (Table 4.1).

Table 4.1 Neurotoxicity of mercury

Myelinopathies
Granule cells in cerebellar cortex
Neuronal swelling
Destruction of astrocytes
Inhibits uptake and release of dopamine, serotonin, norepinephrine

How Does Mercury Affect the Nervous System?

Methylmercury targets and kills neurons in specific areas of the nervous system including the (Fig. 4.1):

Several mechanisms have been proposed to explain how mercury kills neurons:

1. Protein inhibition
2. Disruption of mitochondria function
3. Direct affect on ion exchange in a neuron
4. Disruption of neurotransmitters
5. Destruction of the structural framework of neurons

Methylmercury is especially dangerous to developing babies. This form of mercury is highly toxic and can cross the placenta and the blood–brain barrier. Mercury is concentrated in the brain of the developing fetus because the metal is absorbed quickly and is not excreted efficiently. Children exposed to mercury may be

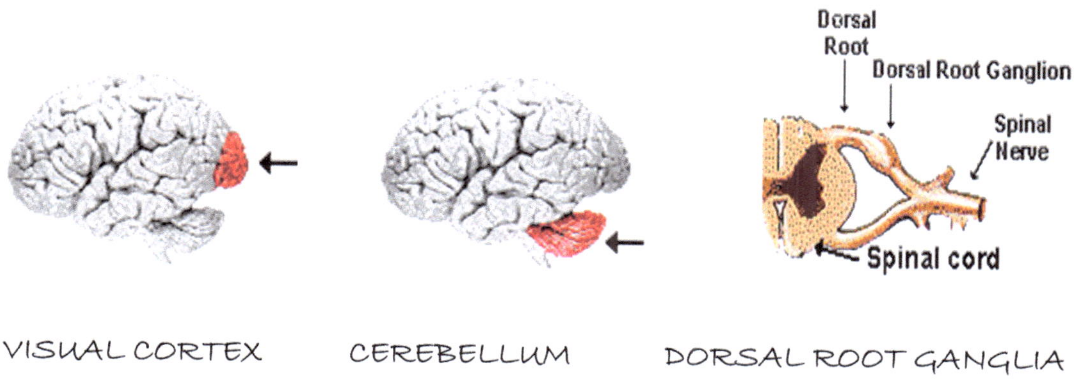

Fig. 4.1 Target areas affected by mercury (Source: English Wikipedia)

born with symptoms resembling cerebral palsy, spasticity and other movement abnormalities, convulsions, visual problems, and abnormal reflexes. The brains of children who have died as a result of mercury poisoning show neuron loss in the cerebellum and throughout the cerebral cortex. Mercury also appears to affect brain development by preventing neurons from finding their appropriate place in the brain.

The adverse effect of mercury on GSH has secondary effects on the levels of Na+, K+, and Mg++ ATPases, all of which are dependent on sulfhydryl compounds. These enzymes, all critical for proper functioning of nervous and other tissues, are all inhibited by various mercurial compounds (Magour et al. 1987). Injection of GSH in animals exposed to methylmercury resulted in the recovery of N+, K+, and Mg++ ATPases (Bapu et al. 1998). In the absence of nutrients to counteract this action, the inhibition of these ATPases results in neurotoxic swelling and destruction of astrocytes (Aschner et al. 1990). Astrocytes are the primary cells responsible for homeostatic control of synaptic pH, Na/K, and glutamate. Mercury is also known to inhibit synaptic uptake of dopamine (Rajanna et al. 1990), serotonin (Oudar et al. 1989), and norepinephrine (Rajanna and Hobson 1985). Mercury apparently has a higher binding affinity for serotonin binding sites. Mercury has also been reported to cause an increase in evoked acetylcholine release followed by a sudden and complete blockade (Cooper and Manalis 1983). Prolonged exposure to methylmercury results in an upregulation of muscarinic cholinergic receptors in the hippocampus and cerebellum and on circulating lymphocytes (Coccini et al. 2000). It also affects the release of neurotransmitters from presynaptic nerve terminals. This may be due to its ability to change the intracellular concentration of Ca^{2+} by disrupting the regulation of Ca^{2+} from intracellular pools and increasing the permeability of plasma membranes to Ca^{2+} (Atchison and Hare 1994). While there is undoubtedly much more to learn about the specific mechanisms of mercury-induced neurotoxicity, the symptoms are fairly clear.

The widespread pollution of Minamata Bay in Japan by methylmercury in 1950s has provided researchers with a clear picture of methylmercury-induced neurotoxicity. Known as Minamata disease, the neurotoxic signs include ataxia, speech impairment, and constriction of visual fields, hypoesthesia, dysarthria, hearing impairment, and sensory disturbances. These neurological problems persisted and were found in other areas of Japan as the mercury contamination spread (Ninomiya et al. 1995) (Tables 4.2 and 4.3).

Neurotoxicity is not only related to methylmercury, as a study of 98 dentists and 54 non-dentist controls revealed. The dentists with an average of 5.5 years of exposure to amalgams performed significantly worse on all of the following neurobehavioral tests: motor speed (finger tapping), visual scanning (trail making), visuomotor coordination and concentration (digit symbol),

Table 4.2 Minamata disease symptoms in male

Stiffness
Dysesthesia
Hand tremor
Dizziness
Loss of pain sensation
Cramping
Atrophy of the upper arm musculature
Arthralgia
Insomnia & Lumbago

Table 4.3 Minamata disease symptoms in female

Leg tremor
Tinnitus
Loss of touch sensation
Leg muscular atrophy & muscular weakness

Table 4.4 Factors that increase mercury release from amalgams

Chewing (food or gum)
Brushing
Bruxism
Hot drinks
Drilling or Polishing

Table 4.5 Psychological symptoms of mercury

Overload
Irritability
Excitability
Temper outbursts
Quarreling
Fearfulness/anxiety
Restlessness
Depression
Insomnia

verbal memory, visual memory, and visuomotor coordination speed (Ngim et al. 1992).

Mercury is also implicated in Alzheimer's disease and other chronic neurological complaints. In 1988, Alzheimer's cadaver studies reported mercury was found in much higher levels in the nucleus basalis of Meynert than in controls (40 ppb vs. 10 ppb) (Thompson et al. 1988). Subsequent studies have shown elevated mercury throughout the brain in individuals with Alzheimer's (Cornett et al. 1998). Furthermore, when rats were exposed to elemental mercury vapor at the same levels as documented in the oral cavity of humans with amalgams, lesions similar to those seen in Alzheimer's disease have occurred (Pendergrass et al. 1997). The same lesions have been demonstrated when rat brains were exposed to EDTA–mercury complex (Pendergrass and Haley 1995).

While amyotrophic lateral sclerosis (ALS) has been associated in some instances with possible cadmium exposure, a published case history revealed a diagnostic case of ALS recovering after amalgam removal (Table 4.4).

Mental health symptoms are also quite common with mercury toxicity. Evidence linking mercury exposure to psychological disorder has been accumulating for 60 years (Table 4.5).

Effects on the Kidney

Many occupational studies indicate that moderate to high exposure to mercury can cause harmful effects on the kidneys. When urine mercury levels are low to moderate, the results are inconclusive with no effects being reported in some studies and mild effects reported in others.

Early indicators of kidney injury include increased levels of protein in the urine (proteinuria) and increased levels of certain enzymes in the blood and urine. Proteinuria is commonly observed in studies reporting kidney effects. Less often, changes to the structure of the kidneys have been shown.

Skin Sensitization

Allergic skin sensitization has been reported in people with occupational exposure to mercury liquid or vapor. Once a person is sensitized to a chemical, contact with even a small amount causes outbreaks of dermatitis with symptoms such as skin redness, itching, rash, and swelling. This can spread from the hands or arms to other parts of the body. Occupational skin sensitization to mercury has been observed in people exposed

Table 4.6 Acute gastrointestinal effects by mercury

Nausea
Vomiting
Severe abdominal cramps & diarrhea
Corrosive ulceration
Bleeding & necrosis of GIT

Table 4.7 Mercury immunotoxicity

Autoimmunity
Decreased cellular immune function
Apoptosis of monocytes & lymphocytes
Decreased phagocytosis
Decreased production of TNF alpha, IL-1
Increased release of ACTH & cortisol

to mercury in dental amalgams, tattoos, or breakage of medical instruments.

Gastrointestinal Tract Toxicity

Limited information suggests that long-term exposure to mercury vapor can cause inflammation and ulceration of the inside of the mouth, sore gums, drooling, diarrhea, and other effects on the digestive system. No exposure information is reported, but presumably the concentrations were high (Table 4.6).

Cardiovascular Toxicity

Both organic and ionic mercury accumulates in the heart and has been associated with elevated blood pressure and abnormal heart rhythms such as tachycardia and ventricular heart rhythms (NASA 2000). It is unknown whether cardiovascular effects of mercury are due to direct cardiac toxicity or to indirect toxicity caused by effects on the neural control of cardiac function (US Environmental Protection Agency 1997).

Immunotoxicity

Mercury increases apoptosis of both monocytes and lymphocytes and reduces the phagocytic ability of monocytes. It has been demonstrated that workers occupationally exposed to mercury vapor exhibited diminished capacity to produce both TNF alpha and IL-1 (Langworth et al. 1993). A number of investigators have reported mercury compounds are capable of immune activation, leading to autoimmunity, while simultaneously reducing the cellular immune response, leading to increased infection (Bigazzi 1992; Druet et al. 1978; Enestrom and Hultman 1984, 1992; Blakley et al. 1980; Dieter et al. 1983; Nordlind 1983; Nakatsuru et al. 1985) which is the classic appearance of immunotoxicity (Vojdani 2007). Simultaneous with immune alterations are changes in the hypothalamic–pituitary–adrenal axis, as exhibited by increased levels of ACTH and corticosterone (Ortega et al. 1997). The increase in corticosterone levels could add to the immunosuppressiveness already present. Not only can mercury cause aberrant responses in both the cellular and humoral immune systems, it may also cause bacteria to become resistant to antibiotics. In a primate study, within 5 weeks of receiving amalgam fillings, the intestinal bacteria of the primates became resistant to penicillin, streptomycin, kanamycin, chloramphenicol, and tetracycline (Kolata 1993) (Table 4.7).

Respiratory Tract

Elemental mercury can enter the body through inhalation of vapor. Inhalation is the most common form of occupational exposure to mercury in United States. Due to high lipid solubility, when inhaled, 75–85 % is rapidly absorbed across the lungs and into the bloodstream. From the bloodstream, it diffuses to all body tissues. Inhalation at high concentrations (acute exposure) causes acute respiratory symptoms including coughing, congestion, dyspnea, bronchitis, pneumonitis, chest pain, reduced vital capacity, pulmonary edema, respiratory failure, and death (US Environmental Protection Agency 1997) (Table 4.8).

Table 4.8 Acute respiratory symptoms

Coughing
Congestion
Dyspnea
Bronchitis
Pneumonitis
Chest pain
Reduced vital activity
Pulmonary edema

Effects on the Eye

Long-term occupational exposure to mercury has caused a grayish-brown or yellow discoloration in the eyes of some people. This haze is not thought to affect vision. A gray band through the cornea (band keratopathy) has also been reported in a few people.

Carcinogenicity

MeHg has been classified by the EPA as a class C, "possible" human carcinogen (NASA 2000). In animal studies, MeHg increases the incidence of renal tumors only in male mice with preexisting tumors. The increase was not noted in female mice. Since the phenomenon was only seen at levels of MeHg that were toxic to the kidneys, the tumorigenic effects were presumed to be "secondary" to cell damage and repair. Therefore, it has been concluded that "in absence of a tumor initiator, long term exposure to sub-toxic doses of MeHg does not appear to increase tumor formation" (NASA 2000).

Genotoxicity

Dose-related aberrations in chromosomes and bone marrow cells were noted when Hg chloride was administered to mice by gavage. The effects of Hg chloride on genetic material are thought to be caused by the ability of Hg to "inhibit the formation of the mitotic spindle which can result in C-mitotic figures" (US Environmental Protection Agency 1997).

Reproductive Toxicity

In mice and rats, methylmercury induces abortions, increases resorption and malformations, and reduces offspring viability. In humans relative to the effects on adult brain, the effects of methylmercury on the developing brain in utero are more diffuse and may involve derangement of cortical cell layers and ectopic neurons. Depolymerization of microtubular structures by methylmercury may be a possible mechanism for these prenatal effects (Clarkson 1989).

Tests for Mercury Exposure

There are two tests available to measure mercury in the body:

The *Mercury Blood Test* measures exposure to all the three types of mercury, but because mercury remains in the bloodstream for only a few days after exposure, the test must be done soon after exposure. Most nonexposed people have mercury levels of 0–2 μg/dl. Levels above 2.8 μg/dl are required to be reported to the health department.

The *Urine Mercury Test* only measures exposure to elemental and inorganic mercury. Organic mercury is not passed out the body in the urine and thus cannot be measured this way. A person with no exposure to mercury would probably have a urine mercury level of 0–20 μg/L. The health department requires reporting of levels above 20.

Safety

Mercury and most of its compounds are extremely toxic and are generally handled with care; in cases of spills involving mercury (such as from certain thermometers or fluorescent light bulbs), specific cleaning procedures are used to avoid toxic exposure (Environmental Protection Agency 2007). It can be inhaled and absorbed through the skin and mucous membranes, so containers of mercury are securely sealed to avoid spills and evaporation. Heating of mercury, or compounds of mercury that may decompose when heated, is always carried out with adequate ventilation in order to avoid exposure to mercury vapor. The most toxic forms of mercury are its organic compounds, such as dimethylmercury and methylmercury. However, inorganic compounds, such as cinnabar, are also highly toxic by ingestion or inhalation of the dust (Oxford University 2009). Mercury can cause both chronic and acute poisoning.

Treatment

Research on the treatment of mercury poisoning is limited. Currently available drugs for acute mercurial poisoning include chelators N-acetyl-D,L-penicillamine (NAP), British anti-Lewisite (BAL), 2,3-dimercapto-1-propanesulfonic acid (DMPS), and dimercaptosuccinic acid (DMSA). In one small study including 11 construction workers exposed to elemental mercury, patients were treated with DMSA and NAP (Bluhm et al. 1992). Chelation therapy with both drugs resulted in the mobilization of a small fraction of the total estimated body mercury. DMSA was able to increase the excretion of mercury to a greater extent than NAP (Bluhm et al. 1992).

Biomagnification of Mercury in Aquatic Fauna

Fish and shellfish have a natural tendency to concentrate mercury in their bodies, often in the form of methylmercury, a highly toxic organic compound of mercury. Species of fish that are high on the food chain, such as shark, swordfish, king mackerel, albacore tuna, and tilefish, contain higher concentrations of mercury than others. As mercury and methylmercury are fat-soluble, they primarily accumulate in the viscera, although they are also found throughout the muscle tissue. When this fish is consumed by a predator, the mercury level is accumulated. Since fish are less efficient at depurating than accumulating methylmercury, fish-tissue concentrations increase over time. Thus, species that are high on the food chain amass body burdens of mercury that can be ten times higher than the species they consume. This process is called biomagnification. Mercury poisoning happened this way in Minamata, Japan, now called Minamata disease. As a result, those consuming high levels of fish should be aware of the symptoms of mercury poisoning.

References

Aschner M, Eberle NB, Miller K, Kimelberg HK (1990) Interactions of methylmercury with rat primary astrocyte cultures: inhibition of rubidium and glutamate uptake and induction of swelling. Brain Res 530: 245–250

Atchison WD, Hare MF (1994) Mechanisms of methylmercury-induced neurotoxicity. FASEB J 8:622–629

Bapu C, Rao P, Sood PP (1998) Restoration of methylmercury inhibited adenosine triphosphatases during vitamin and monothiol therapy. J Environ Pathol Toxicol Oncol 17:75–80

Bigazzi PE (1992) Lessons from animal models: the scope of mercury-induced autoimmunity. Clin Immunol Immunopathol 65:81–84

Blakley BR, Sisodia CS, Mukkur TK (1980) The effect of methylmercury, tetraethyl lead, and sodium arsenite on the humoral immunity response in mice. Toxicol Appl Pharmacol 52:245–254

Bluhm RE, Bobbitt RG, Welch LW, Wood AJ, Bonfiglio JF, Sarzen C, Heath AJ, Branch RA (1992) Elemental mercury vapor toxicity, treatment, and prognosis after acute, intensive exposure in chloralkali plant workers. Part I: history, neuropsychological findings and chelator effects. Hum Exp Toxicol 11:201–210

Chang LW (1997) Neurotoxic effects of mercury: a review. Environ Res 14:329–373

Clarkson TW (1989) Mercury. J Am Coll Toxicol 8:1291–1295

Coccini T, Randine G, Candura SM, Nappi RE, Prockop LD, Manzo L (2000) Low-level exposure to methylmercury modifies muscarinic cholinergic receptor binding characteristics in rat brain and lymphocytes: physiologic implication and new opportunities in biological monitoring. Environ Health Perspect 108: 29–33

Cooper GP, Manalis RS (1983) Influence of heavy metals on synaptic transmission: a review. Neurotoxicology 4:69–83

Cornett CR, Markesbery WR, Ehmann WD (1998) Imbalances of trace elements related to oxidative damage in Alzheimer's disease brain. Neurotoxicology 19:339–346

De Souza Queiroz ML, Pena SC, Salles TS, de Capitani EM, Saad ST (1998) Abnormal antioxidant system in

erythrocytes of mercury exposed workers. Hum Exp Toxicol 17:225–230
Dieter MP, Luster MI, Boorman GA, Charles WJ, Jack HD, Jeffrey WC (1983) Immunological and biochemical responses in mice treated with mercuric chloride. Toxicol Appl Pharmacol 68:218–228
Druet P, Druet E, Potdevin F, Sapin C (1978) Immune type glomerulonephritis induced by $HgCl_2$ in Brown-Norway rat. Ann Immunol 129:777–792
Durham HD, Minotti S, Caporicci E (1997) Sensitivity of platelet microtubules to disassembly by methylmercury. J Toxicol Environ Health 48:57–69
Endo T, Haraguchi K, Sakata M (2003) Renal toxicity in rats after oral administration of mercury-contaminated boiled whale livers marketed for human consumption. Arch Environ Contam Toxicol 44:412–416
Enestrom S, Hultman P (1984) Immune-mediated glomerular nephritis induced by mercuric chloride in mice. Experientia 40:1234–1240
Enestrom S, Hultman P (1992) Dose-response studies in murine mercury-induced autoimmunity and immune-complex disease. Toxicol Appl Pharmacol 113:199–208
Environmental Protection Agency (EPA) (2007) Mercury: spills, disposal and site cleanup
Falconer MM, Vaillant A, Reuhl KR, Laferriere N, Brown DL (1994) The molecular basis of microtubule stability in neurons. Neurotoxicology 15:109–122
Ganther HE (1978) Modification of methylmercury toxicity and metabolism by selenium and vitamin E: possible mechanisms. Environ Health Perspect 25:71–76
Ganther HE, Goudie C, Sunde ML, Kopecky MJ, Wagner P (1972) Selenium: relation to decreased toxicity of methylmercury in diets containing tuna. Science 175:1122–1124
Goldberg RL, Kaplan SR, Fuller GC (1983) Effect of heavy metals on human rheumatoid synovial cell proliferation and collagen synthesis. Biochem Pharmacol 32:2763–2766
InSug O, Datar S, Koch CJ, Shapiro IM, Shenker BJ (1997) Mercuric compounds inhibit human monocyte function by inducing apoptosis: evidence for formation of reactive oxygen species, development of mitochondrial membrane permeability transition and loss of reductive reserve. Toxicology 124:211–224
Kolata G (1993) New suspect in bacterial resistance: amalgam. The New York Times
Langworth S, Elinder CG, Undqvist KG (1993) Minor effects of low exposure to inorganic mercury on the human immune system. Scan J Work Environ Health 19:405–413
Magour S, Maser H, Grein H (1987) The effect of mercury and methylmercury on brain microsomal Na^+, K^+ ATPase after partial delipidisation with Lubrol. Pharmacol Toxicol 60:184–186
Miller OM, Lund BO, Woods JS (1991) Reactivity of Hg (II) with superoxide: evidence for the catalytic dismutation of superoxide by Hg (II). J Biochem Toxicol 6:293–298
Miura K, Koide N, Himeno S, Nakagawa I, Imura N (1999) The involvement of microtubular disruption in methylmercury-induced apoptosis in neuronal and nonneuronal cell lines. Toxicol Appl Pharmacol 160:279–288
Nakatsuru S, Oohashi J, Nozaki H, Nakada S, Imura N (1985) Effect of mercurials on lymphocyte functions in vitro. Toxicology 36:297–305
NASA (2000) Toxicological effects of methylmercury. National Academy of Sciences, Washington, DC
Ngim CH, Foo SC, Boey KW, Jeyaratnam J (1992) Chronic neurobehavioral effects of elemental mercury in dentists. Br J Ind Med 49:782–790
Ninomiya T, Ohmori H, Hashimoto K, Tsuruta K, Ekino S (1995) Expansion of methylmercury poisoning outside of Minamata: an epidemiological study on chronic methylmercury poisoning outside Minamata. Environ Res 70:47–50
Nordlind K (1983) Inhibition of lymphoid-cell DNA synthesis by metal allergens at various concentrations. Effect on short-time cultured non-adherent cell compared to non-separated cells. Int Arch Allergy Appl Immunol 70:191–192
Ortega HG, Lopez M, Takaki A, Huang QH, Arimura A, Salvaggio JE (1997) Neuroimmunological effects of exposure to methylmercury forms in the Sprague-Dawley rat. Activation of the hypothalamic-pituitary adrenal axis and lymphocyte responsiveness. Toxicol Ind Health 13:57–66
Oudar P, Caillard L, Fillon G (1989) In vitro effects of organic and inorganic mercury on the serotonergic system. Pharmacol Toxicol 65:245–248
Oxford University (2009) MSDS. Safety data for mercuric sulphide
Pendergrass JC, Haley BE (1995) Mercury-EDTA complex specifically blocks brain beta-tubulin- GTP interactions: similarity to observations in Alzheimer's disease. In: Friberg LT, Scrauzer GN (eds) Status quo and perspectives of amalgam and other dental materials. Georg Thieme Verlag, Stuttgart, pp 98–105
Pendergrass JC, Haley BE, Vimy MJ, Winfield SA, Lorscheider FL (1997) Mercury vapor inhalation inhibits binding of GTP to tubulin in rat brain: similarity to a molecular lesion in Alzheimer's diseased brain. Neurotoxicology 18:315–324
Rajanna B, Hobson M (1985) Influence of mercury on uptake of dopamine and norepinephrine by rat brain synaptosomes. Toxicol Lett 27:7–14
Rajanna B, Hobson M, Harris L, Ware L, Chetty CS (1990) Effects of cadmium and mercury on Na+, K+ ATPases and the uptake of ^{3}H-dopamine in rat brain synaptosomes. Arch Physiol Biochem 98:291–296
Salonen JT, Seppanen K, Nyyssonen K, Korpela H, Kauhanen J, Kantola M, Tuomilehto J, Esterbauer H, Tatzber F, Salonen R (1995) Intake of mercury from fish, lipid peroxidation, and the risk of myocardial infarction and coronary, cardiovascular, and any death in eastern fisher men. Circulation 91:645–655

Shenker BJ, Guo TL, Shapiro IM (1998) Low-level methylmercury exposure causes human T-cells to undergo apoptosis: evidence of mitochondrial dysfunction. Environ Res 77:149–159

Thompson CM, Markesbery WR, Ehmann WD, Mao YX, Vance DE (1988) Regional brain trace-element studies in Alzheimer's disease. Neurotoxicology 9:1–7

U.S. Environmental Protection Agency (EPA) (1997) Mercury study report to congress, volume II: an inventory of anthropogenic mercury emissions in the United States

Vojdani A (2007) Personal communication

Welsh SO, Soares JH (1976) The protective effect of vitamin E and selenium against methylmercury toxicity in the Japanese quail. Nutr Rep Int 13:43

WHO (2000) Safety evaluation of certain food additives and contaminants. World Health Organisation Food Additives Series, No 44

Zalups RK, Lash LH (1996) Interactions between glutathione and mercury in the kidney, liver and blood. In: Chang LW (ed) Toxicology of metals. CRC Press, Boca Raton, pp 145–163

5 Antioxidant Drugs Mitigating Toxicity

An *antioxidant* is a molecule that inhibits the oxidation of other molecules. Oxidation is a chemical reaction that transfers electrons or hydrogen from a substance to an oxidizing agent. Oxidation reactions can produce free radicals. In turn, these radicals can start chain reactions. When the chain reaction occurs in a cell, it can cause damage or death to the cell. Antioxidants terminate these chain reactions by removing free radical intermediates and inhibit other oxidation reactions. They do this by being oxidized themselves, so antioxidants are often reducing agents such as thiols, ascorbic acid, or polyphenols (Sies 1997).

Antioxidants are important additives in gasoline. These antioxidants prevent the formation of gums that interfere with the operation of internal combustion engines (Dabelstein et al. 2007).

Although oxidation reactions are crucial for life, they can also be damaging; plants and animals maintain complex systems of multiple types of antioxidants, such as glutathione, vitamin C, and vitamin E, as well as enzymes such as catalase, superoxide dismutase, and various peroxidases. Low levels of antioxidants, or inhibition of the antioxidant enzymes, cause oxidative stress and may damage or kill cells.

As oxidative stress appears to be an important part of many human diseases, the use of antioxidants in pharmacology is intensively studied, particularly as treatments for stroke and neurodegenerative diseases. Moreover, oxidative stress is both the cause and the consequence of disease. Antioxidants are widely used in dietary supplements and have been investigated for the prevention of diseases such as cancer, coronary heart disease, and even altitude sickness (Fig. 5.1).

Although initial studies suggested that antioxidant supplements might promote health, later large clinical trials with a limited number of antioxidants detect no benefit and even suggested that excess supplementation with certain putative antioxidants may be harmful (Jha et al. 1995; Baillie et al. 2009; Bjelakovic et al. 2007). Antioxidants also have many industrial uses, such as preservatives in food and cosmetics and to prevent the degradation of rubber and gasoline.

Vitamin E (An Antioxidant Drug)

Description

Vitamin E is a fat-soluble chemical found in the diet in varying amounts. Discovered in 1922, the term vitamin E is used to refer to all tocol and trienol derivatives. The tocols are alpha-, beta-, gamma-, and delta-tocopherols, and the trienols are alpha-, beta-, gamma-, and delta-tocotrienols. All these substances are found in plants and have vitamin E activity, but alpha-tocopherol is the most active form of vitamin E. In the human body, vitamin E is present primarily as alpha-tocopherol. Vitamin E can be isolated from natural sources (plants, vegetables, and meat) or can be made in the laboratory. Therefore, vitamin E is sold commercially as a natural or synthetic preparation. Naturally occurring alpha-tocopherol is now

S. Nabi, *Toxic Effects of Mercury*, DOI 10.1007/978-81-322-1922-4_5,
© Springer India 2014

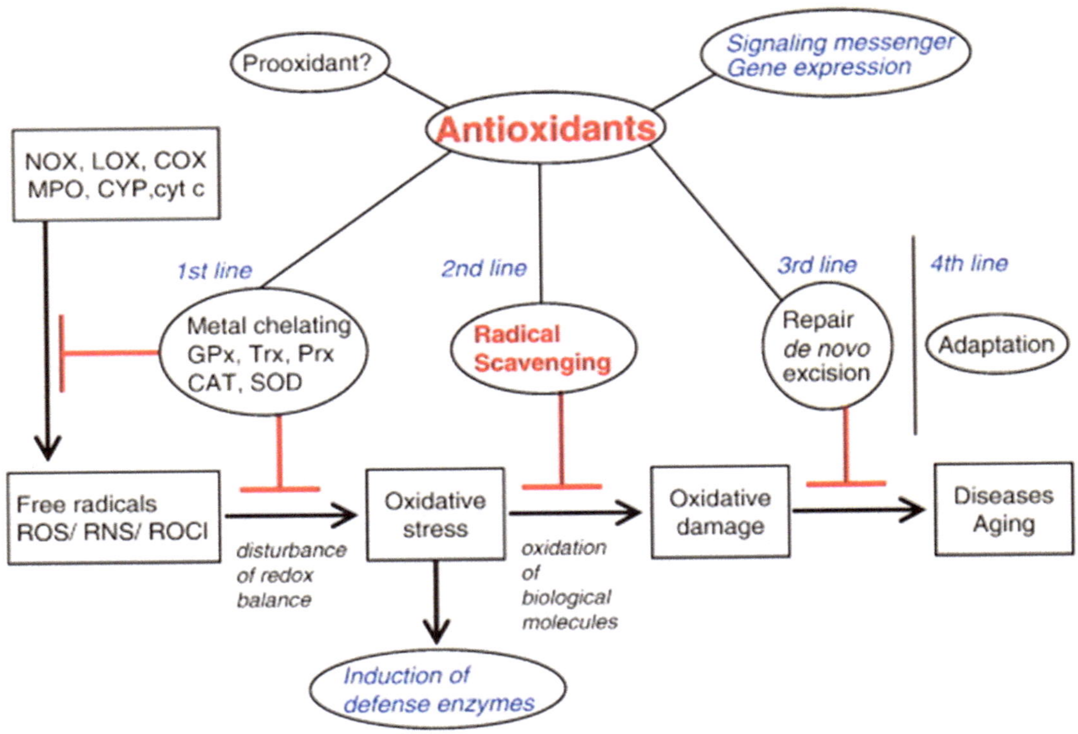

Fig. 5.1 Antioxidants displaying their mode of action (Source: English Wikipedia)

referred to as RRR-alpha-tocopherol (formerly d-alpha-tocopherol), whereas synthetic alpha-tocopherol is referred to as all-rac-alpha-tocopherol (formerly dl-alpha-tocopherol). The esterified forms of vitamin E such as alpha-tocopherol acetate, alpha-tocopherol succinate, and alpha-tocopherol nicotinate are made in the laboratory and are also sold commercially. The alpha form of tocopherol constitutes about 90 % of the tocopherol in animal tissue, originally designated *d-alpha-tocopherol* on the basis of optical activity. Vitamin E was the fifth vitamin discovered when researchers found that a dietary deficiency in laboratory rats produced fetal death in pregnant females. The name "tocopherol" was derived from the Greek words for childbirth (***tos***), to bring forth (***phero***), and the chemical designation for an alcohol (***ol***). Vitamin E acts as a coenzyme in cellular membranes and serves as a scavenger for free radicals that are destructive to the membrane and internal cellular components. Natural sources of vitamin E are vegetable oils, sunflower seeds, almonds, and peanuts. Vitamin E is peanuts. Vitamin E is essential for our growth and survival. However, the human body does not make this vitamin. We depend primarily on diet or supplement for our vitamin E needs. About 20 % of ingested vitamin E is absorbed from the intestine. It is now known that vitamin E undergoes very little degradation in the body. The main degradation products of vitamin E are tocopheryl quinone, tocopheryl hydroquinone, dimers, trimers, and some water-soluble substances. The major route of excretion is through the feces. Adipose tissue, liver, and muscle are major areas for the deposit of vitamin E. The consumption of higher amounts of vitamin E increases its level in all tissues. Vitamin E is present within the cells in its free form as well as bound to proteins. These vitamin E–binding proteins are present in the membrane, cytosol, and nucleus. The role of these vitamin E–binding proteins in the mechanism of action of vitamin E is unknown (Fig. 5.2).

Fig. 5.2 Picture of vitamin E (Source: English Wikipedia)

Method of Action

The chemical formula of alpha-tocopherol is $C_{29}H_{50}O_2$. At least eight compounds having vitamin E activity have been isolated from plant sources. All have a 6-chromonal ring structure and a side chain. The tocols have a phytol side chain, whereas the trienols have a similar structure, with double bonds at the 3′, 7′, and 11′ positions of the side chain. Both tocols and trienols occur as a variety of isomers which differ from one another by the number and location of methyl groups on the chromonal ring. Alpha-tocopherol is the most active form and the side chain is essential for full biological activity of vitamin E.

Vitamin E has more than one mechanism in the body. One of the most well-established mechanisms is its capacity to destroy free radicals generated as a part of the oxidation reaction in the human body or by exogenous agents. This antioxidation mechanism of vitamin E has been demonstrated both in vitro and in vivo. Vitamin E has been shown to stabilize membranes by physiochemical interaction between its phytol side chain and the fatty acid chain of polyunsaturated phospholipids. It inhibits the synthesis of prostaglandins and prevents platelet aggregation in vitro and in vivo. There are some data which show that vitamin E reduces the synthesis of thromboxane and increases the formation of prostacyclin. Thromboxane is considered the most potent platelet-aggregating factor; therefore, further study on the role of vitamin E in regulating the metabolism of arachidonic acid is needed. Currently, the biological activity of vitamin E which attracts most of the interest is the prevention of lipid peroxidation. Alpha-tocopherol is the most active tocopherol against peroxyl radicals ($LOO\cdot$) and delta-tocopherol is the least active (alpha>beta=gamma>delta). The antioxidant activity of vitamin E is based on the ease with which the hydrogen on the hydroxyl group of the chroman ring can be donated to neutralize a free radical (creating a more stable tocopheroxyl radical). As with phospholipids, the polar chroman ring tends to stay near the edges of the membrane, whereas the hydrophobic core will be buried deep into the membrane. When a phospholipid tail becomes peroxidized by a free radical, the tail becomes more polar and migrates to the surfaces where it can meet the tocopherol chroman ring to be neutralized while forming a tocopheroxyl radical. The tocopheroxyl radical can be reduced (restored) to tocopherol directly by ubiquinol or vitamin C and then by glutathione or lipoic acid (via vitamin C), which are in turn reduced by NADH or NADPH.

Properties and Uses

Vitamin E has been shown to interact with some pollutants which are present in the environment and diet. The primary atmospheric pollutants are ozone and nitric oxide which are capable of generating free radicals in the body. Vitamin E has been shown to protect against the harmful effect of ozone and nitric oxide. The major food pollutants are nitrites which are present in fresh fruits and vegetables as well as in bacon, sausage, and cured meat. Nitrites by themselves are not harmful to adults, but they can combine with amines in the stomach to form nitrosamine. Nitrosamines are among the most potent cancer-causing agents for both animals and human beings. The presence of vitamin C or vitamin E in the stomach may prevent the formation of or reduce the levels of nitrosamines In addition to nitrosamines, many other mutagenic substances are formed in the digestive tract. The presence of higher levels of fecal mutagenic substances may increase the risk of some cancers. It has been reported that taking vitamin E or vitamin C reduces the mutagenic substances in the feces (Table 5.1).

Vitamin E protects cells from the toxicity of certain heavy metals. For example, organic mercury is known to cause neurological diseases because of damage to the brain cells. The administration of vitamin E immediately before treatment of animals with vitamin E protects cells from the toxicity of certain heavy metals. For example, organic mercury is known to cause neurological diseases because of damage to the brain cells. Vitamin E may have a role in the management of some neurological diseases. A group of neurologists are initiating a new clinical study using vitamin E in combination with deprenyl in order to slow down the progression of disease in patients with Parkinsonism. It is believed that free radicals are generated by degradation products of dopamine (a chemical which is essential for brain function) and by drugs used in the treatment of Parkinson's disease (Tables 5.2 and 5.3).

Free radicals have been implicated in accelerating the aging processes of organisms as well as individual organs in the body. Therefore, the supplemental use of vitamins on a regular basis should slow down aging processes.

Table 5.1 Recommended dietary allowances of vitamin E for various age groups

RDA for adult male: 10 mg alpha TE
RDA for adult female: 8 mg alpha TE
Children 7–10 years: 7 mg alpha TE
Infants: 4 mg alpha TE
Pregnant and lactating women: 12 mg alpha TE

Table 5.2 Summary of vitamin E deficiency symptoms

Premature hemolysis of red blood cells
Muscle weakness
Increased lung tissue due to smog, cigarette smoking, and air pollution
Shortened life of red blood cells
Anemia
Decreased resistance to bacterial infection
Increased susceptibility to fibrocystic breast disease
Increased tendency to retrolental fibroplasia in babies
Increased tendency to hemolytic anemia in premature babies
Increased susceptibility to thrombosis

Table 5.3 The best sources of dietary vitamin E

Green leafy vegetables and cod liver oil
Coconut and corn oil
Mayonnaise and olive oil
Peanut butter and palm oil
Peanuts and pecans
Soybeans and walnuts

Acetyl-L-Carnitine

L-Carnitine is a trimethylated amino acid that plays essential role in many areas of the body, including fatty acid translocation and muscle function. Carnitine is also acetylated into the ester acetyl-L-carnitine (ALCAR) in the brain, liver, and kidney. ALCAR also plays a variety of roles in the body, including increasing acetylcholine production and stimulation of protein and membrane phospholipid synthesis. Orally administered L-carnitine and ALCAR have profound nutrient repartitioning properties, and the effects of supplementing with them have been extensively researched in many areas.

Carnitine's primary job is in the regulation of cellular metabolism, and it closely interacts with coenzyme A in a variety of reactions. It is required for fatty acid oxidation, and this is the primary theoretical reason for improved exercise performance, as improved fatty acid oxidation will preserve muscle glycogen and improve ATP production (Brass and Hiatt 1998; Brass 2000). It was recently confirmed that supplemental carnitine increases long-chain fatty acid oxidation in healthy individuals without carnitine deficiency (Muller et al. 2002), providing more evidence for an ergogenic benefit. Enhanced fatty acid oxidation and cellular metabolism are also the proposed mechanism of action for the nutrient partitioning benefits (Iossa et al. 2002).

L-Carnitine was able to scavenge superoxide anion to inhibit the lipoperoxidation (Vanella et al. 2000; Rani and Paneerselvam 2002). Carnitine can also act as a chelator by decreasing the concentration of cytosolic iron that plays a very important role in free radical chemistry (Reznick et al. 1992). L-Carnitine has a capacity to enhance nonenzymatic antioxidants, such as vitamin E (Arockia Rani and Paneerselvam 2001).

Acetyl-L-carnitine (ALCAR) plays a strong role in the brain in many ways and has beneficial effects. Studies in aged rodents show markedly improved memory and learning capacities (Ando et al. 2001; Liu et al. 2002a, b; Yasui et al. 2002), while studies in younger rodents show a variety of promising effects as well (De Angelis et al. 1994; Bertoni-Freddari et al. 1996). Other rodent studies have shown that ALCAR significantly protects the brain against a variety of stresses, such as ischemia and reperfusion (Calvani and Arrigoni-Martelli 1999) and mitochondrial uncoupling (Virmani et al. 1995). It also protects against peripheral nerve trauma, "almost eliminating nerve loss" (Hart et al. 2002), and in vitro neuronal apoptosis (Ishii et al. 2000; Bigini et al. 2002). ALCAR also increases levels of dopamine, amino acids, and acetylcholine in the brain, as well as facilitating cholinergic activity (Toth et al. 1993; Ando et al. 2001).

ALCAR supplementation is also accompanied by many positive structural changes in the brain in both the young and the old. It stimulates nerve growth factor (NGF) binding (Hart et al. 2002; Taglialatela et al. 1992), and rodent studies indicate significantly more regenerative elements and reduced degenerative elements (De Angelis et al. 1994; Mckay Hart et al. 2002).

Carnitine has no toxicity, teratogenicity, contraindications, or drug interactions (Benvenga et al. 2001). There are very few side effects associated with carnitine use, and no serious side effects have been reported (Ghidino et al. 1988; Salvioli and Neri 1994; Benvenga et al. 2001). Some users report nausea or stimulation, but these tend to be transient and disappear with time (Benvenga et al. 2001). There are also many anecdotal reports of more vivid dreams.

Both L-carnitine and vitamin E have a primary role as an antioxidant, an anti-inflammatory, and a cytoprotective agent. Because L-carnitine has a capacity to enhance nonenzymatic antioxidants, such as vitamin E, we considered that the use of L-carnitine and vitamin E, in combination, might have additional effects in the prevention of methylmercury-induced toxicities. Therefore, we have decided in this study to use vitamin E and L-carnitine, separately or in combination, in the prevention of methylmercury-induced oxidative stress in rat model.

References

Ando S, Tadenuma T, Tanaka Y, Fukui F, Kobayashi S, Ohashi Y, Kawabata T (2001) Enhancement of learning capacity and cholinergic synaptic function by carnitine in ageing rats. J Neurosci Res 66:266–271

Arockia Rani PJ, Panneerselvam C (2001) Carnitine as a free radical scavenger in ageing. Exp Gerontol 36:1713–1726

Baillie JK, Thompson AAR, Irving JB, Bates MGD, Sutherland AI, MacNee W, Maxwell SRJ, Webb DJ (2009) Oral antioxidant supplementation does not prevent acute mountain sickness: double blind, randomized placebo-controlled trial. QJM 102(5):341–348

Benvenga S, Ruggeri RM, Russo A, Lapa D, Capanni A, Trimarchi F (2001) Usefulness of L-carnitine, a naturally occurring peripheral antagonist of thyroid hormone action, in iatrogenic hyperthyroidism: a randomized, double-blind, placebo-controlled clinical trial. J Endocrinol Metab 86:3579–3594

Bertoni-Freddari C, Fattoretti P, Caselli U, Paoloni R (1996) Acetyl carnitine modulation of the morphology of rat hippocampal synapses. Anal Quant Cytol Histol 18:275–278

Bigini P, Larini S, Pasquali C, Muzio V, Mennini T (2002) Acetyl-L-carnitine shows neuroprotective and neurotrophic activity in primary culture of rat embryo motoneurons. Neurosci Lett 329:334–338

Bjelakovic G, Nikolova D, Gluud LL, Simonetti RG, Gluud C (2007) Mortality in randomized trials of antioxidant supplements for primary and secondary prevention: systematic review and meta-analysis. JAMA 297(8):842–857

Brass EP (2000) Supplemental carnitine and exercise. Am J Clin Nutr 72:618S–623S

Brass EP, Hiatt WR (1998) The role of carnitine and carnitine supplementation during exercise in man and in individuals with special needs. J Am Coll Nutr 17:207–215

Calvani M, Arrigoni-Martelli E (1999) Attenuation by acetyl-L-carnitine of neurological damage and biochemical derangement following brain ischemia and repurfusion. Int J Tissue React 21:1–6

Dabelstein W, Reglitzky A, Schütze A, Reders K (2007) Automotive fuels. In: Ullmann's encyclopedia of industrial chemistry. Wiley-VCH Verlag GmbH & Co. KGaA

De Angelis C, Scarfo C, Falcinelli M, Perna E, Reda E, Ramacci MT, Angelucci L (1994) Acetyl-L-Carnitine prevents age-dependent structural alterations in rat peripheral nerves and promotes regeneration following sciatic nerve injury in young and senescent rats. Exp Neurol 28:103–114

Ghidino O, Azzurro M, Vita G, Sartori G (1988) Evaluation of the therapeutic efficacy of L-Carnitine in congestive heart failure. Int J Clin Pharmacol Ther Toxicol 26:217–220

Hart AM, Wiberg M, Youle M, Terenghi G (2002) Systemic acetyl-L-carnitine eliminates sensory neuronal loss after peripheral axotomy: a new clinical approach in the management of peripheral nerve trauma. Exp Brain Res 145:182–189

Iossa S, Mollica MP, Lionett IL, Crescenzo R, Botta M, Barletta A, Liverini G (2002) Acetyl L-Carnitine supplementation differently influences nutrient partitioning, serum leptin concentration and skeletal muscle mitochondrial in young and old rats. J Nutr 132:636–642

Ishii T, Shimpo Y, Matsuoka Y, Kinoshita K (2000) Anti-apoptotic effect of acetyl-L- carnitine and I-carnitine in primary cultured neurons. Jpn J Pharmacol 83:119–124

Jha P, Flather M, Lonn E, Farkouh M, Yusuf S (1995) The antioxidant vitamins and cardiovascular disease: a critical review of epidemiologic and clinical trial data. Ann Intern Med 123(11):860–872

Liu J, Killilea DW, Ames BN (2002a) Age associated mitochondrial oxidative decay: improvement of carnitine acetyltransferase substrate-binding affinity and activity in brain by feeding old rats acetyl-L-carnitine and/or R-alpha-lipoic acid. Proc Natl Acad Sci 99:1876–1881

Liu J, Head E, Gharib AM, Yuan W, Ingersoll RT, Hagen TM, Cotman CW, Ames BN (2002b) Memory loss in old rats is associated with brain mitochondrial decay and RNA/DNA oxidation: partial reversal by feeding acetyl-L-carnitine and/or R-alpha-lipoic acid. Proc Natl Acad Sci 99:2356–2361

Mckay Hart A, Wiberg M, Terenghi G (2002) Pharmacological enhancement of peripheral nerve regeneration in the rat by systemic acetyl-L-carnitine treatment. Neurosci Lett 334:181–185

Muller DM, Seim H, Kiess W, Loster H, Richter T (2002) Effects of oral L-Carnitine supplementation on in vivo long chain fatty acid oxidation in healthy adults. Metabolism 51:1389–1391

Rani PJ, Panneerselvam C (2002) Effect of L-Carnitine on brain lipid peroxidation and antioxidant enzymes in old rats. J Gerontol A Biol Sci Med Sci 57:B134–B137

Reznick AZ, Kagan VE, Ramsey R, Tsuchiya M, Khwaja S, Serbinova EA, Packer L (1992) Antiradionyl effect of L-propionyl Carnitine protection of the heart against ischemia- repurfusion injury: the possible role of iron chelation. Arch Biochem Biophys 296:394–401

Salvioli G, Neri M (1994) L-acetyl carnitine treatment of mental decline in the elderly. Drugs Exp Clin Res 20:169–176

Sies H (1997) Oxidative stress: oxidants and antioxidants. Exp Physiol 82(2):291–295

Taglialatela G, Angelucci L, Ramacci MT, Werrbach-Perez K, Jackson GR, Perez-polo JR (1992) Stimulation of nerve growth factor receptors in PC12 by acetyl-L-carnitine. Biochem Pharmacol 44:577–585

Toth E, Harsing LG, Sershen H, Ramacci MT, Lajtha A (1993) Effect of acetyl-L-carnitine on extracellular amino acid levels in vivo in rat brain regions. Neurochem Res 18:573–578

Vanella A, Russo A, Acquaviva R, Campisi A, Di Giacomo C, Sorrenti V, Barcellona ML (2000) L-Propionyl-carnitine as superoxide scavenger, antioxidant and DNA cleavage protector. Cell Biol Toxicol 16:99–104

Virmani MA, Biselli R, Spadoni A, Rossi S, Corsico N, Calvani M, Fattorossi A, De Simone C, Arrigoni-Martelli E (1995) Protective actions of L-Carnitine and Acetyl-L-Carnitine on the neurotoxicity evoked by mitochondrial uncoupling or inhibitors. Pharmacol Res 32:383–389

Yasui F, Matsugo S, Ishibashi M, Kajita T, Ezashi Y, Oomura Y, Kojo S, Sasaki K (2002) Effects of chronic acetyl-l-carnitine treatment on brain lipid hydroperoxide level and passive avoidance learningin senescence-accelerated mice. Neurosci Lett 334:177–180

6 Mercury and Cognition

Methylmercury and Behavior Plasticity

Nervous system toxicity is revealed by overt neurological signs that are patently observable, as well as by more subtle effects on behavioral adaptation to ever-changing environments. The latter requires refined testing but is worth the effort, as the impairments that are revealed can be highly informative about the subtle effects of low-level exposure. These effects fall under the umbrella of cognitive effects and are typically reflected in operant or respondent conditioning phenomena as well as "simpler" processes such as habituation and sensitization. They entail manipulation of plastic behavior that has been trained or established during the course of a lifetime. Impairment is reflected in the course of acquisition or in the expression of behavior during a stable baseline. The return for the investment in the extensive testing that is required is the revelation of effects on memory, learning, perception, sensory–motor function, or other subtle effects that significantly impair functioning in an industrialized society (Weiss and Cory-Slechta 1994).

A number of different behavioral domains can be affected by exposure to methylmercury. Sensory function, motor function, sensitivity to certain behaviorally active drugs, and learning have all been identified as toxic endpoints. Behavioral procedures can be designed to emphasize these influences over behavior, but with the caveat that these domains cannot be completely separated from one another or examined in isolation. To characterize the effects of methylmercury on the acquisition of behavior, for example, one must necessarily rely on the perception of stimuli, sensitivity to reinforcement, and motor abilities that permit an animal to respond.

While behavioral analysis of methylmercury toxicity typically involves the use of laboratory models, they address, and can even predict, subtle effects in humans, some of which have economic consequences. For example, estimates of the economic consequences of productivity loss associated with methylmercury exposure in United States begin with decrements of scores on IQ tests (Trasande et al. 2005), a highly refined form of testing. Animal models of human IQ tests do not exist, but correlations between exposure levels that impair operant behavior in nonhuman species and those that affect scores on IQ tests (Cory-Slechta 1990; Paule et al. 1990, 1999) support the use of such testing as a crucial component in the characterization of methylmercury neurotoxicity.

Operant behavior is any behavior that is sensitive to its consequences, thus underlying a broad range of behavioral phenomena. While operants (i.e., voluntary responses) can be extraordinary complex and extensive over both time and space (Marr 1979; Schick 1971), they often are studied with simple responses such as lever presses or traversal through a maze. The complexity can be reduced by identifying three important terms in the "three-term contingency" of operant behavior (Fig. 6.1). These are responses, consequences, and the context in which

S. Nabi, *Toxic Effects of Mercury*, DOI 10.1007/978-81-322-1922-4_6,
© Springer India 2014

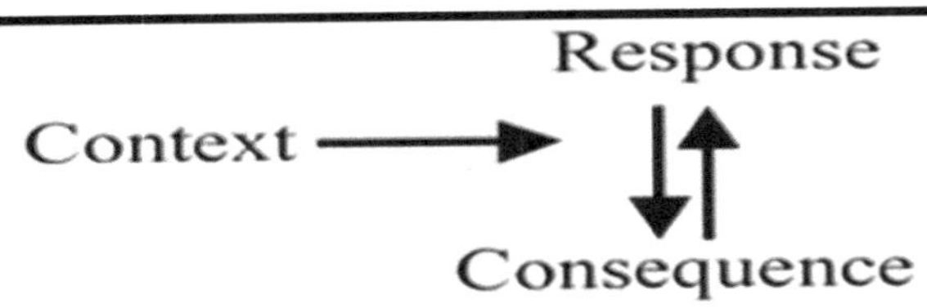

Element	Examples	Function
Antecedents (context)	Current context Past context Genetic history	Discrimination Remembering Sensory properties of chemicals.
Behavior	Force, topography precision	Motor effects Tremor
Consequences	Reinforcing, or aversive properties	Self-administration Avoidance
Combination	Schedule-controlled operant	Changes in rate or pattern. Choice

Fig. 6.1 Three-term contingency describing the control of operant behavior. Simply put, consequences act on responses by changing their probability of occurrence. This response–reinforcer relationship always takes place in some context. The application of these different elements to the study of the behavioral effects of drugs and toxicants is described in the columns on the right (Source: English Wikipedia)

behavior occurs. A consequence (food, drug, shock, or irritant) may follow a response. If that contingency between the response and the consequence increases the response rate over baseline levels, then it is called a reinforcer and the response is called an operant. Positive and negative reinforcement and punishment are defined according to how response rate changes as a function of the consequences (Catania 1991). All of this takes place in a context. Thus, for example, the stimulus context in which a left lever press is reinforced can differ from one in which a right lever press is reinforced. Discrimination describes the extent to which response rates in the presence of these stimuli differ.

As noted in Fig. 6.1, the experimental arrangement can be structured so as to emphasize the role of consequences and thereby examine the reinforcing efficacy of a drug or food (Griffiths et al. 1977), the irritancy of a chemical (Wood 1981), or the sensitivity to the richness of different sources of reinforcers (Davison and McCarthy 1988). Patterns of lever pressing or the response's physical characteristics (force, duration, displacement) can be emphasized to study motor function (Newland 1994, 1995). The context can be exploited to examine sensory function (Maurissen 1995; Rice 1994) as well as memory, discrimination, or generalization (Miller and Eckerman 1986). Schedule-controlled operant behavior, generally reflected in behavior supported by complex arrangements of contingencies, often involves complex behavioral patterns that can be quite sensitive to chemical exposure (Cory-Slechta 1994; Rice 1988; Weiss 1981). One important advantage of operant testing is that it reveals the extent to which impairment occurs in an intact, behaving organism. Behavior can be quite sensitive to toxicant exposure, and behavior analyses can test hypotheses generated in in vitro experiments as well as generate hypotheses for pursuit in further studies (Kulig and Jaspers 1999; Weiss and Cory-Slechta 1994).

Identifying Effects of MeHg Exposure on Learning

Newland and Paletz (2000) proposed that MeHg's effects are dominated by disruptions in the relationship between a response and its consequences and that the role of stimulus control processes (discrimination, memory) is relatively minimal. These determinants of operant responses, their context, and their consequences are inseparable influences over the behavior of the whole animal, making it impossible to eliminate any single term in a behavioral analysis of a neurotoxicant (Davison and Nevin 1999). It is possible, however, to design procedures that minimize the influence of a particular element.

If MeHg's effects are primarily on the response–reinforcer relationship, then it will be necessary to use procedures that exploit this relationship in order to identify these effects. These procedures should also minimize the role of memorial or discrimination processes. This runs counter to many approaches that are commonly used in behavioral toxicology because they emphasize discrimination or, especially, memory processes. For example, procedures such as radial arm mazes, delayed match to sample, and delayed discrimination procedures, which are commonly used in identifying "cognitive" effects of toxicant exposure, should be less sensitive to developmental MeHg if such exposure does not disrupt discrimination or contextual control processes.

Motor Function

Behavior always entails a motor act, so it is necessary to eliminate motor deficits as confounds in experiments on cognitive function. Sensory–motor function has been shown to be disrupted by MeHg exposure in humans, nonhuman primates, and laboratory rodents and pigeons, so the possibility that motor deficits confound effects on behavioral plasticity must be addressed (Beuter et al. 1999; National Research Council 2000; Watanabe and Satoh 1996). With MeHg, the pattern of sensory–motor deficit differs, depending on whether exposure is developmental or chronic and beginning in adulthood.

A direct comparison of the effects of developmental and chronic, adult-onset exposure on motor skills showed dose-related effects of MeHg only after extensive, chronic exposure. Rats were exposed either chronically or prenatally (via maternal drinking water) to 0, 0.5, or 5.0 ppm of MeHg in drinking water, approximating 0, 40, or 400 μg/kg/day (Day et al. 2005). Half of the rats were maintained on a coconut oil diet, while the other half ate a fish oil diet high in docosahexaenoic acid (DHA), but diet did not influence methylmercury's effects, so will not be considered further here. Chronic, adult-onset exposure produced dose-related increases in grip strength, hind-limb crossing when the rat is held by the tail (a marker of chronic, high-dose exposure), gait abnormalities, and diminished running wheel activity. Similar effects have been reported in other experiments, but with higher exposure levels and, consequently, more rapid onset of effects (Sakamoto et al. 1998). Pigeons are also susceptible to sensory–motor deficits after chronic exposure (Evans et al. 1982).

Developmental exposure produced none of these effects (Day et al. 2005). Since the effects on choice, fixed-ratio acquisition, or high-rate behavior were all associated with developmental exposure, it can be concluded that they did not reflect such pronounced motor deficits. There remains the possibility that subtler motor deficits associated with developmental exposure influence the effects seen, but this seems unlikely. The elimination of subtler deficits that might be associated with developmental exposure (Rice 1989) requires the more fine-grained analyses described elsewhere. This issue has been addressed in experiments described above by noting distinctions between MeHg's effects on choice and overall response rate or between pauses that separate high-rate bursts of responding and the initiation of such bursts. For example, it is difficult to argue that an effect on the acquisition of fixed-ratio responding is a motor deficit when the effect is an increase in overall response rates.

Sensory Deficits

Deficits in visual, auditory, and tactile function have been reported in animal models (Newland and Paletz 2000; Rice 1996) and in human exposures (Amin-Zaki and Majeed 1981; Bakir et al. 1980; National Research Council 2000; Watanabe and Satoh 1996). As with motor deficits, MeHg affects sensory function, and some of these effects could reflect MeHg's disruption of cortical development (O'Kusky 1985). A large portion of the cortex is devoted to the processing and integration of sensory information, and these cortical regions accumulate MeHg during developmental exposure. Cortical deficits would be reflected in higher-order sensory functions such as integration of the visual stimuli. These include clumsiness and deficient tactile sensitivity (Rice and Gilbert 1995), impaired contrast-sensitivity function (Rice and Gilbert 1990), and high-frequency hearing (Rice and Gilbert 1992). While these effects could reflect cortical function, they could, and probably do, also result from toxicity in sensory pathways afferent to the cortex.

Cortical damage could produce agnosia, and the range of possible effects is broad. This is an area that has not been pursued aggressively in the animal literature, although the study of contrast-sensitivity contours in exposed monkeys does represent an examination of complex visual function (Rice and Gilbert 1990).

Sensory effects could result in deficits on tests of human cognitive function. Language disorders in children, for example, have been linked to deficits in higher-order auditory processing (Bishop et al. 1999; Fitch et al. 1997). Dyslexia and other disorders of language development have been traced, in some cases, to the initial processing of auditory information in the primary auditory cortex. For example, difficulties in distinguishing stop consonants can delay language development because of resulting difficulties in distinguishing between, say, the phonemes /b/ and /t/, with the consequence that, say, the syllables "ba" and "ta" sound the same. It is not known at present whether sensory deficits associated with developmental exposure to MeHg have such consequences.

Discrimination and Memory Processes

Discrimination and memory processes are disrupted by exposure to numerous drugs and toxic substances (Miller and Eckerman 1986). It is natural, therefore, to expect that MeHg exposure would produce similar effects. However, several studies have failed to show effects of developmental MeHg exposure on commonly used tests of discrimination or memory.

Another discrimination-based procedure is the visual discrimination reversal, which can be arranged in a T-maze or in an operant chamber. In the traditional operant chamber, it is arranged as follows. When the key light above the right lever is illuminated, responding on that lever is reinforced and responding on the left lever is on extinction (EXT, no reinforcement for lever pressing is available). When the light above the left lever is illuminated, the opposite is true; responding on the left lever is reinforced and responding on the right lever is on EXT. A trial begins when a light is illuminated and ends following a response or some specified period of time. The active lever varies pseudorandomly from trial to trial. This preparation arranges for a perfect correlation between the location of a visual stimulus and the location of the lever paired with reinforcement.

In a recently published experiment, rats were exposed developmentally to 0.5 ppm of MeHg in drinking water and trained, as adults, on a spatial discrimination reversal task as well as several alternation tasks (Widholm et al. 2004). There was no effect on the spatial discrimination reversal procedure, but developmental MeHg exposure resulted in a small but statistically significant increase in the number of errors on both delayed and nondelayed (actually a very short delay) spatial alternation tasks. MeHg's effects were unrelated to the duration of the delay, suggesting that memory processes were not affected by MeHg. The pattern of errors indicated a tendency to perseverate on a lever and especially on the lever that did not produce a reinforcer, i.e., a "lose–stay" pattern. The authors concluded that the effects reflected a diminished sensitivity to the consequences of the animals' behavior, an interpretation consistent with the hypothesis that MeHg selectively targets the response–reinforcer relationship while sparing discrimination processes.

Neural and Behavioral Mechanisms

In the present review, an argument has been presented that MeHg affects a response–reinforcer relationship and leaves relatively intact contextual (or stimulus) control processes such as discrimination and memory processes. In this section, we summarize the approach used to define a behavioral mechanism and then link it to the neural consequences of MeHg exposure. Specifically, we argue that the behavioral consequences of developmental exposure to MeHg (adult-onset exposure is quite different) can be traced to MeHg's disruption of cortical development and, possibly, of dopamine and GABA pathways. To do this, it is necessary to show (a) that a response–reinforcer relationship can be isolated experimentally and affected specifically in behavioral procedures used to examine MeHg's neurotoxicity, (b) that the neural regions affected have effects similar to what has been seen with MeHg exposure, and (c) that developmental MeHg exposure affects development of the appropriate neurochemical pathways as well as of the cortex.

MeHg and the Response–Reinforcer Relationship

Operant behavior, which is any behavior that is sensitive to its consequences, reflects both discrimination (also called stimulus control or, in the more recent literature, contextual control) and reinforcement processes. It is impossible to eliminate completely any category (context, consequence, or response) and still have behavior. Contextual control and reinforcement processes can be isolated, however, by arranging conditions such that one set of processes does not play a differential role, thereby allowing the other's effects to be seen in greater relief (Davison and Nevin 1999). For example, in the fixed-ratio acquisition and progressive-ratio experiments, the number of responses required for a reinforcer changes frequently but the context remains the same. Therefore, contextual processes cannot exert differential control over behavior. On the other hand, if stimulus conditions change frequently but reinforcement rates remain constant, then contextual control processes are likely to play a major role relative to reinforcement processes.

Under a concurrent schedule, two levers are available simultaneously and lever pressing on each one is reinforced. However, responding on one lever is usually reinforced at a higher rate than the other. This presents a more complex case than the acquisition of FR responding. Evidence suggests that concurrent-schedule procedures nevertheless tap reinforcement rather than discrimination processes. The evidence derives from comparisons of this standard implementation with modifications that have been made to examine memory (Alsop and Davison 1991; McCarthy and Davison 1991), sensory (Davison and McCarthy 1988; White and McKenzie 1982; White and Wixted 1999), or discrimination (Alsop and Davison 1991; Nevin et al. 1993) processes. In those experiments, concurrent schedules were arranged to highlight contextual control processes and demonstrate the minimal degree to which these processes act in the standard arrangement.

To see this, compare a two-lever concurrent schedule with a two-lever discrimination procedure. For example, under the typical concurrent schedule, the animal is presented with a left and a right lever that, when pressed, produce reinforcers at rates of one and four reinforcers/min, respectively. Contextual stimuli are limited to the spatial location of the levers; after time, right becomes the context for richly reinforced lever presses, and left becomes the context for leanly reinforced presses. In a subsequent condition, the locations of the levers remain unchanged, but the left and right levers produce reinforcers at, say, four and one reinforcers/min respectively. The key here is that the context barely changes but the relationship between a response and a reinforcer does. Therefore, reinforcement processes should have relatively greater, and therefore primary, impact on behavior (Davison and Nevin 1999).

Now consider a common discrimination procedure. The animal faces the same two levers, but stimulus conditions are explicitly arranged to play a larger role in behavior. For example, the left lever may be rich when the houselight flickers, but the right lever is rich when the houselight is steady. As these two houselight conditions alternate, the established response–reinforcer relationship between lever pressing and food will depend on the houselight. Accordingly, discrimination or contextual control processes are allowed to play a larger role in response rates on the two levers as the two response–reinforcer relationships (left and right lever pressing) are brought under the control of the context. If the houselight stimulus is presented and then removed before the animal can respond, then the procedure has changed to a memory task, and time and discrimination both become factors, but still associated with contextual control (Alsop and Davison 1991; McCarthy and Davison 1991). This approach may be too complicated to study discrimination or memory processes per se but has been used successfully to examine competing control between discrimination and reinforcement processes (Alsop and Davison 1991; Davison and McCarthy 1988; Nevin et al. 1993; White and McKenzie 1982; White and Wixted 1999).

Note that in a "spatial discrimination reversal" procedure, an analysis similar to the one used with simple concurrent schedules applies, except that the difference in reinforcement rates on the two levers is more extreme: one reinforcer–response and extinction for one set of sessions and the reverse for a second set. This is a large difference between reinforcement rates that could increase the role of discrimination processes, but the dominant influence still seems to be the relationship between pressing a lever and receiving a reinforcer. After all, the largest change is in the response–reinforcer relationship (left lever produces reinforcement, then the right lever produces reinforcement); the animal always faces the same two levers.

This stands in contrast to a "visual discrimination reversal" which, although it sounds similar to a spatial discrimination reversal, may functionally be quite different. In the visual discrimination procedure, a light is presented over the lever that will produce a reinforcer. As usually implemented, this will be the left lever for half of the trials. Thus, the stimulus conditions change from trial to trial: sometimes the light is over the left lever and sometimes it is over the right. The discrimination is more complex because it enlists multiple sensory systems and ignores spatial position. The animal still faces two levers, as in the spatial discrimination reversal, but left lever presses are reinforced only if the light is on over the left lever (and lights can be aversive to nocturnal rats), and they are not reinforced if the light is over the right lever. We have conducted visual and spatial discrimination reversals with Long–Evans rats, and they are quite different procedures. The spatial discrimination can be acquired in 3–10 sessions, depending on the rat. The visual discrimination can take as few as 10 and as many as 90 sessions to acquire. This could represent a case in which it might be assumed that behavior is under the control of something simple, like which lever the light is over, but instead the controlling context is more complex and could be the entire configuration of stimuli (Carter and Werner 1978).

The idea that discrimination and reinforcer processes are distinct, independent, and experimentally separable has been examined rigorously and quantitatively in empirical and theoretical treatments (Alsop and Davison 1991; Davison and Nevin 1999; McCarthy and Davison 1980). Thus, while it is impossible to remove the role of context or the role of consequences from studies of operant behavior, it is possible to arrange conditions so that the role of one process is minimal across all opportunities to behave while the role of the other is exaggerated. This minimizes or eliminates its role in differentiating behavior and permits the other set of processes to play a large role.

Neural Effects of MeHg Exposure

The profile of MeHg's developmental behavioral toxicity profile is consistent with the sensitivity of laminar neural structures, such as the cortex, to developmental MeHg exposure. Some of the earliest animal models of MeHg neurotoxicity as well as postmortem examination of exposed humans implicate disruptions in cell migration and associated derangement of the layers characteristic of the neocortex or cerebellum (Choi et al. 1978; Sager et al. 1982). In other experiments with nonhuman primates, MeHg induces the presence of reactive glia in the visual cortex (Charleston et al. 1994, 1995).

The sensitivity of the neocortex has been noted also with developmental exposure to low-to-moderate levels of MeHg (Barone et al. 1998). In that study, the effects were especially pronounced in the occipital neocortex but were also noted in basal forebrain nuclei, hippocampus, and brain stem nuclei. In the cortex, MeHg exposure resulted in distortions of the morphology and size of cortical lamina of 10- and 21-day-old rats. This exposure interfered with the nerve growth factor (NGF) signal cascade (i.e., NGF transduction), resulting in a dramatic loss of neurites on neurons as well as decreased cell density and cell size in the occipital neocortex of rats (Parran et al. 2001, 2003, 2004). In in vitro models (PC12 and cortical cells), MeHg's effects occurred after differentiation had taken place and were specific to the important developmental processes of dendritic elaboration and axogenesis. These are the processes by which neurons contact and communicate with one another and by which the activity of one portion of the nervous system influences or modulates activity in another.

Neurite outgrowth was impaired at mercury concentrations as low as about 30 n*M*, or about 0.006 ppm, in the medium, concentrations that were 40-fold lower than those that resulted in cytotoxicity. At first, this concentration sounds extraordinarily low, close to background levels, but the concentration in tissue is substantially higher than that in medium (Meacham et al. 2005). In the studies showing impaired neurite formation at 0.006 ppm, the concentration of mercury inside cortical cells was on the order of 0.3 to about 3 ppm. This compares with brain concentration of 0.5–9.5 ppm in neonates using the exposure protocol reported by Newland and colleagues, a protocol that altered choice in transition, high-rate behavior, and sensitivity to amphetamine (Newland and Reile 1999; Newland et al. 2004).

Neurochemically, methylmercury exposure targets catecholamines, including dopamine, as well as GABA systems. Both dopamine and GABA neurotransmitter systems are closely linked to reinforcement processes. In adult rats, the presence of MeHg in striatum produces dose-dependent increases in dopamine in this region, one of the regions receiving dopamine projections that originate in the midbrain (Faro et al. 1998, 2002, 1997). In in vitro striatal punches, the presence of MeHg increases the concentration of dopamine in extracellular media and decreases its concentration in the neuron, an effect consistent with enhanced release of dopamine (Bemis and Seegal 1999). Postnatal developmental exposure produced dose-dependent increases in the turnover of catecholamines (dopamine and norepinephrine) at 20–40 days of age. With developmental exposure, GABA systems in the cortex are also especially sensitive (O'Kusky and McGeer 1989), although norepinephrine systems seem to be more sensitive in the caudate–putamen (O'Kusky et al. 1988). The aforementioned neurochemical effects were seen in the tissue of adults after developmental MeHg exposure or from tissue exposed in vitro, so they are relevant to developmental exposure.

The behavioral significance of these effects has been demonstrated using drug challenges in behaving animals. Developmental exposure to methylmercury (relatively high levels compared with contemporary exposure protocols) affects sensitivity to amphetamine, a dopamine and noradrenergic agonist in relatively young animals (Buelke-Sam et al. 1985; Cagiano et al. 1990; Eccles and Annau 1982; Hughes and Sparber 1978; Rossi et al. 1997). Moreover, these effects are specific to certain drug classes, extend into fully adult animals, and arise with dosing protocols involving relatively low exposure levels, suggesting that they represent permanent changes in the neurochemical tone of the intact nervous system under physiologically relevant conditions (Rasmussen and Newland 2001). Adult rats exposed during gestation to MeHg (0.5 and 6 ppm, resulting in 40 and 500 μg/kg/day) were especially sensitive to amphetamine when tested as adults (Rasmussen and Newland 2001). They were also relatively insensitive to pentobarbital, a GABA agonist. No specific sensitivity was demonstrated to a cholinergic or glutamatergic antagonist. An unpublished study in our lab shows no sensitivity to cholinergic or serotonergic agonists.

Conclusion

Methylmercury is a neurotoxicant that is well known to disrupt sensory–motor function with adult-onset exposure and, at high exposure levels, to produce cerebral palsy-like signs and mental retardation after developmental exposure. It is becoming evident that alterations in cognitive or intellectual function appear with low-level exposures. While these effects can be subtle, they can carry a significant economic cost (Trasande et al. 2005).

The pattern of behavioral effects seen in animal studies of developmental MeHg exposure suggests a disruption in the acquisition or maintenance of a response–reinforcer relationship, with minimal contribution from contextual (or stimulus) control processes such as discrimination or even memory. The effects include disruption of choice, the acquisition of choice, persistent or perseverative responding on changing fixed-ratio schedules, and effects on discrimination reversal procedures suggestive of perseveration or diminished sensitivity to the consequences of behavior. Animal studies even suggest behavioral interventions to ameliorate MeHg's effects, interventions that entail the exaggeration of the discrepancy in the relative rate of reinforcement from concurrently available response alternatives. Behavioral procedures that tap memory processes are relatively insensitive to MeHg exposure. Where memorial processes do apply, they seem to involve what has been called "working memory," which entails stimuli that the animal must remember but that can change from task to task.

Studies of cortical function indicate that these are the very behavioral processes that affect the response–reinforcer relationship, choice, and perseveration. Drugs that act on dopamine systems have behavioral effects that include alterations in choice, in delay discounting, and in perseveration. The activity of neurons in dopamine-rich midbrain areas and of neurons in selected cortical areas is sensitive not just to reinforcer deliveries but also to the relative value of different reinforcing consequences as compared with others that are available concurrently. This pattern of effects is, therefore, consistent with the neuroanatomical and neurochemical effects of developmental MeHg exposure.

Finally, in postmortem examination of methylmercury-exposed people as well as in experimental methylmercury exposures in laboratory rodents, the size and interconnectivity of the neocortex appear to be especially sensitive to developmental exposure. Psychopharmacological challenges have implicated both dopamine and GABAergic systems as being permanently altered by developmental methylmercury exposure. Taken together, these raise the possibility that developmental methylmercury exposure is a model of abnormal cortical development with effects that implicate the sensitivity of behavior to reinforcing consequences.

References

Alsop B, Davison M (1991) Effects of varying stimulus disparity and the reinforcer ratio in concurrent schedule and signal-detection procedures. J Exp Anal Behav 5(6):67

Amin-Zaki L, Majeed MA (1981) Methylmercury poisoning in the Iraqi suckling infant: a longitudinal study over five years. J Appl Toxicol 1:210–214

Bakir F, Rustam H, Tikriti S, Al-Damluji SF, Shihristani H (1980) Clinical and epidemiological aspects of methylmercury poisoning. Postgrad Med J 5(6):1

Barone S Jr, Haykal-Coates N, Parran DK, Tilson HA (1998) Gestational exposure to methylmercury alters the developmental pattern of trk-like immunoreactivity in the rat brain and results in cortical dysmorphology. Brain Res Dev Brain Res 10(9):13

Bemis JC, Seegal RF (1999) Polychlorinated biphenyls and methylmercury act synergistically to reduce rat brain dopamine content *in vitro*. Environ Health Perspect 10(7):879

Beuter A, De Geoffroy A, Edwards R (1999) Analysis of rapid alternating movements in Cree subjects exposed to methylmercury and in subjects with neurological deficits. Environ Res 80:64

Bishop DV, Bishop SJ, Bright P, James C, Delaney T, Tallal P (1999) Different origin of auditory and phonological processing problems in children with language impairment: evidence from a twin study. J Speech Lang Hear Res 42:155

Buelke-Sam J, Kimmel CA, Adams J, Nelson CJ, Vorhees CV, Wright DC, St Omer V, Korol BA, Butcher RE, Geyer MA (1985) Collaborative behavioral teratology study: results. Neurobehav Toxicol Teratol 7:591–624

Cagiano R, De Salvia MA, Persichella M, Renna G, Tattoli M, Cuomo V (1990) Behavioral changes in the offspring of rats exposed to diazepam during gestation. Eur J Pharmacol 177:67–74

Carter DE, Werner TJ (1978) Complex learning and information processing by pigeons: a critical analysis. J Exp Anal Behav 2(9):565

Catania AC (1991) Glossary. In: Iversen IH, Lattal KA (eds) Experimental analysis of behavior: part 2. Elsevier, Amsterdam, pp G1–G44

Charleston JS, Bolender RP, Mottet NK, Body RL, Vahter ME, Burbacher TM (1994) Increases in the number of reactive glia in the visual cortex of *Macaca fascicularis* following subclinical long-term methyl mercury exposure. Toxicol Appl Pharmacol 12(9):196

Charleston JS, Body RL, Mottet NK, Vahter ME, Burbacher TM (1995) Autometallographic determination of inorganic mercury distribution in the cortex of the calcarine sulcus of the monkey *Macaca fascicularis* following long-term subclinical exposure to methylmercury and mercuric chloride. Toxicol Appl Pharmacol 13(2):325

Choi BH, Lapham LW, Amin-Zaki L, Saleem T (1978) Abnormal neuronal migration, deranged cerebral cortical organization, and diffuse white matter astrocytosis of human fetal brain: a major effect of methylmercury poisoning *in utero*. J Neuropathol Exp Neurol 3(7):719

Cory-Slechta DA (1990) Behavioral measures of neurotoxicity. National Academy Press, Washington, DC. Bridging experimental animal and human behavioral toxicity studies. p 137–158

Cory-Slechta DA (1994) Implications of changes in schedule-controlled operant behavior. In: Weiss B, O'Donoghue J (eds) Neurobehavioral toxicity: analysis and interpretation. Raven Press, New York, pp 195–214

Davison M, McCarthy D (1988) Matching models of signal detection: the matching law. Erlbaum, Hillsdale, pp 216–248

Davison M, Nevin JA (1999) Stimuli, reinforcers, and behavior: an integration. J Exp Anal Behav 7(1):439

Day J, Reed M, Newland MC (2005) Neuromotor deficits in aging rats exposed to methylmercury and n-3 fatty acids. Neurotoxicol Teratol 2(7):629

Eccles CU, Annau Z (1982) Prenatal methyl mercury exposure, II: alterations in learning and psychotropic drug sensitivity in adult offspring. Neurobehav Toxicol Teratol 4:377–382

Evans HL, Garman RH, Laties VG (1982) Neurotoxicity of methylmercury in the pigeon. Neurotoxicology 3:21–36

Faro LR, Duran R, Do Nascimento JL, Alfonso M, Picanco-Diniz CW (1997) Effects of methyl mercury on the *in vivo* release of dopamine and its acidic metabolites dopac and hva from striatum of rats. Ecotoxicol Environ Saf 3(8):95

Faro LR, Do Nascimento JL, Alfonso M, Duran R (1998) Acute administration of methylmercury changes *in vivo* dopamine release from rat striatum. Bull Environ Contam Toxicol 6(0):632

Faro LR, Do Nascimento JL, Alfonso M, Duran R (2002) Mechanism of action of methylmercury on *in vivo* striatal dopamine release: possible involvement of dopamine transporter. Neurochem Int 4:455

Fitch RH, Miller S, Tallal P (1997) Neurobiology of speech perception. Annu Rev Neurosci 20:331

Griffiths RR, Brady JV, Bradford LD (1977) Advances in behavioral pharmacology. In: Predicting the abuse liability of drugs with animal drug self-administration procedures: psychomotor stimulants and hallucinogens. Academic Press, New York, pp 164–208

Hughes JA, Sparber SB (1978) D-amphetamine unmasks postnatal consequences of exposure to methylmercury *in utero*: methods for studying behavioral teratogenesis. Pharmacol Biochem Behav 8:365–375

Kulig BM, Jaspers RMA (1999) Assessment techniques for detecting neurobehavioral toxicity. In: Niesink RJM, Jaspers RMA, Kornet LMW, van Ree JM, Tilson HA (eds) Introduction to neurobehavioral toxicology. CRC Press, Boca Raton, pp 70–113

Marr MJ (1979) Second-order schedules and the generation of unitary response sequences. Advances in the analysis of behavior. In: Zeiler MD, Harzem P (eds) Reinforcement and the organization of behavior, vol 1. Wiley, New York, pp 223–260

Maurissen JPJ (1995) Neurotoxicology: approaches and methods. Academic Press, San Diego. Neurobehavioral methods for the evaluation of sensory functions, pp 239–264

McCarthy D, Davison MC (1980) Independence of sensitivity to relative reinforcement rate and discriminability in signal detection. J Exp Anal Behav 3(4):273

McCarthy DC, Davison MD (1991) The interaction between stimulus and reinforcer control on remembering. J Exp Anal Behav 8(6):51

Meacham CA, Freudenrich TM, Anderson WL, Sui L, Lyons-Darden T, Barone S Jr, Gilbert ME, Mundy WR, Shafer TJ (2005) Accumulation of methylmercury or polychlorinated biphenyls in *in vitro* models of rat neuronal tissue. Toxicol Appl Pharmacol 20(5):177

Miller DB, Eckerman DA (1986) Learning and memory measures. In: Annau Z (ed) Neurobehavioral toxicology. Johns Hopkins University Press, Baltimore, pp 94–149

National Research Council (2000) Toxicological effects of methylmercury. National Academy Press, Washington, DC, Committee on the Toxicological Effects of Methylmercury

Nevin JA, Cate H, Alsop B (1993) Effects of differences between stimuli, responses, and reinforcer rates on conditional discrimination performance. J Exp Anal Behav 5(9):147

Newland MC (1994) Operant behavior and the measurement of motor dysfunction. In: Weiss B, O'Donoghue JL (eds) Neurobehavioral toxicity: analysis and interpretation. Raven Press, New York, pp 273–297

Newland MC (1995) Motor function and the physical properties of the operant: applications to screening and advanced techniques. In: Chang LW, Slikker W (eds) Neurotoxicology: approaches and methods. Academic, San Diego, pp 265–299

Newland MC, Paletz EM (2000) Animal studies of methylmercury and PCBs: what do they tell us about expected effects in humans. Neurotoxicology 21:1003–1027

Newland MC, Reile PA (1999) Blood and brain mercury levels after chronic gestational exposure to methylmercury in rats. Toxicol Sci 5(0):106

Newland MC, Reile PA, Langston JL (2004) Gestational exposure to methylmercury retards choice in transition in aging rats. Neurotoxicol Teratol 2(6):179

O'Kusky J (1985) Synaptic degeneration in rat visual cortex after neonatal administration of methylmercury. Exp Neurol 8(9):32

O'Kusky JR, McGeer EG (1989) Methylmercury-induced movements and postural disorders in developing rat: high-affinity uptake of choline, glutamate, and gamma-aminobutyric acid in the cerebral cortex and caudate-putamen. Neurochemistry 5(3):999

O'Kusky JR, Boyes BE, McGeer EG (1988) Methylmercury-induced movement and postural disorders in developing rat: regional analysis of brain catecholamines and indoleamines. Brain Res 43(9):138

Parran DK, Mundy WR, Barone S Jr (2001) Effects of methylmercury and mercuric chloride on differentiation and cell viability in pc12 cells. Toxicol Sci 5(9):278

Parran DK, Barone S Jr, Mundy WR (2003) Methylmercury decreases ngf-induced trka autophosphorylation and neurite outgrowth in pc12 cells. Brain Res Dev Brain Res 14(1):71

Parran DK, Barone S Jr, Mundy WR (2004) Methylmercury inhibits trka signaling through the erk1/2 cascade after ngf stimulation of pc12 cells. Brain Res Dev Brain Res 14(9):53

Paule MG, Forrester TM, Maher MA, Cranmer JM, Allen RR (1990) Monkey versus human performance in the nctr operant test battery. Neurotoxicol Teratol 1:503–507

Paule MG, Chelonis JJ, Buffalo EA, Blake DJ, Casey PH (1999) Operant test battery performance in children: correlation with IQ. Neurotoxicol Teratol 2:223–230

Rasmussen EB, Newland MC (2001) Developmental exposure to methylmercury alters behavioral sensitivity to d-amphetamine and pentobarbital in adult rats. Neurotoxicol Teratol 23:45–55

Rice DC (1988) Quantification of operant behavior. Toxicol Lett 4(3):361

Rice DC (1989) Delayed neurotoxicity in monkeys exposed developmentally to methyl-mercury. Neurotoxicology 1(0):645

Rice DC (1994) Neurobehavioral toxicity: analysis and interpretation. Raven Press, New York. Testing effects of toxicants on sensory system function by operant methodology,. pp 299–318

Rice DC (1996) Sensory and cognitive effects of developmental methylmercury exposure in monkeys, and a comparison to effects in rodents. Neurotoxicology 1(7):139

Rice DC, Gilbert SG (1990) Effects of developmental exposure to methyl mercury on spatial and temporal visual function in monkeys. Toxicol Appl Pharmacol 10(2):151

Rice DC, Gilbert SG (1992) Exposure to methyl mercury from birth to adulthood impairs high-frequency hearing in monkeys. Toxicol Appl Pharmacol 11(5):6

Rice DC, Gilbert SG (1995) Effects of developmental methylmercury exposure or lifetime lead exposure on vibration sensitivity function in monkeys. Toxicol Appl Pharmacol 13(4):161

Rossi AD, Ahlbom E, Ogren SO, Nicotera P (1997) Prenatal exposure to methylmercury alters locomotor activity of male but not female rats. Exp Brain Res 117:428–436

Sager PR, Doherty RA, Rodier PM (1982) Effects of methylmercury on developing mouse cerebellar cortex. Exp Neurol 7(7):179

Sakamoto M, Wakabayashi K, Kakita A, Hitoshi T, Adachi T, Nakano A (1998) Widespread neuronal degeneration in rats following oral administration of methylmercury during the postnatal developing phase: a model of fetal-type Minamata disease. Brain Res 78(4):351

Schick K (1971) Operants. J Exp Anal Behav 1(5):413

Trasande L, Landrigan PJ, Schecter C (2005) Public health and economic consequences of methylmercury

toxicity to the developing brain. Environ Health Perspect 11(3):590
Watanabe C, Satoh H (1996) Evolution of our understanding of methylmercury as a health threat. Environ Health Perspect 104(Suppl 2):367–379
Weiss B (1981) Microproperties of operant behavior as aspects of toxicity: quantification of steady-state operant behavior. Elsevier, Amsterdam, pp 249–265
Weiss B, Cory-Slechta DA (1994) Principles and methods of toxicology. Raven Press, New York. Assessment of behavioral toxicity, pp 1091–1155
White KG, McKenzie J (1982) Delayed stimulus control: recall for single and relational stimuli. J Exp Anal Behav 3(8):305
White KG, Wixted JT (1999) Psychophysics of remembering. J Exp Anal Behav 7(1):91
Widholm JJ, Villareal S, Seegal RF, Schantz SL (2004) Spatial alternation deficits following developmental exposure to Aroclor 1254 and/or methylmercury in rats. Toxicol Sci 82:577
Wood RW (1981) Determinants of irritant termination behavior. Toxicol Appl Pharmacol 6(1):260

Part II

Review of Literature

7 Exposure, Sources, and Intoxication

The leperous distilment; whose effect
Holds such an enmity with blood of man
That swift as quicksilver it courses through
The natural gates and alleys of the body,
And with a sudden vigor it doth posset
And curd, like eager droppings into milk,
The thin and wholesome blood: so did it mine;

(W. Shakespeare: Hamlet, Prince of Denmark. Act I, Scene 5 (1600))

The use of mercury in manufacturing and medical purposes has been recorded since classical times in China, Egypt, Greece, and Rome. Concomitantly, poisoning by this metal has also been reported since 2,000 years ago, such as in Pliny the Elder's (23–79 AD) *Naturae Historiarum Libri*, which refers to cinnabar (HgS) poisoning among miners at Almaden, Spain (Rackham 1952). Mercury as a poison has been documented for many centuries.

Historically, mercury poisoning has been mainly occupational and iatrogenic. In the eighteenth century, Ramazinni described the occupational diseases developed by workers exposed to mercury (Goldwater 1936). Elemental and inorganic mercury both continue to be widely used in industrial applications. In the sixteenth century, calomel (Hg_2Cl_2, mercurous chloride) was introduced as a treatment for syphilis (De Laguna 1955; Sigerest 1996). Medical administration of mercury was largely practiced until the twentieth century. It was present in cathartic, antisyphilitic, antihelminthic, diuretic, and many other preparations. It is still used in Chinese herbal medicines, in the form of calomel or even cinnabar, according to the traditional pharmacopoeia (Ernst and Coon 2001). Some of these preparations which have become popular in western countries exceed the maximum concentrations permitted by regulatory bodies (World Health Organization 1991). Dental mercury amalgam, which releases low amounts of mercury (elemental mercury vapor and inorganic ions), was first recorded in China in 600 AD. The safety of mercury amalgam has long been a source of controversy (Dodes 2001; Clarkson 2002).

Mercury is a ubiquitous contaminant, and a range of chemical species is generated by human activity and natural environmental change. Elemental mercury and its inorganic and organic compounds have different toxic properties, but all of them are considered hazardous in human exposure. In an equimolecular exposure basis, organomercurials with a short aliphatic chain are the most harmful compounds and they may cause irreversible damage to the nervous system. Methylmercury (CH_3Hg+) is the most studied following the neurotoxic outbreaks identified as Minamata disease and the Iraq poisoning.

S. Nabi, *Toxic Effects of Mercury*, DOI 10.1007/978-81-322-1922-4_7,
© Springer India 2014

The first description of CNS pathology dates from 1954. Since then, the clinical neurology, the neuropathology, and the mechanisms of neurotoxicity of organomercurials have been widely studied. The high thiol reactivity of CH_3Hg^+, as well as all mercury compounds, has been suggested to be the basis of their harmful biological effects.

> Thiol poisons, especially mercury and its compounds, reacting with SH groups of proteins lead to the lowered activity of various enzymes containing sulfhydryl groups. This produces a series of disruptions in the functional activity of many organs and tissues of the organism. (Trakhtenberg 1964)

Methylmercury Exposure

Methylmercury is a commonly encountered form of environmental mercury due both to its widespread use and to biomethylation by aquatic organisms. Exposure to methylmercury in the food chain has led to catastrophic episodes of intoxication (Takeuchi et al. 1962; Bakir et al. 1973), and exposure to inorganic and organomercurials still poses a significant toxicological problem (Adams et al. 1983; Hansen 1990). Poisoning has occurred after both acute and chronic exposure to MeHg. Chronic poisoning with MeHg typically results in ataxia, disturbances of sensory and visual function, and extremity weakness (Chang 1980). Mercury leads the pack in the potency of its toxicity and in the pervasiveness of its presence in the environment, medicine, and dentistry. Doctors who administer mercury-laden vaccines and dentists who plant highly toxic mercury in people's mouths in the form of dental amalgam cannot seem to see the forest from the trees and curb their use of it. It is reasonable to assume a direct correlation between rising environmental mercury levels, mercury exposure through dental amalgam, heavy fish consumption, and exposure to mercury in vaccines with the rapidly expanding diabetic pandemic, not to mention the host of drugs and even chemicals put into foods that are part of the diabetic equation.

> Because glycemic regulation is one of the body's most central homeostatic mechanisms, mercury's attack is most problematic, even at low concentrations, and indicates that it is playing a great role in the dramatic rise in diabetes. (IMVA 2006)

In contrast to the historical exposure to mercury, in the mid-twentieth century, there appeared a new and unexpected form of mercury poisoning resulting from the environmental exposure to short-chain alkyl mercury compounds. Although organomercurials had been known since the nineteenth century, important poisoning outbreaks occurred in the twentieth century. These organic forms of mercury were widely used as anti-fungicides for seed and cereal crop preservation and affected the general population mainly through contaminated food. Furthermore, organomercury compounds synthesized for various purposes have also exposed the population to these new agents (World Health Organization 1989, 1990, 1991).

Sources of Exposure

According to the Agency for Toxic Substances and Disease Registry (ATSDR) of the U.S. Department of Health Services, mercury is listed as the third most frequently found (lead and arsenic are first and second) and the most toxic substance in the United States (ATSDR 2001). This figure originates from the US Government's Priority List of Hazardous Substances. This list includes, in order of priority, substances that have been at hazardous waste sites on the National Priorities List (Superfund sites) that "pose the most significant potential threat to human health due to their known or suspected toxicity and the frequency of exposure." Of 1,467 hazardous waste sites listed on the National Priorities List in 1998, toxic levels of mercury were identified in 714. Mercury toxicity is also considered the second most common cause of acute heavy metal poisoning, with 3,596 cases reported in 1997 by the American Association of Poison Control Centers (Ozuah 2000).

Annual worldwide emissions of mercury into the atmosphere have been estimated at 2,200 metric tons (Ferrara et al. 2000). One-third of these emissions are estimated to originate from natural sources (volcanic eruptions and decay of mercury-containing sediments) and two-thirds from man-made sources. Twenty-five percent of worldwide emissions come from fossil fuel combustion. In the United States, 26 % (64.7 tons/year) of atmospheric mercury emissions come from medical waste incineration, such as cremation (ATSDR 1999).

There are currently 1,782 advisories (one per body of water) issued by the US Environmental Protection Agency (EPA) in 41 states in the United States restricting the consumption of any locally caught fish or shellfish due to their mercury content. Sixteen states have issued statewide or statewide–coastal advisories recommending restricting the consumption of fish caught in the state or along the coastline due to methylmercury contamination (ATSDR 1999). The Environmental Working Group, in a presentation to the Food Advisory Committee of the U.S. Food and Drug Administration (FDA), recently presented data warning of the consequences for fetuses of women who follow the current FDA's fish consumption advisory and eat 12 oz of "safe" fish per week. The Environmental Working Group estimates that more than 25 % of children in utero in the United States would be exposed to levels of mercury above the EPA safe reference dose (0.1 μg methylmercury/kg body weight/day) for at least 30 days during gestation and would have an increased risk for neurological damage (EWG 2001).

The ATSDR considers anyone who lives in close proximity to a former mercury mining site, recycling facility, municipal or medical incinerator, or coal-fired electric generating plant to be at risk for mercury toxicity. Anyone who routinely consumes contaminated fish, subsistence hunters who consume meat or organ tissues of marine mammals or feral wildlife, individuals with a "large number" of dental amalgams, pregnant or nursing women (and their developing fetuses and breast-fed babies), those who use consumer products containing mercury (skin-lightening creams or antiseptic facial products, mercury-containing diuretics or laxatives, and teething powders), or those living or working in buildings painted with mercury-containing latex paint are also considered at significant risk. Mercury-containing latex paint was removed from paint manufacturing in 1991 but may still be available in the reserve inventories of contractors and warehouses (ATSDR 1999).

> Inspite of manifold privilege of consistent fish gobbling, it oddments a grave informant of mercury exposure to man leading to cataclysm happenings of intoxication. (Rizvi et al. 2005)

Methylmercury in the Environment

Mercury is found in the environment in three basic states: elemental mercury or mercury vapor, inorganic mercury, and organic mercury (ethyl-, methyl-, alkyl-, or phenylmercury). Each form has an individual toxicological profile and metabolic fate. The most frequent sources of mercury exposure are open to debate. On an individual exposure basis, the estimated intake and retention of elemental mercury vapor (from dental amalgams and atmospheric pollution) in non-occupationally exposed individuals has a much broader range (3.9–21.0 μg/day) than either inorganic (4.3 μg/day) or methylmercury (1–6 μg/day) exposure (National Research Council 2000).

Mercury in Sediments

When mercury is first deposited in sediment, it is rapidly and strongly complexed to various components of the sediment. Mercury is most strongly bound to sulfur-containing organic and inorganic particles. In surface sediments, up to 62 % of the mercury present is bound to these types of particles (Walters and Wolery 1974). To a lesser extent, mercury is also bound strongly to clays, mineral sediments containing iron and manganese oxides, and fine sands (Reimers and Krenkel 1974). Only a small portion of mercury in sediments is released into the pore water. In this interstitial water, mercury appears to be associated primarily with organic acids such as fulvates and humates with little or none of the mercury in the unbound form (Fitzgerald and Lyons 1973). Of mercury present in deeper sediments, 65–75 % is also bound to organic acids (Walters and Wolery 1974). With or without agitation, the rate of release of mercury from sediments is hardly measurable (Reimers and Krenkel 1974).

Methylation

Because of the greater toxicity of methylmercury as compared to non-alkylmercury compounds, great attention has been directed toward the formation and passage of methylmercury in aquatic sediments. Organisms present in many types of sediments are able to methylate inorganic mercury under ideal laboratory conditions (Jernelov 1969; Gillespie 1972). Methylating organisms that have been isolated grow only under very strict conditions: They are microaerophilic, being killed if the sediment is agitated; they grow only in a narrow pH range; and, even under ideal conditions, they are slow growers (Spangler et al. 1972). Methylation appears to occur only in the top 1–2 cm of sediment. Burrowing sediment organisms, however, can expose mercury present at deeper layers to the methylating process (Jernelov 1970). When the pH of sediment is raised, mercury is bound less tightly to organic acids and sulfide complexes and is more readily available for methylation (Matsumura et al. 1972). When mercury is bound to sulfides, there is little demonstrable methylation under anaerobic conditions. Even under aerobic conditions the rate of methylation is only about 0.001 % that for mercuric chloride under the same conditions (Fagerstrom and Jernelov 1971). Methylation even under ideal conditions can at best convert less than 1.5 % of the inorganic mercury present per month (Jensen and Jernelov 1969; Jacobs and Keeney 1974).

Demethylation

Little or no methylmercury, however, is found in sediments (Andren and Harriss 1973). This might be explained by methylmercury's lesser tendency to be absorbed by sediment constituents and greater tendency to be desorbed than inorganic mercury. Methylmercury tends to be bound only to sulfur-containing sediment particles, and, even in sulfur-containing sediments, the rate of absorption is one-half to one-third that for inorganic mercury salts. The rate of desorption of methylmercury from any type of sediment is from 10 to 1,000 times that for inorganic mercury (Reimers and Krenkel 1974).

Another possible explanation for not finding appreciable amounts of methylmercury in sediments is that there is a greater tendency for sediments to demethylate than to methylate mercury compounds. As much as 15 % of bacterial isolates from mercury-containing sediments have been found to demethylate mercury (Spangler et al. 1973). These organisms are hardly being able to demethylate both aerobically and anaerobically. The demethylation process is rapid, with 100 % of any methylmercury added to the cultures being demethylated within 4 days and, in some cases, within 1 day (Spangler et al. 1972, 1973). A majority of the organisms isolated that demethylate mercury have been identified as belonging to the *Pseudomonas species* (Spangler et al. 1972).

Methylmercury in Aquatic Food Chains

Mercury is avidly taken up by lower biologic orders in aquatic ecosystems (Huckabee and Goldstein 1973; Fagerstrom and Jernelov 1974; Biesinger 1974). Greater than 75 % of methylmercury present in these lower orders is taken up directly from water. Even in higher orders, such as predatory fish, as much as 60 % of methylmercury present is taken up from water (Jernelov and Lann 1971). At each higher trophic level, the concentration of methylmercury usually increases (Matida and Kumada 1969; Jernelov and Lann 1971). In fish, this might be explained by methylmercury's prolonged half-life. Methylmercury is rapidly cleared from the aquatic environment and bound mostly to muscle tissue. When exposed to similar concentrations of methylmercury and inorganic mercury, fish are able to absorb methylmercury from water 100 times as fast as the inorganic mercury and are able to absorb five times as much methylmercury from food as compared to inorganic mercury (De Freitas et al. 1974). Once absorbed, methylmercury is retained two to five times as long as inorganic mercury. With increased fish size, both the uptake of methylmercury from the environment and the clearance of methylmercury from the fish are decreased. Because, however, methylmercury is strongly bound to muscle, methylmercury does accumulate appreciably with increased muscle mass and increased duration of exposure. With fish of the same size and with similar conditions of exposure, the rate of uptake and clearance of methylmercury is approximately the same in all species (De Freitas et al. 1974).

Direct methylation of inorganic mercury by members of higher biologic orders has been postulated to account for the higher methylmercury levels found in these orders. For instance, liver homogenates of certain species of tuna and trout have been found to methylate mercury (Imura et al. 1972). In vivo experiments, however, where fish and rats have been exposed to methylmercury, have suggested the occurrence of demethylation, with a larger fraction of the mercury in the liver and kidneys being in the form of inorganic mercury (Burrows et al. 1974; Norseth and Clarkson 1970).

Methylmercury Toxicity

The studies about methylmercury toxicity became ubiquitous and diversified since the outbreak of environmental catastrophes such as those in Minamata (1950s) and Niigata (1960s). In such episodes, as a consequence of methylmercury exposure, the exposed individuals exhibit severe forms of neurological disease which include a collection of cognitive, sensory, and motor disturbance (Eto 2000; Takeuchi et al. 1979).

The studies on methylmercury toxicity have tried to evaluate its impact on several ecosystems around the world including places in Japan, Iraq, Canada, Africa, Brazilian Amazon, and India (Malm 1998; Harada et al. 2001; Agarwal et al. 2007), as well as to understand its toxicological effect on biological systems. Methylmercury was firstly recognized as a potent neurotoxicant for the adult nervous system in studies performed on exposed workers of a chemical factory in England (Hunter et al. 1940; Hunter and Russell 1954). Later, its importance as a neurotoxicant for the nervous system during development was recognized in the Minamata's outbreak (Eto 2000; Takeuchi et al. 1979). Since then, several studies of exposed human populations as well as experiments with laboratory animals demonstrated that exposure to toxic levels of methylmercury during pre- and postnatal life causes neurological abnormalities, cognitive impairment, and behavioral disturbance (Steuerwald et al. 2000; Cordier et al. 2002). Methylmercury vulnerability of the developing brain reflects the ability of lipophilic methylmercury to cross the placenta and to concentrate in the central nervous system (CNS) once the blood–brain barrier is not fully developed in the prenatal period (Castoldi et al. 2001; Lepham et al. 1995).

Membrane Interactions and Transporter-Mediated Methylmercury Toxicity

Methylmercury quickly diffuses across membranes without significant partitioning in lipid bilayers. Thus, it has been proposed that methylmercury toxicity is mediated by methylmercury membrane leakage (Lakowicz and Anderson 1980). However, it has also been suggested that the potential of methylmercury to increase oxidative events leading to cell damage is controlled by methylmercury binding to membrane transporters. Methylmercury absorption, distribution, and excretion are commonly mediated by plasma membrane protein transporters (Sekine et al. 2000). In addition, it has been possible to investigate at molecular level the mechanisms of methylmercury transport through membrane transporters with broad substrate selectivity. These transporters are known as "multispecific." The main route for methylmercury transmembrane transport seems to be the amino acid transport system L, which transports large amino acids (Aschner and Aschner 1990).

It has been proposed that methylmercury–cysteine conjugate is the pathway whereby methylmercury exerts its toxicity (Morkzan et al. 1995). Once the presence of such transporters is crucial for toxicity to occur at least through this mechanism, transporter inhibition is expected to be beneficial to prevent disorders caused by methylmercury toxicity.

Cellular Mechanism to Methylmercury Intoxication

Calcium Homeostasis: Calcium ion (Ca^{2+}) plays a critical role in CNS cell death. Ca^{2+} increase beyond physiological levels activates catabolic enzymes such as phospholipases, proteases, and endonucleases, causes mitochondrial dysfunction, and disturbs cytoskeletal organization. Several lines of evidence indicate that at low concentrations MeHg disrupts Ca^{2+} homeostasis, increasing its intracellular level in a number of experimental situations, including primary culture of cerebellar granule cells (Limke et al. 2003). This effect has all the potential to disrupt the synaptic function and impair the neural development (Marty and Atchison 1998).

Mitochondrial Damage Induced by Methylmercury

Mercury can induce apoptosis in human T lymphocytes. The target organelle was the mitochondrion and that induction of Oxidative stress led to activation of death-signaling pathways. (Shenker et al. 1999)

Mitochondria are the main intracellular sites for reactive oxygen production and one of the most susceptible targets for radical species to exert their actions. Importance of mitochondria for methylmercury toxicity was recognized from studies performed both in vivo and in vitro. In vivo exposure to methylmercury causes its accumulation inside mitochondria followed by a series of biochemical changes in these organelles (Denny and Atchison 1994). These effects are similar to those observed in studies of mitochondrial respiratory chain inhibition (Mori et al. 2007).

Rats exposed to methylmercury in vivo display neurological symptoms after a latent period. Mitochondrial function (as measured by oxygen consumption of brain slices) is impaired during the symptomatic phase but not during the latent phase (Yoshino et al. 1966b). Although MeHg concentrations are maximal during the latent phase, the effects of MeHg on mitochondria may be indirect as they are preceded by inhibition of protein synthesis (Yoshino et al. 1966a, b).

Synaptosomes from rats treated with MeHg and from naive rats exposed to MeHg in vitro have reduced rates of respiration (Verity et al. 1975). This effect is blocked by removal of K^+, suggesting that there is an increase in the K^+ permeability of the inner mitochondrial membrane. Alterations in respiration are also observed in guinea pig brain slices at slightly higher concentrations of MeHg (Fox et al. 1975). The decrease in respiratory rates may be due to MeHg- induced inhibition of the tricarboxylic acid cycle. This is consistent with earlier work in which in vivo methylmercury exposure reported decreased succinate dehydrogenase activity (Yoshino et al. 1966b). High MeHg levels cause impairment of mitochondrial function as the organelle exhibits a membrane permeability transition state. MeHg exposure induces a decrease in the activity of enzymes of the mitochondrial energy metabolism such as cytochrome C oxidase (CCO), superoxide dismutase (SOD), and succinate dehydrogenase (SDH) Yoshino et al. 1966b. This is probably due to the decrease in the respiratory rate caused by MeHg-induced inhibition of the tricarboxylic acid cycle. This is consistent with previous work showing that MeHg exposure decreases succinate dehydrogenase activity (Naganuma et al. 1998). In vitro MeHg exposure of isolated mitochondria from rat liver inhibits electron transport and phosphorylation, increases K^+ permeability, and dissipates the mitochondrial membrane potential (MMP) Sone et al. 1977. Loss of MMP results in efflux of mitochondrial Ca^{+2} and inhibition of mitochondrial Ca^{+2} uptakes (Levesque and Atchison 1991). In addition, MeHg exposure in isolated rat brain mitochondria causes

ATP-dependent and ATP-independent decrease in Ca^{+2} uptake and increase in Ca^{+2} effluxes from mitochondria (Denny et al. 1993). Although mitochondria participate in Ca^{+2} buffering at relatively elevated Ca^{+2}, the affinity of the uniport carrier for Ca^{+2} is low, and mitochondria may play only a minor role in buffering Ca^{+2} under normal conditions (Levesque and Atchison 1991).

Microtubule Network

MeHg seems to interact with cytoplasmic cytoskeletal components, including microtubules (Sager et al. 1983). In vitro studies demonstrated that MeHg presents high affinity for tubulin sulfhydryl groups (-SH), depolymerizing cerebral microtubules and directly inhibiting their assembly (Sager et al. 1983; Vogel et al. 1989). In addition, several works reported that MeHg promotes microtubule disruption in a number of cell models, including human fibroblasts (Sager et al. 1983), neuroblastoma, and glioma cells (Prasad et al. 1979; Miura et al. 1984).

Methylmercury and Neurotransmitter System

Several metal compounds have been shown to interfere with neurotransmission. MeHg directly affects the mechanisms of neurotransmission, including release and uptake of neurotransmitters, enzymatic neurotransmitter metabolic inactivation, and postsynaptic events associated with receptor activation (Atchison 2005). Some neurotoxicants indirectly interfere with neurotransmission by interacting, for example, with energy metabolism, sodium channels, or ATPases. Furthermore, changes of any parameter of neurotransmission can be the result of neuronal death due to cytotoxic effects of the neurotoxicants (Orrenius and Nicotera 1994).

The rising of extracellular glutamate levels is responsible for the constant activation of metabotropic and ionotropic glutamate receptors, thus elevating Na^{+} influx and Ca^{2+} release from intracellular organelles that may trigger a biochemical cascade which increases the production of ROS (Orrenius and Nicotera 1994). Oxidative stress by itself inhibits the astrocytic glutamate uptake through a direct action on the transporter proteins (Park et al. 1996; Volterra et al. 1994).

Although the toxic damage caused by MeHg might be intrinsically prevalent in neurons, many of the published evidences suggest that neuronal damage in response to MeHg most likely represents aberrant control of the extracellular milieu by the astrocytes (Shanker et al. 2001). In line with this argument, it should be remarked that the neurotoxic effect of MeHg could be reverted with antagonists of N-methyl-D-aspartate (NMDA) receptor (Park et al. 1996).

Moreover, MeHg has been described to produce increases in the spontaneous release of other neurotransmitters such as dopamine, GABA, acetylcholine, and serotonin from rat brain synaptosomes (Komulainen and Tuomisto 1981; Minema et al. 1989; Juarez et al. 2002). MeHg also inhibits astrocytic uptake of cystine and cysteine, the key precursors for glutathione biosynthesis (Shanker et al. 2001).

Methylmercury and Metallothioneins

Metallothioneins (MTs) constitute a family of proteins characterized by unusual cysteine abundance (Hidalgo et al. 2001). Under physiological conditions, MTs are unusually rich in multiple cysteine residues allowing their binding to metal centers and enabling them to serve as a heavy metal detoxification system (Gonzalez-Duarte 2003). MTs are predominantly expressed in the central nervous system, and it is important to gain new insight into how MTs are regulated in the brain in pathological injury, such as that produced by MeHg intoxication.

Some studies have reported the potential role of MTs in attenuating the cytotoxicity induced by MeHg (Hidalgo et al. 2001; Aschner et al. 1997). Although the interaction of MTs with MeHg ions has long been established, elucidation of the binding features of MeHg–MT species has been hampered by the inherent difficulties of MeHg–thiolate chemistry, which mainly arise from the diverse coordination preferences of Hg (II) and the various ligation modes of the thiolate ligands (Wright et al. 1990). Nevertheless, the analysis of MeHg binding to MTs has been intensively studied. In contrast, the chemistry of MeHg–MT complexes has attracted much less attention. Earlier reports demonstrated the inability of MT in the detoxification of MeHg and that it is unable to bind to MeHg either in vivo or in vitro (Wright et al. 1990; Rising et al. 1995). Subsequent attempts to induce brain MT by exposure to MeHg+ gave inconsistent results: MT concentrations remained unchanged in rats, whereas MT and mRNA concentrations increased in MeHg-treated rat neonatal astrocyte cultures (Rising et al. 1995). However, there is increasing evidence that induction of MTs in astrocytes attenuates and even reverses the cytotoxicity caused by MeHg, indicating binding of MeHg by an astrocyte-specific MT isoform, MT1 (Yao et al. 1999).

Methylmercury in Human Body

Methylmercury is almost completely absorbed (95–100 %) in the human gastrointestinal tract (Ozuah 2000; Clarkson 2002), 90 % of which is eventually eliminated through the feces. Methylmercury is present in the body as a water-soluble complex, mainly with the sulfur atom of thiol ligands (Clarkson 2002), and crosses the blood–brain barrier complexed with L-cysteine in a molecule resembling methionine. Methylmercury is absorbed into the placenta and stored in the fetal brain in concentrations that exceed maternal blood levels (Cernichiari et al. 1995). After being released from cells in a complex with reduced glutathione, methylmercury is degraded in the bile duct to an L-cysteine complex. Only 10 % of methylmercury is eliminated through the kidneys. The rest either undergoes enterohepatic recycling or demethylation by microflora in the intestine and immune system and eventual elimination through the feces.

Most methylmercury in animal exposure studies is degraded to, and eliminated as, inorganic mercury at the rate of 1 % per day (Clarkson 2002). At least one study has demonstrated the capacity of two common forms of gastrointestinal yeast to convert inorganic mercury to methylmercury (Yannai et al. 1991). Demethylation by intestinal microflora is a crucial step in the elimination of methylmercury from the body, but research has not yet identified the mechanisms or the microbes responsible for this detoxification system (Clarkson 2002). Enterohepatic reabsorption is also a significant event in the metabolism of methylmercury; more than 70 % is reabsorbed from the gut and returned to the liver (Clarkson 2002; Alexander and Aaseth 1982).

Inorganic mercury has been found as the major form of mercury in brain tissue in humans fatally exposed to methylmercury (Davis et al. 1994). The conversion of methylmercury to inorganic mercury is thought to take place in phagocytic cells in the liver or in the astroglial cells of the brain (Clarkson 2002).

Methylmercury Toxicity in Man

Recently, more has been elucidated about the toxicity of methylmercury in man. Although chromosomal breaks have been found in onion root tips exposed to concentrations of methylmercury that cause neurotoxicity in animals (Ramel 1969), no genetic defects or excessive chromosomal abnormalities have been found in children with congenital methylmercury poisoning (RCMD 1974). With severe long-term methylmercury poisoning, brain atrophy with associated presenile dementia and atrophy of the islets of Langerhans of the pancreas with associated diabetes mellitus have resulted (RCMD 1974). Methylmercury poisoning has yet to be demonstrated in human populations not exposed directly to methylmercury or to food contaminated with methylmercury. Recent studies of populations who subsist mainly on seafood that is naturally high in methylmercury have failed to demonstrate any evidence of methylmercury poisoning, even with whole blood methylmercury levels that average three to eight times that a comparative non-fish-eating population (Turner et al. 1980; Marsh et al. 1974).

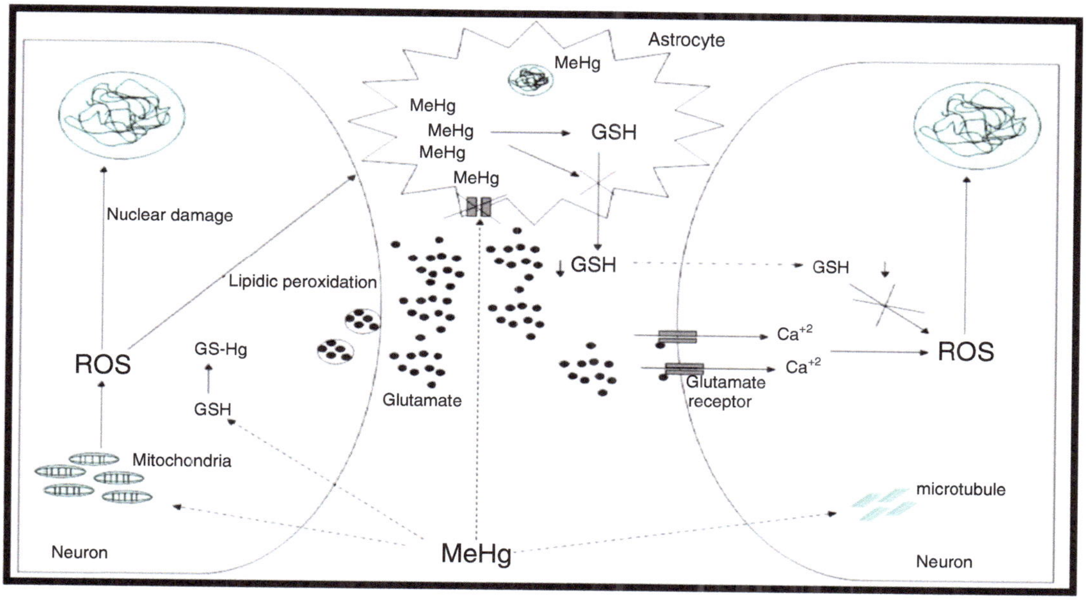

Fig. 7.1 A schematic model of some of the currently proposed mechanism for cellular damage induced by MeHg in the CNS. In the extracellular environment, MeHg inhibits glutamate uptake, as well as a number of the amino acids that are associated with the synthesis of astrocytic glutathione (*GSH*). Accumulation of glutamate in the extracellular space and the resulting excessive activation of NMDA receptors can result in excitotoxicity and, ultimately, cell death. Other proposed mechanisms are related with mitochondrial MeHg-associated dysfunction, including impaired cytoplasmic Ca2+ homeostasis and release of ROS, metabolic inhibition that leads to impaired ATP production, lipid peroxidation, and nuclear damage. MeHg also can provoke microtubule chain disruption decreasing vesicular migration or genotoxicity (do nascimento et al. 2008)

CNS Damage Induced by Methylmercury

The majority of toxicity due to methylmercury exposure involves the central nervous system. Methylmercury can cause demyelination, autonomic dysfunction, sensory nerve conduction delay, abnormal neuronal migration, and abnormal central nervous system cell division. Chronic toxicity symptoms include paresthesia, peripheral neuropathy, cerebellar ataxia, akathisia, spasticity, memory loss, dementia, constricted vision, dysarthria, impaired hearing, smell and taste, tremors, and depression (Ozuah 2000; Clarkson 2002).

Cardiovascular Damage Induced by Methylmercury

Methylmercury exposure also appears to increase risk for cardiovascular disease. In a long-term prospective study, both intake of nonfatty freshwater fish and hair mercury content demonstrated a statistically significant correlation with increased risk for acute myocardial infarction (Salonen et al. 1995). Men with the highest hair mercury had a 2.9-fold increased risk for cardiovascular death. An examination of the same cohort found a significant correlation between hair mercury and increased risk for progression of carotid atherosclerosis (Salonen et al. 2000). Prenatal exposure to methylmercury has been correlated with significant blood pressure elevations in 7-year-old children as a result of maternal fish intake (Sorensen et al. 1999) (Fig. 7.1).

Mercury Toxicity in Food Chain Organisms

In aquatic ecosystems, mercury is quite toxic to lower biological orders and to juveniles of certain species. At concentration of less than 0.1 ppb, methylmercury causes a decrease in the growth rate of phytoplankton and a decreased reproduction of daphnia (Biesinger 1974; Harada et al. 1970). At similar levels, inorganic mercury causes a decreased long-term survival of fiddler crab larvae (Matida et al. 1971). In fish, toxicity has been noted at 3 ppb for both methylmercury and mercuric chloride (Matida et al. 1971; Weir and Hine 1970). These toxic levels of mercury compare with normal methylmercury levels in surface water of less than 0.001 ppb and inorganic mercury levels of less than 0.05 ppb (Andren and Harriss 1973; Hartung 1973).

Mercury in Vaccines

Mercury is currently mixed with DTaP, HIB, and hepatitis B vaccines or is used in the manufacturing process for vaccines, with resultant trace amounts being present in the final product. Based on existing Centers for Disease Control (CDC) recommendations for vaccinations, a typical 6-month-old child, if receiving all thimerosal (49.6 % ethylmercury) containing vaccines, could potentially be injected with as much as 187.5–200 μg of methylmercury, the equivalent of more than 1.0 μg/day. This amount exceeds the reference limits for exposure to mercury set by the EPA of 0.1 μg/kg/day (Halsey 1999). In the United States, at the FDA's request, all the vaccines are currently being produced as thimerosal-free or thimerosol-reduced (>95 % reduction) vaccines. Thimerosol-preserved vaccines are still available and used in clinical practice.

> Medical and health officials seem to live in an unconscious fog when it comes to mercury even though Methyl mercury induces oxidative stress and cell cytotoxicity through mitochondrial apoptosis pathways. (IMVA 2006)

Someday it will dawn on both dentists and doctors who use mercury that they are actually poisoning children and people.

Ethylmercury (fungicides, thimerosal in vaccines, and gamma-globulin) also causes renal and central nervous system toxicity and is deposited in the liver, kidneys, skin, brain, spleen, and plasma (Clarkson 2002). Ethylmercury, like methylmercury, is metabolized to the inorganic form and accounts for 50 % of the mercury eliminated in urine. Ethylmercury may actually be converted to inorganic mercury in the tissues in greater amounts and more rapidly than methylmercury (Clarkson 2002). As with methylmercury, the feces are the main natural route of elimination.

References

Adams CR, Zeigler DK, Lin JT (1983) Mercury intoxication simulating amyotrophic lateral sclerosis. J Am Med Assoc 250:642–643

Agarwal R, Kumar R, Behari JR (2007) Mercury and lead content in fish species from the river Gomti, Lucknow, India, as biomarkers of contamination. Bull Environ Contam Toxicol 78:118–122

Alexander J, Aaseth J (1982) Organ distribution and cellular uptake of methylmercury in the rat as influenced by the intra and extracellular glutathione concentration. Biochem Pharmacol 31:685–690

Andren AW, Harriss RC (1973) Methylmercury in estuarine sediments. Nature 245:256–257

Aschner M, Aschner JL (1990) Mercury neurotoxicity: mechanisms of blood-brain barrier transport. Neurosci Biobehav Res 14:169–176

Aschner M, Cherian MG, Klaassen CD, Palmiter RD, Erickson JC, Bush AI (1997) Metallothioneins in brain: the role in physiology and pathology. Toxicol Appl Pharmacol 142:229–242

Atchison WD (2005) Is chemical neurotransmission altered specifically during methylmercury-induced cerebellar dysfunction? Trends Pharmacol Sci 26:549–557

ATSDR (1999) Toxicological profile for mercury. Agency for Toxic Substances and Disease Registry. U.S. Department of Health and Human Services. Public Health Service, Atlanta

ATSDR (2001) 2001 CERCLA priority list of hazardous substances. Agency for Toxic Substances and Disease Registry. U.S. Department of Health and Human Services. Public Health Service, Atlanta

Bakir F, Damluji SF, Amin-Zaki L, Murtadha M, Khalidi A, Al-Rawi NY, Tikriti S, Dhahir HI, Clarkson TW, Smith JC, Doherty RA (1973) Methyl mercury poisoning in Iraq. Science 181:230–241

Biesinger KE (1974) The chronic toxicity of mercury to Daphnia magna. National water quality laboratory report. U.S. Environmental Protection Agency, Duluth

Burrows WD, Taimi KI, Krenkel PA (1974) The uptake and loss of methylmercury by freshwater fish. In: Congreso Internacional Del Mercurio, Barcelona

Castoldi AF, Coccini T, Ceccatelli S, Manzo L (2001) Neurotoxicity and molecular effects of methylmercury. Brain Res Bull 55:197–203

Cernichiari E, Brewer R, Myers GJ, Marsh DO, Lapham LW, Cox C, Shamlaye CF, Berlin M, Davidson PW, Clarkson TW (1995) Monitoring methylmercury during pregnancy; maternal hair predicts fetal brain exposure. Neurotoxicology 16:705–710

Chang LW (1980) Mercury. In: Spencer PS, Schaumburg HH (eds) Experimental and clinical neurotoxicology. The Williams and Wilkins Co., New York, pp 508–526

Clarkson RW (2002) The three modern faces of mercury. Environ Health Perspect 110:11–23

Cordier S, Garel M, Mandereau L, Morcel H, Doineau P, Gosme-Seguret S, Josse D, White R, Amiel-Tison C (2002) Neurodevelopmental investigations among methylmercury-exposed children in French Guiana. Environ Res 89:1–11

Davis LE, Kornfield M, Mooney HS, Fiedler KJ, Haaland KY, Orrison WW, Cernichiari E, Clarkson TW (1994) Methylmercury poisoning: long-term clinical, radiological, toxicological and pathological studies of an affected family. Ann Neurol 35:680–688

De Freitas ASW, Qadri SU, Case BE (1974) Origins and fate of mercury compounds in fish. In: Proceedings of the international conference on transport of persistent chemicals in aquatic ecosystems, Ottawa. p III-31–III-36

De Laguna A (1955) Pedacio Dioscorides Anazarbeo, Acerca de la Materia Medicinal y de los venenos mortiferos. Libro V, Cap. LXIX: Del Azogue, Juan Latio, Anvers, Belgium. p 540–542

Denny MF, Atchison WD (1994) Elevations in the free intrasynaptosomal concentration of endogenous zinc by methylmercury. J Neurochem 14:290

Denny MF, Hare MF, Atchison WD (1993) Methylmercury alters intrasynaptosomal concentrations of endogenous polyvalent cations. Toxicol Appl Pharmacol 122:222–232

Do Nascimento JLM, Oliveira KRM, Crespo-Lopez ME, Macchi BM, Maues LAL, Pinheiro MCN, Silveira LCL, Herculano AM (2008) Methylmercury neurotoxicity and antioxidant defenses. Indian J Med Res 128:373–382

Dodes JE (2001) The amalgam controversy: an evidence-based analysis. JADA 132:348–356

Environmental Working Group (EWG) (2001) What women should know about mercury contamination of fish? Environmental Working Group, Washington, DC, pp 1–4

Ernst E, Coon CT (2001) Heavy metals in traditional Chinese medicines: a systemic review. Clin Pharmacol Ther 70:497–504

Eto K (2000) Minamata disease. Neuropathology 20:S14–S19

Fagerstrom T, Jernelov A (1971) Formation of methylmercury from pure mercuric sulfide in aerobic sediment. Water Res 5:121–122

Fagerstrom T, Jernelov A (1974) Biological methylation of mercury, food chain accumulations, ecological effects and routes of exposure to man. CRC Crit Rev Environ Control 4:296

Ferrara R, Mazzolai B, Lanzillotta E, Nucaro E, Pirrone N (2000) Temporal trends in gaseous mercury evasion from the Mediterranean seawaters. Sci Total Environ 259:183–190

Fitzgerald WF, Lyons WB (1973) Organic mercury compounds in coastal waters. Nature 242:452–453

Fox JH, Patel-Mandlik K, Cohen MM (1975) Comparative effects of organic and inorganic mercury on brain slice respiration and metabolism. J Neurochem 24:757–762

Gillespie DC (1972) Mobilization of mercury from sediments into guppies (Poecilia reticulata). J Fish Res Bd Can 29:1035–1041

Goldwater LJ (1936) From Hippocrates to Ramazzini: early history of industrial medicine. Ann Med Hist 8:27–35

Gonzalez-Duarte P (2003) Metallothioneins. In: McCleverty J, Meyer TJ (eds) Comprehensive coordination chemistry II, 8th edn. Elsevier-Pergamon, Amsterdam, pp 213–228

Halsey NA (1999) Limiting infant exposure to thimerosal in vaccines and other sources of mercury. JAMA 282:1763–1766

Hansen JC (1990) Human exposure to metals through consumption of marine foods: a case study of exceptionally high intake among Greenlanders. In: Furness RW, Rainbow PS (eds) Heavy metals in the marine environment. CRC Press, Boca Raton, pp 227–243

Harada M, Nakanishi J, Yasoda E, Pinheiro MC, Oikawa R, de Assis Guimaraes, Harriss RC, White D, MacFarlane R (1970) Mercury compounds reduce photosynthesis in plankton. Science 170:736

Harada M, Nakanishi J, Yasoda E, Pinheiro MC, Oikawa T, de Assis Guimaraes G, da Silva Cardoso B, Kizaki T, Ohno H (2001) Mercury pollution in the Tapajos River basin, Amazon: mercury level of head hair and health effects. Environ Int 27:285–290

Hartung R (1973) A survey of heavy metals in the lower Mississippi River. University of Michigan, Ann Arbor

Hidalgo J, Aschner M, Zatta P, Vasak M (2001) Roles of the metallothionein family of proteins in the central nervous system. Brain Res Bull 55:133–145

Huckabee JW, Goldstein RA (1973) Dynamic redistribution of methylmercury in a pond ecosystem. In: Proceedings of the first annual NSF trace contaminants conference. National Science Foundation

Hunter D, Russell DS (1954) Focal cerebral and cerebellar atrophy in a human subject due to organic mercury compounds. J Neurol Neurosurg Psychiatry 17:235–241

Hunter D, Bomfond RR, Russell DS (1940) Poisoning by methylmercury compounds. Q J Med 35:193–213

Imura N, Pan S, Ukita T (1972) Methylation of inorganic mercury with liver homogenate of tuna fish. Chemosphere 5:197

International Medical Veritas Association (IMVA) (2006) Diabetes and mercury poisoning. Medical news

Jacobs LW, Keeney DR (1974) Methylmercury formation in mercury treated river sediments during *in situ* equilibration. J Environ Qual 3:121–126

Jensen S, Jernelov A (1969) Biological methylation of mercury in aquatic organisms. Nature 223:753–754

Jernelov A (1969) Are there any difference between "biologically" and "chemically" synthesized methylmercury? Vatten 3:304

Jernelov A (1970) Release of methylmercury from sediments with layers containing inorganic mercury at different depths. Limnol Oceanogr 15:958–960

Jernelov A, Lann H (1971) Mercury accumulation in food chains. Oikos 22:403

Juarez BI, Martinez ML, Montante M, Dufour L, García E, Jimenez-Capdeville ME (2002) Methylmercury increases glutamate extracellular levels in frontal cortex of awake rats. Neurotoxicol Teratol 24:767–771

Komulainen H, Tuomisto J (1981) Interference of methylmercury with monoamine uptake and release in rat brain synaptosomes. Acta Pharmacol Toxicol 48:214–222

Lakowicz JR, Anderson CJ (1980) Permeability of lipid bilayers to methylmercuric chloride: quantification by fluorescence quenching of a carbazole-labeled phospholipids. Chem Biol Interact 30:309–323

Lepham LW, Cermichiari E, Cox C, Meyers GJ, Baggs RB, Brewer R, Shamlaye CF, Davidson PW, Clarkson TW (1995) An analysis of autopsy of brain tissue from infants prematurely exposed to methylmercury. Neurotoxicology 16:689–704

Levesque PC, Atchison WD (1991) Disruption of brain mitochondrial calcium sequestration by methylmercury. J Pharmacol Exp Ther 256:236–242

Limke TL, Otero-Montanez JK, Atchison WD (2003) Evidence for interactions between intracellular calcium stores during methylmercury induced intracellular calcium dysregulation in rat cerebellar granule neurons. J Pharmacol Exp Ther 304:949–958

Malm O (1998) Gold mining as a source of mercury exposure in the Brazilian Amazon. Environ Res 77:73–78

Marsh DO, Turner MD, Crispin Smith J, Wun Choi J, Clarkson TW (1974) Methylmercury (MeHg) in human population eating large quantities of marine fish. II. American Samoa: cannery workers and fishermen. In: Proceedings of the first international mercury conference, Barcelona, 2. Fabrica de Moneda y Timbre, Madrid, pp 235–239

Marty MS, Atchison WD (1998) Elevations of intracellular Ca^{2+} as a probable contributor to decreased viability in cerebellar granule cells following acute exposure to methylmercury. Toxicol Appl Pharmacol 150:98–105

Matida Y, Kumada H (1969) Distribution of mercury in water, bottom mud and aquatic organisms of Minamata Bay, the River Agano and other water bodies in Japan. Bull Freshw Fish Res Lab 19:73

Matida Y, Kumada H, Kimura S, Saiga Y, Nose T, Yokote M, Kawatsu H (1971) Toxicity of mercury compounds to aquatic organisms and accumulation of the compounds by the organisms. Bull Freshw Fish Res Lab 21:197–227

Matsumura F, Gotoh Y, Boush GM (1972) Factors influencing the translocation and transformation of mercury in river sediment. Bull Environ Contam Toxicol 8:267–272

Minema DJ, Cooper GP, Greeland RD (1989) Effects of methylmercury on neurotransmitters release from rat brain synaptosomes. Toxicol Appl Pharmacol 99:510–521

Miura K, Inokawa M, Imura N (1984) Effects of methylmercury and some metal ions on microtubule networks in mouse glioma cells and in vitro tubulin polymerization. Toxicol Appl Pharmacol 73:218–231

Mori N, Yasutake A, Hirayama K (2007) Comparative study of activities in reactive oxygen species production/defense system in mitochondria of rat brain and liver, and their susceptibility to methylmercury toxicity. Arch Toxicol 81:769–776

Morkzan EM, Kerper LE, Ballatori N, Clarkson TW (1995) Methylmercury-thiol uptake into cultured brain capillary endothelial cells on amino acid system L. J Pharmacol Exp Ther 272:1277–1284

Naganuma A, Miura K, Tanaka-Kagawa T, Kitahara J, Seko Y, Toyoda H, Imura N (1998) Overexpression of manganese-superoxide dismutase prevents methylmercury toxicity in HeLa cells. Life Sci 62:L157–L161

National Research Council (2000) Toxicological effects of methylmercury. National Academy Press, Washington, DC, Committee on the Toxicological Effects of Methylmercury

Norseth T, Clarkson TW (1970) Studies on the biotransformation of 203 Hg-labelled methylmercury chloride in rats. Arch Environ Health 21:717–727

Orrenius S, Nicotera P (1994) The calcium ion and cell death. J Neural Transm 43:1–11

Ozuah PO (2000) Mercury poisoning. Curr Probl Pediatr 30:91–99

Park ST, Lim KT, Chung YT, Kim SU (1996) Methylmercury induced neurotoxicity in cerebral neuron culture is blocked by antioxidants and NMDA receptor antagonists. Neurotoxicology 17:37–45

Prasad KN, Nobles E, Ramanujam M (1979) Differential sensitivity of glioma cells and neuroblastoma cells to methylmercury toxicity in cultures. Environ Res 19:189–201

Rackham H (1952) Pliny: natural history (English trans.), Loeb Classical ser. W. Heinemann, London

Ramel C (1969) Genetic effects of organic mercury compounds: I. cytological investigations on allium roots. Hereditas 61:208

Reimers RS, Krenkel PA (1974) Sediment sorption phenomena. CRC Crit Revs Environ Control 4:265

Research Committee on Minamata Disease (RCMD) (1974) Pathological, clinical and epidemiological research about Minamata disease; 10 years after (trans: Ishizaki R). Interuniversity Consortium for Environmental Studies, Durham

Rising L, Vitarella D, Kimelberg HK, Aschner M (1995) Metallothionein induction in neonatal rat primary astrocyte cultures protects against methylmercury cytotoxicity. J Neurochem 65:1562–1568

Rizvi SJ, Zahir F, Jairajpuri DS (2005) Mercury: the lurking danger. Anil Aggrawal's Internet J Forensic Med Toxicol 6

Sager PR, Doherty RA, Olmsted JB (1983) Interaction of methylmercury with microtubules in cultured cells and in vitro. Exp Cell Res 146:127–137

Salonen JT, Seppanen K, Nyyssonen K, Korpela H, Kauhanen J, Kantola M, Tuomilehto J, Esterbauer H, Tatzber F, Salonen R (1995) Intake of mercury from fish, lipid peroxidation, and the risk of myocardial infarction and coronary, cardiovascular, and any death in eastern fisher men. Circulation 91:645–655

Salonen JT, Seppanen K, Lakka TA, Salonen R, Kaplan GA (2000) Mercury accumulation and accelerated progression of carotid atherosclerosis: a population-based prospective 4 year follows up study in men in eastern Finland. Atherosclerosis 148:265–273

Sekine T, Cha SH, Endou H (2000) The multispecific organic anion transporter (OAT) family. Pflugers Arch 440:337–350

Shanker G, Allen JW, Mutkus LA (2001) Methylmercury inhibits cysteine uptake in cultured primary astrocytes, but not in neurons. Brain Res 914:159–165

Shenker BJ, Guo TL, Shapiro IM (1999) Induction of apoptosis in human T-cells by methyl mercury: temporal relationship between mitochondrial dysfunction and loss of reductive reserve. Toxicol Appl Pharmacol 157:23–35

Sigerest HE (1996) Four treatises of Theophrastus von Hohenheim called Paracelsus. The Johns Hopkins University Press, Baltimore

Sone N, Margareta K, Larsstuvold, Kagawa Y (1977) Effect of methyl mercury on phosphorylation, transport, and oxidation in mammalian mitochondria. J Biochem 82:859–868

Sorensen N, Murata K, Budtz-Jorgensen E, Weihe P, Grandjean P (1999) Prenatal methylmercury exposure as a cardiovascular risk factor at seven years of age. Epidemiology 10:370–375

Spangler WJ, Sigarelli JL, Miller HM (1972) Studies on the biomethylation of mercury. Midwest Research Institute, Kansus City

Spangler WJ, Spigarelli JL, Rose JM, Flippin RS, Miller HH (1973) Degradation of methylmercury by bacteria isolated from environmental samples. Appl Microbiol 25:488–493

Steuerwald U, Weihe P, Jorgensen PJ, Bjerve K, Brock J, Heinzow B, Budtz-Jorgensen E, Grandjean P (2000) Maternal seafood diet, methyl-mercury exposure and neonatal neurological function. J Pediatr 136:599–605

Takeuchi T, Norikawa N, Matsumoto H, Shiraishi Y (1962) Pathological study of Minamata disease in Japan. Acta Neuropathol 2:40–57

Takeuchi T, Eto N, Eto K (1979) Neuropathology of childhood cases of methylmercury poisoning (Minamata disease) with prolonged symptoms, with particular reference to the decortication syndrome. Neurotoxicology 1:1–20

Trakhtenberg IM (1964) Chronic effects of mercury on organisms: the micromercurialism phenomenon on mercury handlers. Ch VI: 109-34, DHEW Publ. No. (NIH) 74-473, and mercury: cardiovascular adverse effects

Turner MD, Marsh DO, Smith JC, Inglis JB, Clarkson TW, Rubio CE, Chiriboga J, Chiriboga CC (1980) Methylmercury in populations eating large quantities of marine fish. Arch Environ Health 35:367–378

Verity MA, Brown WJ, Cheung M (1975) Organic mercurial encephalopathy: in vivo and in vitro effects of methylmercury on synaptosomal respiration. J Neurochem 25:759–766

Vogel DG, Margolis R, Mottet NK (1989) Analysis of methylmercury binding sites on tubulin subunits and microtubules. Pharmacol Toxicol 64:196–201

Volterra A, Trotti D, Racagni G (1994) Glutamate uptake is inhibited by arachidonic acid and oxygen radicals via two distinct and additive mechanisms. Mol Pharmacol 46:986–992

Walters IJ, Wolery TJ (1974) Transfer of heavy metal pollutants from Lake Erie bottom sediments to the overlying water. Water Resources Center, Engineering Experiment Station, The Ohio State University Columbus, Ohio, 43210

Weir PA, Hine CH (1970) Effects of various metals on behavior of conditioned goldfish. Arch Environ Health 20:45

World Health Organization (1989) Environmental health criteria 86: environmental aspects of mercury. International programme of chemical safety. World Health Organization, Geneva, p 115

World Health Organization (1990) Methyl mercury, vol 101. Distribution and Sales Service, International Programme on Chemical Safety, Geneva, p 144

World Health Organization (1991) Inorganic mercury, vol 118. Distribution and Sales Service, International Programme on Chemical Safety, Geneva

Wright JG, Natan MJ, MacDonnell FM, Ralston DM, O'Halloran TV (1990) Mercury (II) thiolate chemistry and the mechanism of the heavy metal biosensor MerR. Prog Inorg Chem 38:323–412

Yannai S, Berdicevsky I, Duek L (1991) Transformations of inorganic mercury by Candida albicans and saccharomyces cerevisiae. Appl Environ Microbiol 57:245–247

Yao CP, Allen JW, Conklin DR, Aschner M (1999) Transfection and overexpression of metallothionein-I in neonatal rat primary astrocyte cultures and in astrocytoma cells increases their resistance to methylmercury-induced cytotoxicity. Brain Res 818:414–420

Yoshino Y, Mozai T, Nakao K (1966a) Biochemical changes in the brain in rats poisoned with an alkylmercury compound, with special reference to the inhibition of protein synthesis in brain cortex slices. J Neurochem 13:1223–1230

Yoshino Y, Mozai T, Nakao K (1966b) Distribution of mercury in the brain and its subcellular units in experimental organic mercury poisoning. J Neurochem 13:397–406

8 Methylmercury and Oxidative Stress

MeHg triggers ROS production, suppresses insulin secretion, and induces apoptosis in-cell-derived HIT-T15 cells and isolated mouse pancreatic islets

(Chen et al. 2006)

ROS (reactive oxygen species) are generally very small molecules and are highly reactive due to the presence of unpaired valence shell electrons. ROS form as a natural by-product of the normal oxygen metabolism and have important roles in cell signaling. These molecules are generated continuously during oxidative metabolism and consist of inorganic molecules, such as superoxide radical anion (O_2^-), hydrogen peroxide (H_2O_2), hydroxyl radicals (OH^-), as well as organic molecules such as alkoxyl and peroxyl radicals (Schulz et al. 2000). Some evidences suggest that the disturbance in the balance between oxidative and reductive cell processes is involved in the pathogenesis of many neurodegenerative conditions such as Alzheimer's disease, amyotrophic lateral sclerosis (ALS), and Parkinson's disease. Other conditions such as autoimmune and inflammatory diseases, cancer, and diabetes mellitus also seemed to be related to this disturbance (Schulz et al. 2000).

MeHg has been thought to induce ROS and generation of oxidative events leading to cell damage. Previous studies have suggested that there is a relationship between these events with dysfunction of the cellular energetic metabolism and disruption of the electron transport chain. These phenomena generate oxidative stress (Clarkson 1997; Shanker et al. 2002). MeHg exposure increases the rate of ROS in the cerebellum (in vivo) and in the brain synaptosomes as well as in the cerebellum neuronal cultures, hypothalamic neuronal cell line, and mixed reaggregating cell cultures (Ali et al. 1992; Sarafian et al. 1994; Sarafian 1999). The formation of these species was critical to determine the damage and the cell death in distinct cell types such as astrocytes and neurons.

It seems that the intensity of MeHg exposure is a crucial factor to establish whether the neuronal death occurs by necrosis or apoptosis (Kunimoto 1994; Castoldi et al. 2000). However, the mechanism of cell death induced by oxidative stress via MeHg has not been well characterized.

Inhibition of Protein Synthesis

Disruption of protein synthesis may be an early manifestation of MeHg toxicity in vitro and in vivo and has been proposed to be the proximal event and primary mechanism of action of MeHg in the nervous system (Yoshino et al. 1966; Verity et al. 1977). However, no direct relationship between the inhibition of protein synthesis and neuropathological changes in MeHg poisoning has been established.

MeHg alters protein phosphorylation, although the patterns of alteration differ somewhat from study to study. Sarafian and Verity (1990) reported

S. Nabi, *Toxic Effects of Mercury*, DOI 10.1007/978-81-322-1922-4_8,
© Springer India 2014

stimulation of protein phospholabeling in primary cultures of cerebellar granule cells exposed to low concentrations of MeHg 24 h, whereas cerebellar glial cells had decreased in protein phosphorylation under identical exposure conditions.

Methylmercury and Antioxidant Defenses

Many studies have already established that MeHg neurotoxicity evokes oxidative stress with formation of ROS in the CNS and that the increase of ROS induces cell damage and death in the CNS. In order to avoid the damage caused by ROS, such as DNA strand breaks, lipid peroxidation, and protein modification, mechanisms have been developed during evolution which dispose or prevent the generation of ROS (Dringen 2000). However, the underlying mechanisms responsible for the protection of CNS against MeHg neurotoxicity are still poorly understood.

It is well known that cell defenses against free radicals such as ROS include scavenger compounds such as glutathione, cysteine, melatonin, and enzymes with antioxidant activities as superoxide dismutase, catalase, and glutathione peroxidase (Olivieri et al. 2000).

It was demonstrated that MeHg induces a concentration-dependent increase in ROS formation in rat neonatal neuronal culture and astrocyte culture (Park et al. 1996; Sorg et al. 1998; Shanker and Aschner 2003). It was also shown that this effect can be reverted by the use of n-propyl gallate (PG), a free radical scavenger; superoxide dismutase (SOD), an antioxidant enzyme; and α-phenyl-tert-butyl nitrone (PBN), a lipophilic hydroxyl radical spin-trapping agent (Shanker and Aschner 2003; Gasso et al. 2001).

Endogenous glutathione (GSH) is one of the most abundant and essential thiol tripeptide present in mammalian cells for scavenging reactive oxygen species (Dringen 2000). The involvement of GSH in the neurotoxicity of MeHg was also evaluated, showing that the increased oxidative stress is related with the depleted intracellular GSH levels (Lee et al. 2001; Shanker et al. 2005). The excessive formation of ROS induced by MeHg exposure can be reverted under treatment with L-2-oxothiazolidine-4-carboxylic acid (OTC), which increases the amount of intracellular GSH, as well as the depletion of GSH by treatment with buthionine-L-sulphoxane (BSO) can potentiate the production of ROS induced by MeHg in rat primary cerebral astrocytes (Do Nascimento et al. 2008).

Recently a human population study in the Amazon correlated the MeHg exposure with the levels of glutathione and catalase activity. Surprisingly, it was demonstrated that high blood levels of glutathione in women exposed to high concentrations of MeHg may be explained by the increase of glutathione peroxidase activity (Pinheiro et al. 2007). In the same population, the inhibition of catalase activity was also observed. These changes likely reflect adaptive responses of the Amazonian population to oxidative stress induced by MeHg.

Other studies revealed that the GSH content may vary in different regions of the CNS, demonstrating that the GSH amount is higher in cerebral cells than in cerebellar cells (Kaur et al. 2007; Adachi and Kunimoto 2005). This may explain the higher susceptibility of cerebellar cells to MeHg toxicity in comparison with cerebral cells, but the reason why certain areas of CNS showed different sensitivity to MeHg toxicity remains unclear.

In addition, MeHg poisoning can induce sympathetic ganglia toxicity and neurite outgrowth inhibition (Soderstrom and Ebendal 1995; Miura et al. 2000). Compounds that possess sulfhydryl (-SH) groups attenuate MeHg neurotoxicity, once at least part of MeHg effects occurs through interaction with -SH groups in cellular proteins (Mullaney et al. 1994). In this context, primary neuronal cultures from avian sympathetic ganglion were used to evaluate the protective role of antioxidant agents with – SH group such as L-cysteine against MeHg toxicity. It was reported that MeHg induces massive cell death (neurite death) and that L-cysteine could fully protect (nearly 100 %) the sympathetic neuron against this damage. The effect of GSH was also tested showing the same properties of cysteine (De Melo Reis et al. 2007).

The use of methionine, an antioxidant agent which does not possess – SH groups, fails to promote cell protection against MeHg intoxication, proving the relevance of – SH groups to this effect (De Melo Reis et al. 2007). Another antioxidant that protects the brain from oxidative stress is vitamin E, which maintains the integrity of membrane by inhibiting lipid peroxidation (Ricciarelli et al. 2001). Recent findings reported the protective effect of the antioxidants tocopherols and tocotrienols (analogs to vitamin E) against MeHg neurotoxicity (Osakada et al. 2004; Khanna et al. 2006). In cerebellar granule cells (CGC), these compounds effectively prevent cell death caused by MeHg intoxication as well as cell migration (Shichiri et al. 2007).

Evidences also suggest that the treatment with trolox (6-hydroxy-2, 5, 7, 8,-tetramethylchroman-2-carboxylic acid), other antioxidant derivative from vitamin E, might provide prevention against oxidative stress. In MeHg-treated rats, it detected many apoptotic cells in the cerebellar granule layers and the treatment with trolox clearly repressed the appearance of these apoptotic processes (Usuki et al. 2001).

A number of different hypotheses have been suggested to explain the mechanism by which the antioxidant defenses protect CNS against MeHg neurotoxicity, which include scavenging and removal of free radicals, reversal of glutamate uptake impairment, inhibition of cytochrome c release, and caspase activation (Aschner et al. 2007).

It has been established that MeHg inhibits glutamate transport by astrocytes by an unknown mechanism which leads to the increase of ROS generation (Aschner et al. 2007). It was demonstrated that a variety of antioxidants can prevent the overproduction of ROS, in this way attenuating MeHg neurotoxicity. Some workers have focused on the effect of antioxidant agents in the impairment of EAA transport elicited by MeHg (Do Nascimento et al. 2008).

Mechanisms of Mercury Toxicity

Mercury can cause biochemical damage to tissues and genes through diverse mechanisms, such as interrupting intracellular calcium homeostasis, disrupting membrane potential, altering protein synthesis, and interrupting excitatory amino acid pathways in the central nervous system (Yee and Choi 1996). Mitochondrial damage, lipid peroxidation, microtubule destruction (National Research Council 2000), and the neurotoxic accumulation of serotonin, aspartate, and glutamate are all mechanisms of methylmercury neurotoxicity (Yee and Choi 1996).

One of the major mechanisms behind MeHg-induced toxicity is via generation of reactive oxygen species (ROS) and depletion of glutathione (GSH). The balance between the oxidative and reductive cellular processes is critical for MeHg-induced neurotoxicity. Over time, both methylmercury and elemental mercury vapor in the brain are transformed to inorganic mercury and become firmly bound to sulfhydryl-containing macromolecules (National Research Council 2000). Both methylmercury and inorganic mercury bind to various molecular weight thiol-containing proteins (glutathione, cysteine, albumin, etc.). The binding and dissociation of these mercury–thiol complexes are believed to control the movement of mercury and its toxic effects in the body (Clarkson 2002).

Mitochondrial damage from oxidative stress may be the earliest sign of neurotoxicity with methylmercury. A study in neural tissue indicates the electron transport chain appears to be the site where free radicals are generated, leading to oxidative damage induced by methylmercury (Yee and Choi 1996).

Concluding Remarks

The molecular mechanisms of MeHg damage in both adult and developing CNS are not fully understood. Early reports have described a number of possible cellular mechanisms to explain the neurotoxicity induced by MeHg. Most of these studies reported the high affinity of MeHg for thiol groups (-SH) which are present in cytoskeletal proteins, enzymes, and peptides that contain the amino acid cysteine (Kaur et al. 2007).

The effects of MeHg on the normal functioning of the nervous system are numerous. It is unlikely that any single event is responsible for the neurotoxicity of MeHg. Rather, MeHg likely causes disruptions in cellular processes including synaptic function, excitability, ion regulation, and protein synthesis.

There have been discrepancies in the outcomes of epidemiological studies estimating the effect of MeHg from fish diet. The availability of nutritional factors such as docosahaxaenoic acid (DHA) might influence MeHg toxicity and may explain the discrepancies from the different studies.

Antioxidants (both enzymatic and nonenzymatic) provide protection against deleterious metal-mediated free radical attacks. Vitamin E and melatonin can prevent the majority of MeHg-mediated damage both in vitro systems and in metal-loaded animals. Toxicity produced by MeHg showed that the protective effect of vitamin E against lipid peroxidation may be associated rather with the level of nonenzymatic antioxidants than the activity of enzymatic antioxidants. Molecular and cellular approaches can be a strategy to critically examine the possibility of therapeutic actions such as antioxidants or chelating agents in the treatment of neurodegeneration produced by MeHg. The rising tonnage (approximately 20 tons a day) put into the air everyday by human activity.

> Because mercury is increasingly becoming elevated in all forms of life, we can assume that more people will have some defects in pancreatic function. Pancreatic support is increasingly necessary for optimal health. (IMVA 2006)

In short, the challenges posed by methylmercury poisoning nowadays can be summarized in the following points:

- The need for a definition and description of the effects of low-level exposure to methylmercury through seafood and freshwater fish, particularly in relation to neurodevelopment, i.e., in vitro and prenatal exposure. This would require a definition of the non-effect range, *lowest observed adverse effect level* (LOAEL) for MeHg concentrations in the most vulnerable populations. Furthermore, even if the fetal brain has been identified as the most vulnerable target, many sensitivity factors still remain to be investigated.
- The need for an improved understanding of MeHg neurotoxicity mechanisms. Although a wealth of information is available on the subject, what is required is a clear explanation that takes the phenomena underlying specific clinical and neuropathological toxicity manifestations fully into account.
- A better knowledge of the above-mentioned points will undoubtedly lead to more effective preventative public health measures.
- At last four courses of action now seem warranted. First, toxic mercurial materials in agriculture and industry should be replaced by less toxic substitutes. Second, controls should be applied at the point of origin to prevent the discharge of potentially harmful Hg wastes. Third, continued periodic monitoring of Hg in fish and wildlife is needed for identification of potential problem areas and for evaluation of ongoing mercury curtailment programs. And fourth, additional research is merited on mechanisms of mercury accumulation and detoxification in comparatively pristine ecosystems.

> DON'T TREAT THE SYMPTOMS. TREAT THE CAUSE. DISCOVER WHY HEAVY METALS MAY BE CAUSING YOUR UNEXPLAINED HEALTH PROBLEMS.
>
> [IAN M. SOLLEY]

References

Adachi T, Kunimoto M (2005) Acute cytotoxic effects of mercuric compounds in cultured astrocytes prepared from cerebral hemisphere and cerebellum of newborn rats. Biol Pharm Bull 28:2308–2311

Ali SF, LeBel CP, Bondy SC (1992) Reactive oxygen species formation as a biomarker of methylmercury and trimethyltin neurotoxicity. Neurotoxicology 13:637–648

Aschner M, Syversen T, Souza DO, Rocha JBT, Farina M (2007) Involvement of glutamate and reactive oxygen species in methylmercury neurotoxicity. Braz J Med Biol Res 40:285–291

Castoldi AF, Barni S, Turin I, Gandini C, Manzo L (2000) Early acute necrosis, delayed apoptosis and cytoskeletal breakdown in cultured cerebellar granule neurons exposed to methylmercury. J Neurosci Res 59: 775–787

Chen YW, Huang CF, Tsai KS, Yang RS, Yen CC, Yang CY, Lin-Shiau SY, Liu SH (2006) Methylmercury induces pancreatic beta-cell apoptosis and dysfunction. Chem Res Toxicol 19:1080–1085

Clarkson TW (1997) The toxicity of mercury. Crit Rev Clin Lab Sci 34:369–403

Clarkson RW (2002) The three modern faces of mercury. Environ Health Perspect 110:11–23

De Melo Reis RA, Herculano AM, da Silva MC, dos Santos RM, de Nascimento JL (2007) In vitro toxicity induced by methylmercury on sympathetic neurons is reverted by L-cysteine or glutathione. Neurosci Res 58:278–284

Do Nascimento JLM, Oliveira KRM, Crespo-Lopez ME, Macchi BM, Maues LAL, Pinheiro MCN, Silveira LCL, Herculano AM (2008) Methylmercury neurotoxicity and antioxidant defenses. Indian J Med Res 128:373–382

Dringen R (2000) Metabolism and functions of glutathione in brain. Prog Neurobiol 62:649–671

Gasso S, Cristofol RM, Selema G, Rosa R, Rodriguez-Farre E, Sanfeliu C (2001) Antioxidant compounds and Ca_2^+ pathway blockers differentially protect against methylmercury and mercuric chloride neurotoxicity. J Neurosci Res 66:135–145

International Medical Veritas Association (IMVA) (2006) Diabetes and mercury poisoning. Medical News

Kaur P, Aschner M, Syversen T (2007) Role of glutathione in determining the differential sensitivity between the cortical and cerebellar regions towards mercury-induced oxidative stress. Toxicology 230:164–177

Khanna S, Roy S, Parinandi NL, Maurer M, Sen CK (2006) Characterization of the potent neuroprotective properties of the natural vitamin E alphatocotrienol. J Neurochem 98:1474–1486

Kunimoto M (1994) Methylmercury induces apoptosis of rat cerebellar neurons in primary culture. Biochem Biophys Res Commun 204:310–317

Lee YW, Ha MS, Kim YK (2001) Role of reactive oxygen species and glutathione in inorganic mercury-induced injury in human glioma cells. Neurochem Res 26:1187–1193

Miura K, Himeno S, Koide N, Imura N (2000) Effects of methylmercury and inorganic mercury on the growth of nerve fibers in cultured chick dorsal root ganglia. Tohoku J Exp Med 192:195–210

Mullaney KJ, Fehm MN, Vitarella D, Wagoner DE, Aschner M (1994) The role of -SH groups in methyl mercuric chloride-induced D-aspartate and rubidium release from rat primary astrocyte cultures. Brain Res 641:1–9

National Research Council (2000) Toxicological effects of methylmercury. National Academy Press, Washington, DC, Committee on the Toxicological Effects of Methylmercury

Olivieri G, Brack C, Muller-Spahn F, Stahelin HB, Herrmann M, Renard P, Brockhaus M, Hock C (2000) Mercury induces cell cytotoxicity and oxidative stress and increases beta-amyloid secretion and tau phosphorylation in SHSY5Y neuroblastoma cells. J Neurochem 74:231–236

Osakada F, Hashino A, Kume T, Katsuki H, Kaneko S, Akaike A (2004) Alpha-tocotrienol provides the most potent neuroprotection among vitamin E analogs on cultured striatal neurons. Neuropharmacology 47:904–915

Park ST, Lim KT, Chung YT, Kim SU (1996) Methylmercury induced neurotoxicity in cerebral neuron culture is blocked by antioxidants and NMDA receptor antagonists. Neurotoxicology 17:37–45

Pinheiro MCM, Crespo-Lopez ME, Vieira JLF, Oikawa T, Guimaraes GA, Araujo CC, Amoras WW, Ribeiro DR, Herculano AM, do Nascimento JLM, Silveira LCL (2007) Mercury pollution and childhood in Amazon riverside villages. Environ Int 33:56–61

Ricciarelli R, Zingg JM, Azzi A (2001) Vitamin E: protective role of a Janus molecule. FASEB J 15: 2314–2325

Sarafian TA (1999) Methylmercury-induced generation of free radical: biological implications. Met Ions Biol Syst 36:415–444

Sarafian T, Verity MA (1990) Altered patterns of protein phosphorylation and synthesis caused by methylmercury in cerebellar granule cell culture. J Neurochem 55:922–929

Sarafian TA, Vartavarian L, Kane DJ, Bredesen DE, Verity MA (1994) Bcl-2 expression decreases methyl mercury-induced free radical generation and cell killing in a neural cell line. Toxicol Lett 74:149–155

Schulz JB, Lindenau J, Seyfried J, Dichgans J (2000) Glutathione, oxidative stress and neurodegeneration. Eur J Biochem 267:4904–4911

Shanker G, Aschner A (2003) Methylmercury-induced reactive oxygen species formation in neonatal cerebral astrocytic cultures is attenuated by antioxidants. Mol Brain Res 110:85–91

Shanker G, Mutkus LA, Walker SJ, Aschner M (2002) Methylmercury enhances arachidonic acid release and cytosolic phospholipase A2 expression in primary cultures of neonatal astrocytes. Brain Res Mol Brain Res 106:1–11

Shanker G, Syversen T, Aschner JL, Aschner M (2005) Modulatory effect of glutathione status and antioxidants on methylmercury induced free radical formation in primary cultures of cerebral astrocytes. Mol Brain Res 37:11–22

Shichiri M, Takanezawa Y, Uchida K, Tamai H, Arai H (2007) Protection of cerebellar granule cells by tocopherols and tocotrienols against methylmercury toxicity. Brain Res 1182:106–115

Soderstrom S, Ebendal T (1995) *In vitro* toxicity of methyl mercury: effects on nerve growth factor (NGF)-responsive neurons and on NGF synthesis in fibroblasts. Toxicol Lett 75:133–144

Sorg O, Schilter B, Honegger P, Monnet-Tschudi F (1998) Increased vulnerability of neurones and glial cells to low concentrations of methylmercury in a prooxidant situation. Acta Neuropathol 96:621–627

Usuki F, Yasutake A, Umehara F, Tokunaga H, Matsumoto M, Eto K, Ishiura S, Higuchi I (2001) *In vivo* protection of a water-soluble derivative of vitamin E, Trolox, against methylmercury-intoxication in the rat. Neurosci Lett 304:199–203

Verity MA, Brown WJ, Cheung M, Czer G (1977) Methylmercury inhibition of synaptosomes and brain slice protein synthesis. In vivo and in vitro studies. J Neurochem 29:673–679

Yee S, Choi BH (1996) Oxidative stress in neurotoxic effects of methylmercury poisoning. Neurotoxicology 17:17–26

Yoshino Y, Mozai T, Nakao K (1966) Distribution of mercury in the brain and its subcellular units in experimental organic mercury poisoning. J Neurochem 13:397–406

Part III

Toxicity Evaluation Methodology

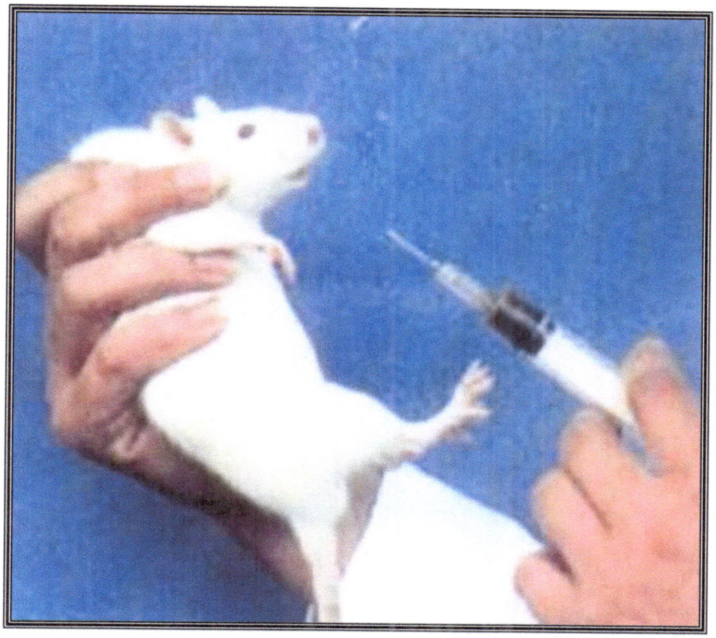

9 Materials and Analytical Procedures

Animals and Administration Procedure

Three-month-old healthy male albino rats of Wistar strain weighing 150±50 g supplied by Central Animal House, J.N. Medical College, AMU, Aligarh, were used throughout present investigation. The animals were kept in highly hygienic conditions using polypropylene cages maintained at 28±5 °C with lighting conditions (photoperiod) of 12 h of light and 12 h of darkness. They were kept on pellet diet (Hindustan Lever Ltd., Mumbai, India) measuring up to the nutritional standards recommended by the U.S. National Research Council Publication No. 990, titled "Nutritional Requirements of Laboratory Animals" and water was available ad libitum. The animals were acclimatized for 1 week after being transferred from their colonies to the working laboratory. The animal room and cages were cleaned daily to keep them dust-free. The rats showing signs of disease were removed immediately after identification. Female rats were excluded from the study because of their cyclic hormonal variations.

The rats were randomly assigned to five treatment groups.

Normal Control (NC)

Ten rats were given 0.2 ml of isotonic saline orally by intragastric intubation for 28 days.

Methylmercury Chloride Treated (MMC)

In this group, ten rats were given 2 mg/kg body weight of methylmercury chloride orally by intragastric intubation for 14 days and were kept without toxicant for the next 14 days.

Methylmercury Chloride and Vitamin E Treated (MMC+Vit E)

Rats of this group, ten in number, were administered with 2 mg/kg body weight of methylmercury chloride orally by intragastric intubation for 14 days and then treated with 100 mg/kg body weight of vitamin E orally by intragastric intubation for the next 14 days.

Methylmercury Chloride and Acetyl-L-Carnitine Treated (MMC+ALCAR)

In this group, ten rats were given 2 mg/kg body weight of methylmercury chloride orally by gavage for 14 days, and then for the next 14 days, they were treated with 100 mg/kg body weight of L-carnitine orally.

S. Nabi, *Toxic Effects of Mercury*, DOI 10.1007/978-81-322-1922-4_9,
© Springer India 2014

Methylmercury Chloride, Vitamin E, and Acetyl-L-Carnitine Treated (MMC+Vit E+ALCAR)

Ten rats of this group were treated with methylmercury chloride orally for 14 days, and for the next 14 days, they were given vitamin E and acetyl-L-carnitine in combination at respective doses mentioned above by oral administration at a gap of 30 min between the compounds.

Methylmercury chloride (CAS: 115-09-03) was purchased from Sigma-Aldrich (St. Louis, MO, USA). All other chemicals used were of analytical grade or purest quality purchased from Merck, Fluka, Himedia, or Loba.

Analytical Procedures

For biochemical estimations, all the above-mentioned groups were treated orally once a day for 28 consecutive days. All the animals were sacrificed on the 29th day after being tested for cognitive deficits in the early hours of the day. The solution of MMC was prepared by dissolving 25 mg of MMC in 20 ml physiological saline (1.25 mg/ml). The dose of MMC was selected based on recent estimate of daily ingestion in an environmentally exposed population (Passos et al. 2008).

Cytogenetic study was done in first two groups only, i.e., the rats which were given MMC only and the rats of control group. Groups treated with vitamins were skipped for cytogenetic assays as handful of work has been done on these antioxidants earlier by a number of authors, and every time, positive results were analyzed whether in MNT or in chromosomal aberrations. Keeping those results in consideration, we only mentioned methylmercury chloride results of cytogenetic parameters in our study. Ten rats were included in each group.

Dissection of Rats and Collection of Different Organs

At the end of the treatment, overnight-fasted rats in each group were killed by cervical dislocation (one of the most acceptable methods of euthanasia) on the scheduled days between 10:30 and 11:30 am without using anesthesia.

The controls as well as experimental rats were grasped at their neck near the base of skull, with the thumb and forefinger of one hand, and hind limbs and tail with the other. A swift but controlled motion separated the cervical vertebrae from the base of skull. This resulted in instantaneous loss of consciousness and loss of all vital signs within a few minutes.

The head of rats (killed by decapitation) was skinned. The skeletal coverings over the skull were removed with the use of a small pair of bone forceps, scalpel, scissors, and nail clippers. Great care was taken to avoid any laceration of brain tissue. After exposing the brain surface from the top and sides, it was gently detached from the base of skull and removed.

For all biochemical investigations, fresh unfixed brain was used. The brain and spinal cord were removed rapidly from individual rat and dissected out on an ice plate. The blood clots adhering to the brain were removed by washing with the cold normal saline. Thereafter, the cerebrum, cerebellum, brain stem, and spinal cord were rapidly dissected out and weighed to the nearest milligram on an electrical balance.

The animal body was cut ventrally with the help of sharp scissors, and the lung, heart, and pancreas were immediately excised. Whole tissues were washed with chilled saline (4 °C), blotted, and weighed.

Collection of Blood for Plasma "Total Antioxidant Power" and "Xanthine Oxidase Activity"

The blood from rat in a given group was drawn by cardiac puncture. It was collected in centrifuge tubes using heparin as anticoagulant. Blood was mixed gently by inversion two to three times and immediately cooled to 4 °C in a refrigerator. The samples were centrifuged at 2,500 rpm for 30 min. Plasma was aliquoted and either stored at 4 °C or frozen at −20 °C for future use.

Collection of Bone Marrow for Cytogenetic Study

One 20 min prior to sacrifice of rats, colchicine (0.4 mg/100 g body wt., i.p.) was injected into the animals. The time of sacrifice was decided on preliminary observations for scoring the optimal metaphase plates.

Slide Preparation for Chromosomal Studies

Slide preparation and staining were done according to the established protocol of Preston et al. (1987). Briefly, both the femurs of each rat were dissected and bone marrow was flushed out by aspirating with a syringe containing 0.56 % KCl. The cells were treated with a hypotonic solution of 0.56 % KCl and incubated at 37 °C for 30 min following centrifugation at 1,500 rpm for 10 min. Contents were fixed in glacial acetic acid: methanol (1:3 v/v). The cells were then prepared for microscopical examination by adding three to four drops of cell suspension to precleaned chilled ethanol-dipped slides. Slides were air-dried and stained with 5 % Giemsa for 5 min.

Slide Preparation for Bone Marrow Micronucleus Test

The MNT from animals was carried out according to Schmid (1975). BMCs from both the femurs were flushed out as a fine suspension into a tube containing 0.5 ml fetal bovine serum (FBS) and centrifuged at 1,000 rpm for 10 min. The pellet was resuspended in FBS. The suspension was smeared on to the precleaned air-dried slides following fixing in 100 % methanol for 5 min. and stained permanently with May–Grünwald and Giemsa. Slides for both the categories were cleared in xylene and permanently mounted by DPX (Schmid 1975). Slides were selected on the basis of staining quality, coded randomly and scored blindly. In each group, a maximum of 4,000 cells were examined at a magnification of 10 × 100X. The presence of micronucleated polychromatic erythrocytes was visually scored by optical microscopy using a Nikon 80i field microscope. Cells were considered to be micronucleated when they contained neatly defined chromatin corpuscles with a diameter of less than one-third the diameter of the cell nucleus and stained equal or lighter than the nucleus of the cell from which the micronucleated cell had developed (Schmid 1975).

Peripheral Blood Micronucleus Assay

This test was performed to investigate the mutagenicity in circulating erythrocytes. If this test is conducted along with bone marrow micronucleus test and chromosome aberration test, the accuracy of genotoxicity evaluation is 100 %. This assay was conducted in the same animals used for bone marrow micronucleus assay (four animals/group). The blood was collected from heart puncture and smears were prepared immediately on clean glass slides, air-dried, fixed in methanol for 10 min, and stained with May–Grünwald–Giemsa.

References

Passos CJS, Sampaio DS, Lemire M, Fillion M, Guimaraes JRD, Lucotte M, Mergler D (2008) Daily mercury intake in fish-eating populations in the Brazilian Amazon. Exp Sci Environ Epidemiol 18:76–87

Preston RJ, Dean BJ, Galloway S, Holden H, McFee AF, Shelby M (1987) Mammalian in vivo cytogenetic assay: analysis of chromosomal aberrations in bone marrow cells. Mutat Res 189:157–165

Schmid W (1975) The micronucleus test. Mutat Res 31:9–15

10 Biochemical Estimation Techniques

Biochemical Studies

Tissue Preparation

Tissues were minced with sharp scissors and then thawed and homogenized with the aid of York's homogenizer fitted with Teflon plunger in a proportion of 1:10(w/v) ice-cold phosphate buffer (50 mM; PH 7.4). Aliquots of homogenates were used for the determination of malondialdehyde (MDA), hydroperoxides, reduced glutathione (GSH), and total sulfhydryl (TSH) levels. The homogenates were centrifuged for 60 min at 10,000×g at 4 °C, and determination of superoxide dismutase (SOD) and catalase (CAT) activities were carried out with the resultant supernatant. Protein concentration was determined according to the method of Lowry et al. (1951) using purified bovine serum albumin as standard.

Estimation of Lipid Peroxide

Lipid peroxide content was estimated according to the method of Ohkawa et al. (1979).

Principle

Acetic acid detaches the lipid and protein of the tissue. The protein in the reaction mixture is dissolved by the addition of sodium dodecyl sulfate. 2-Thiobarbituric acid (TBA) reacts with lipid peroxide, hydroperoxide, and oxygen labile double bond to form the color products with absorption maxima at 532 nm.

Reagents

(I) 0.15 M potassium chloride
(II) Sodium dodecyl sulfate (SDS) 8 %
(III) Acetic acid (20 %)
(IV) 2-Thiobarbituric acid (0.8 %)
800 mg of TBA was suspended in 20 ml double distilled water; the pH was adjusted to 7.0 by 0.1 N NaOH. This TBA was dissolved and volume was adjusted to 100 ml with distilled water.
(V) n-Butanol

Procedure

0.2 ml of tissue homogenate was mixed with 1.0 ml of 20 % acetic acid. Subsequently, 0.2 ml of 8 % aqueous SDS was mixed in the above reaction mixture; the pH of the mixture was adjusted at 4.0 using concentrated NaOH solution if needed. After adjusting the pH of the reaction mixture, 1.5 ml of 0.8 % TBA solution and sufficient amount of distilled water were added to a final volume of 4.0 ml. Then the reaction mixture was incubated in a boiling water bath for 1 h. After cooling to room temperature, 3 ml of n-butanol was mixed. The reaction mixture was then centrifuged at 10,000×g for 15 min. A clear

S. Nabi, *Toxic Effects of Mercury*, DOI 10.1007/978-81-322-1922-4_10,
© Springer India 2014

butanol fraction obtained after centrifugation was used for measuring the absorbance at 532 nm in Beckman DU-640 spectrophotometer. An appropriate standard made up of malondialdehyde (MDA) 2.5 nmol was run simultaneously.

Standard absorbance of malondialdehyde (2.5 nmol) was used to calculate the amount of lipid peroxide in the samples, and results were expressed as nmol of MDA/g tissue weight.

Estimation of Lipid Hydroperoxide

Lipid hydroperoxide was estimated by the method of Haldebrandt and Roots (1975).

Principle

The principle of this method is the formation of Fe $(SCN)_3$ from ferrous ammonium sulfate and potassium thiocyanate on peroxidation of Fe^{2+} –Fe^{3+} by H_2O_2 which results into the development of an intense pale color that can be recorded at 480 nm.

Reagents

(I) 0.15 KCl
(II) Trichloroacetic acid (TCA, 20 % w/v)
(III) 10 mM ferrous ammonium sulfate
97.72 mg of ferrous ammonium sulfate was dissolved in double distilled water and the volume was made up to 25 ml.
(IV) 2.5 M potassium thiocyanate
6.11 g of KSCN was dissolved in double distilled water and the volume was made up to 25 ml in a volumetric flask.

Procedure

Two milliliter of each homogenate (10 %, w/v) was treated with 1 ml of 20 % TCA. It was incubated at 0 °C for 30 min and centrifuged at 14,000×g for 30 min in cold. The above TCA soluble extract was used for the estimation of hydroperoxide. One milliliter of the clear supernatant was carefully mixed with 0.2 ml of 10 mM Fe $(NH_4)_2$ SO_4 and 0.1 ml of 2.5 M KSCN. The tubes were shaken thoroughly and kept for 10 min and then absorbance of the pink color formed was read at 480 nm in spectrophotometer against a reagent blank. An appropriate standard of cumin hydroperoxide (4 nmol) was simultaneously run.

Standard absorbance of cumin hydroperoxide (4 nmol) was used to calculate the amount of lipid hydroperoxide in the samples, and results were expressed as nmol of cumin hydroperoxide per g tissue.

Estimation of Total Sulfhydryl Group (TSH)

The estimation of sulfhydryl group was done by the method of Ellman (1959) as modified by Sedlak and Lindsay (1968).

Principle

5-5′-Dithiobis-2-nitrobenzoic acid (DTNB) is reduced by –SH groups of glutathione (GSH) in alkaline medium to produce one mole of 2-nitro-5-mercaptobenzoic acid per mole of –SH group. Since the anion (2-nitro-5-mercaptobenzoic acid) has an intense yellow color, it can be used to measure –SH group at 412 nm.

Reagents

(I) Standard solution
A standard solution of 2×10^{-3} M of GSH was prepared by dissolving 6.146 mg GSH in 10 ml of 0.02 M EDTA.
(II) 0.02 M EDTA
(III) 0.2 M tris buffer in 0.2 M EDTA, pH 8.2
(IV) 0.01 M DTNB
0.01 M solution of DTNB was prepared by dissolving 99 mg DTNB in 25 ml of absolute methanol.
(V) Absolute methanol

Procedure

Various parts of the brain and different organs were homogenized in chilled 50 mM, pH 7.4 phosphate buffer, and the volume was adjusted to give a 10 % (w/v) homogenate. In 0.1 m of tissue homogenate (10 %), 1.5 ml of 0.2 M tris buffer (pH 8.2) and 0.1 ml of DTNB were added. The mixture was shaken and made to 10 ml with 8.3 ml of absolute methanol. The reaction mixture was centrifuged at 6,000×g for 5 min in cold. The absorbance of the clear supernatant was read at 412 nm. A calibration curve with different concentrations of GSH (200–1,600 µmoles) was obtained according to the same procedure as described above. The values were plotted by least square method.

Total –SH group in the samples were calculated using the standard curve and the results were expressed as µmoles/g tissue.

Estimation of Free Sulfhydryl Group (GSH)

Free sulfhydryl group was estimated by the method of Ellman (1959) as modified by Sedlak and Lindsay (1968).

Principle

Same as for total sulfhydryl group estimation.

Reagents

(I) Standard solution
A standard solution of 2×10^{-3} M of GSH was prepared by dissolving 6.146 mg GSH in 10 ml of 0.02 M EDTA.
(II) 0.15 M KCl
(III) 10 % TCA
(IV) 0.4 M tris buffer in 0.2 M EDTA, pH 8.9
(V) 0.01 M DTNB

Procedure

Various tissues were homogenized (10 %, w/v) in chilled 50 mM, pH 7.4 phosphate buffer. 1 ml tissue homogenate was deproteinized by adding 1 ml of 10 % TCA and centrifuged at 6,000×g for 5 min. 0.5 ml aliquot from clear supernatant was mixed with 0.5 ml double distilled water. Thereafter, 2 ml of 0.4 M tris buffer and 0.1 ml DTNB were added to it with proper stirring. The absorbance was read at 412 nm within 5 min of the addition of DTNB. A calibration curve with different concentrations of GSH (200–1,600 µmoles) was drawn by the same procedure as described above. The values were plotted with least square method.

Free –SH (GSH reduced) in the samples were calculated using the standard curve and the results were expressed as µmoles/g tissue.

Estimation of Superoxide Dismutase (SOD)

SOD activity was measured by the method of Marklund and Marklund (1974).

Principle

SOD principle depends upon auto-oxidation of pyrogallol:

$$\text{Pyrogallol} + O_2 \xrightarrow[\text{SOD}]{\text{Auto-Oxidation}} \text{oxidation product} + O_2 \quad (10.1)$$

$$2O_2^- + 2H^+ \longrightarrow O_{2+}H_2O_2 \quad (10.2)$$

Procedure

Different rat tissues were homogenized in chilled 50 mM, pH 7.4 phosphate buffer (10 w/v). Homogenate was centrifuged in cold at 10,000×g for 60 min. 0.05 ml of clear supernatant was added to 2.85 ml of 0.05 M tris succinate buffer (pH 8.2), mixed well, and incubated at 25 °C for 20 min. The reaction was started by adding 0.1 ml of 8 mM of pyrogallol solution. The contents were shaken well, and change in OD/min was immediately recorded for 3 min at 420 nm. A reference set consisting 0.05 ml of double distilled water instead of the sample solution (clear supernatant) was also run similarly.

Calculation

$$\text{SOD} = \frac{(\text{A/min. ref.} - \text{A/min. sample}) \times 30}{(\text{A/min. ref.}/2 \times 0.05 \times 1)} \text{Units/10mg tissue}$$

where

A/min. ref. = change of OD/min in ref. set
A/min. sample = change of OD in sample set

Activity Unit

One unit of the enzyme is defined as the amount of enzyme which causes a 50 % inhibition of pyrogallol auto-oxidation under assay conditions.

Estimation of Catalase (H_2O_2: H_2O_2 Oxidoreductase, EC 1.11.1.6)

Catalase activity was assayed according to the method of Aebi (1984).

Principle

It catalyzes the following reaction:

$$2H_2O_2 \xrightarrow{\text{Catalase}} 2H_2O + O_2$$

In the UV range, H_2O_2 shows a continual increase in absorption with decreasing wavelength and maximum at 240 nm. The decomposition of H_2O_2 can be followed directly by the decrease in extinction at 240 nm (E 240 = 40 cm^2/μmol). The difference in extinction (E 240) per unit time is the measure of the catalase activity.

Reagents

(I) 50 mM phosphate buffer (pH 7.0)
(II) 30 mM hydrogen peroxide
0.34 ml of 30 % H_2O_2 was diluted to 100 ml with phosphate buffer.

Procedure

Three milliliter of H_2O_2–phosphate buffer was pipetted into the cuvette, required amount of tissue supernatant (cytosolic fraction) was added as enzyme source, and the contents were mixed thoroughly. The decrease in absorbance at 240 nm was recorded after every 30 s for 3 min. The results were expressed as units/mg protein.

Protein Estimation

Protein estimation in tissue homogenates was done by the method of Lowry et al. (1951).

Principle

This method is based on color reactions of amino acids tryptophan and tyrosine with Folin's phenol reagent. By the reaction of these amino acids with phosphomolybdic acid and phosphotungstic acid (present in Folin's reagent), a blue color is formed which is colorimetrically estimated at 625 nm. The color is the result of reduction of phosphomolybdic acid and biuret reaction of proteins with Cu^{++} ions in alkaline medium.

Reagents

(I) 2.0 % (w/v) sodium carbonate in 0.1 N NaOH.
(II) 1.0 % (w/v) copper sulfate
(III) 2.0 % (w/v) sodium potassium tartrate
(IV) Alkaline copper sulfate solution
1.0 ml of reagent (II) + 1.0 ml of reagent (III) + 48 ml of reagent (I) in the same sequence

(V) Folin–Ciocalteu phenol reagent
2 N solution obtained commercially was diluted 1:1 with double distilled water before use.

(VI) Standard bovine serum albumin (1 mg/ml)
Stock standard was diluted ten times to get the working standard of 100 μg/ml.

Procedure

0.9 ml normal saline and 1.5 ml of 10 % of TCA were added to 0.1 ml of tissue or subcellular fraction. The contents were kept in cold for 4 h and protein precipitate was recovered by centrifugation. Protein was dissolved in 1.0 ml 0.1 N NaOH. 5.0 ml of alkaline copper sulfate solution was then added to the diluted sample and incubated at 37 °C. After 30 min, 0.5 ml of Folin–Ciocalteu reagent was added at the same temperature. Optical density of the blue color developed was read at 625 nm exactly after 30 min. Standard protein solution (BSA, 20–100 μg) and reagent blank were also run simultaneously.

Standard curve of BSA (20–100 μg) was plotted and the concentration of protein in the samples was calculated. The results were expressed as mg protein/100 g tissue weight.

Measurement of FRAP of Plasma (Total Antioxidant Power)

The ferric reducing ability of plasma, the FRAP assay, which estimates "total antioxidant power" was measured as described by Benzie and Strain (1996) with minor modifications.

Principle

Ferric to ferrous ion reduction at low pH results in the formation of a colored ferrous tripyridyltriazine complex.

Reagents

(I) Acetate buffer: 300 mM, pH 3.6
3.1 g sodium acetate and 16 ml glacial acetic acid were dissolved in 1 l of distilled water. pH was maintained and contents were stored at 4 °C.

(II) Dilute HCl: 40 mM
1.46 ml concentration HCl (11 M) was dissolved in 1 l of distilled water and stored at room temperature.

(III) TPTZ (2, 4, 6-tri [2-pyridyl]-s-triazine): 10 mM
0.031 g of TPTZ was dissolved in 10 ml of 40 mM HCl at 50 °C in water bath. It was freshly made on the day of assay in a new corning tube

(IV) Ferric chloride: 20 mM
0.054 g of $FeCl_3.6H_2O$ was dissolved in 10 ml of distilled water and freshly prepared on the day of assay.

(V) Standard solution of ferrous sulfate (1 mM) was prepared by dissolving 0.278 g of $FeSO_4.7H_2O$ in 1 l of distilled water.

Procedure

The assay was carried out in a total volume of 1.0 ml containing a suitable aliquot of plasma in 0.1 ml and 900 μl of freshly prepared FRAP reagent, prepared by mixing 10.0 ml of 300 mM sodium acetate buffer, pH 3.62, 1.0 ml of 0 mM ferric chloride, and 1.0 ml of 10 mM 2, 4, 6-tripyridyl-s-triazine solution prepared in 40 mM HCl. Both FRAP reagent and plasma samples were preincubated for 5 min at 30 °C before starting the reaction. Incubation was done for 5 min at 30 °C and absorbance was recorded at 593 nm against a reagent blank in a Beckman DU-640 spectrophotometer.

The "total antioxidant power" (FRAP value) was calculated by utilizing a standard ferrous sulfate. Results were expressed as μmol of ferrous ion formed/min/dl.

Assay of Xanthine Oxidase Activity

The activity of xanthine oxidase in plasma and tissue homogenate was assayed by the method of Noro et al. (1983) with suitable modification.

Procedure

The reaction mixture consisted of 70 mM sodium phosphate buffer, pH 7.4, 10–40 µl of plasma or 100–200 µl of 10 % tissue homogenate, and 8 mM xanthine. After a 5-min preincubation, the reaction was started by the addition of xanthine and incubation was carried out for 15 min at 37 °C. The reaction was terminated by the addition of 0.5 ml ice-cold 10 % perchloric acid, followed by incubation for 10 min at room temperature. Samples were then centrifuged at 4,000×g for 10 min. The absorbance of the supernatant was recorded at 290 nm against a reagent blank in a Beckman DU 640 spectrophotometer.

Reagents

(I) 70 mM sodium phosphate buffer, pH 7.4
(II) 8 mM xanthine
(III) 10 % perchloric acid
(IV) Standard solution, 30–360 µM uric acid
Xanthine oxidase activity was calculated by using a standard uric acid. One unit of enzyme activity is defined as µmols of uric acid formed/min/ml for plasma and µmols of uric acid formed/min/mg protein for tissue.

Glutathione Peroxidase (GSH-Px) Assay in Plasma

Glutathione peroxidase (GSH-Px) activity in plasma was measured by the method of Paglia and Valentine (1967).

Procedure

Antioxidant enzyme glutathione peroxidase (GSH-Px) activity was determined using glutathione reductase and NADPH. The method is based on the oxidation of NADPH at 25 °C, which is indicated by the decrease in absorbance at 340 nm, according to Paglia and Valentine (1967). Data were expressed as units of NADPH oxidized/min/ml. In short, reaction mixture was 0.2 mM NADPH, 4 mM GSH, 4 mM EDTA, 4 mM NaN_3, 1 i.u. glutathione reductase, 0.2 mM H_2O_2, and the sample in 100 mM phosphate buffer, pH 7.0. Plasma was added to the assay mixture and the reaction was started by the addition of H_2O_2. Oxidation of NADPH to $NADP^+$ was monitored continuously at 340 nm for 10 min.

Reagents

(I) 0.2 mM NADPH
(II) 4 mM GSH
(III) 4 mM EDTA
(IV) 4 mM NaN_3
(V) 1 i.u. Glutathione reductase
(VI) 0.2 mM H_2O_2
(VII) 100 mM phosphate buffer, pH 7.0
Glutathione peroxidase activity was calculated by using a standard NADPH. One unit of enzyme activity is defined as µmol NADPH oxidized/min/ml for plasma and µmol NADPH oxidized/min/mg protein for tissue.

Observation of Body Weight, Food Intake, Water Intake, and Tissue Weight

Body weight gain, food intake, and water intake changes were checked daily by weighing animals, observing the amount of pellet taken by them per day and the volume of water drunk daily. After decapitation, animal nervous and non-nervous tissues were weighed, and change in weight between treated and control group rats was estimated manually and calculated by applying analysis of variance.

References

Aebi H (1984) Catalase in vitro. Methods Enzymol 105:121–126

Benzie IFF, Strain JJ (1996) The ferric reducing ability of plasma (FRAP) as a measure of "antioxidant power": the FRAP assay. Anal Biochem 239:70–76

Ellman GL (1959) Tissue sulfhydryl groups. Arch Biochem Biophys 82:70

Haldebrandt AG, Roots I (1975) Reduced nicotinamide adenine dinucleotide phosphate (NADPH) dependent formation and breakdown of hydrogen peroxide during mixed function oxidation reaction in liver microsome. Arch Biochem Biophys 171:385–397

Lowry OH, Rosebrough NJ, Farr AL, Randall RJ (1951) Protein measurement with folin phenol reagent. J Biol Chem 193:265–275

Marklund S, Marklund G (1974) Involvement of superoxide anion radical in the autoxidation of pyrogallol and a convenient assay for superoxide dismutase. Eur J Biochem 47:469–474

Noro T, Oda Y, Miyase T, Ueno A, Fukushima S (1983) Inhibitors of xanthine oxidase from the flowers and buds of Daphne genkwa. Chem Pharm Bull 31:3984–3987

Ohkawa H, Ohishi N, Yagi K (1979) Assay for lipid peroxides in animal tissue by thiobarbituric reaction. Anal Biochem 95:351–358

Paglia DE, Valentine WN (1967) Studies on the quantitative and qualitative characterization of erythrocyte glutathione peroxidase. J Lab Clin Med 70:158–169

Sedlak J, Lindsay RH (1968) Estimation of total, protein-bound and nonprotein sulfhydryl groups in tissue with Ellman's reagent. Anal Biochem 25:192–205

11 Neurobehavioral Assessment Models

Behavioral Experiments

Elevated Plus Maze Test

The elevated plus maze (EPM) is a rodent model of anxiety that is used as a screening test for putative anxiolytic or anxiogenic compounds and as a general research tool in neurobiological anxiety research (Fig. 11.1).

Method

The test setting consists of a plus-shaped apparatus with two open and two enclosed arms, each with an open roof, elevated 40–70 cm from the floor. The model is based on rodents' aversion of open spaces. This aversion leads to the behavior termed thigmotaxis, which involves avoidance of open areas by confining movements to enclosed spaces or to the edges of a bounded space. In EPM this translates into a restriction of movement to the enclosed arms (Pellow et al. 1985; Treit et al. 1993; Rodgers 1997; Carobrez and Bertoglio 2005).

Anxiety reduction in the plus maze is indicated by an increase in the proportion of time spent in the open arms (time in open arms/total time in open or closed arms) and an increase in the proportion of entries into the open arms (entries into open arms/total entries into open or closed arms). Total number of arm entries and number of closed-arm entries are usually employed as measures of general activity (Hogg 1996). The relation between the EPM and other tests of exploratory activity (open field and emergence) has been analyzed in two mouse strains (Lalonde and Strazielle 2008).

Validity

While EPM is the most commonly employed behavioral animal anxiety model, there are several issues concerning the validity of the model. Classical clinical anxiolytics, such as benzodiazepines (e.g., Valium), do reduce measures of anxiety in EPM. However, more novel compounds, such as 5-HT_{1A} agonists (e.g., BuSpar), give mixed results. Selective serotonin reuptake inhibitors and tricyclic antidepressants, which are commonly employed in clinical settings to treat anxiety disorders, also do not lead to a stable anxiolytic effect on EPM (Carobrez and Bertoglio 2005). This raises the possibility that EPM is a suitable model for testing GABA-related compounds, such as benzodiazepines or direct $GABA_A$ agonists, but not for other drugs. Despite this, the model is commonly employed for screening putative anxiolytics and for general research into the brain mechanisms of anxiety (Engin and Treit 2008) because of the ease of employment and the vast number of studies already in the literature. An alternative explanation to the lack of correlation between this model and treatments for the purportedly related DSM-IV conditions is that the psychopharmacological paradigm relating to these conditions is flawed.

S. Nabi, *Toxic Effects of Mercury*, DOI 10.1007/978-81-322-1922-4_11,
© Springer India 2014

Fig. 11.1 Elevated plus maze (Source: English Wikipedia)

Depression

Major depressive disorder, or a general term "depression," is a common and highly heterogeneous psychiatric disorder. It is a long lasting and even life-threatening disorder with symptoms including deficits of cognitive, psychomotor, and emotional processes. The illness can be referred to a wide variety of abnormal variations in an individual's mood, which is characterized by periods of depressed mood, profound sadness, or loss of interest in activities (anhedonia). The negative moods caused by depression significantly interfere with the normal functional ability of affected people, and the symptoms include a persistent sad or empty mood, feeling of hopelessness and worthlessness, changes in sleep and appetite, difficulty of concentrating and making decisions, and recurring thoughts of death or suicide (Fava and Kendler 2000).

There are about one in six people in the United States who will succumb to depression at some point during their life span (Kessler et al. 2005), and according to the World Health Organization, depression is projected to reach second place as leading contributor to the global burden of disease by the year 2020 (Murray and Lopez 1997). The current antidepressants have late drug efficacy about 3–6 weeks; besides, many patients' illness cannot be alleviated (Hua-Cheng Yan et al. 2010). Novel and effective depression treatments are needed indeed.

Modeling Depression in Animals

It is difficult to develop an animal model that perfectly reproduces the symptoms of depression in patients. Animals lack self-consciousness, self-reflection, and consideration; moreover, hallmarks of the disorder such as depressed mood, low self-esteem, or suicidality are hardly accessible in nonhumans. However, depression, as other mental disorders, constitutes of endophenotypes (Hasler et al. 2004) that can be reproduced independently and evaluated in animals. An ideal animal model offers an opportunity to understand molecular, genetic, and epigenetic factors that may lead to depression. By using animal models, the underlying molecular alterations and the causal relationship between genetic or environmental alterations and depression can be examined, which would afford a better insight into pathology of depression. In addition, animal models of depression are indispensable for identifying novel therapies for depression.

Endophenotypes in Animal Model of Depression

The following endophenotypes have been described (Hua-Cheng Yan et al. 2010):

- Anhedonia: The loss of interest is a core symptom of depression. Anhedonia in rodents can be assessed by sucrose preference or by intracranial self-stimulation.
- Behavioral despair: Behavioral despair might be assessed with tests such as the forced swimming test or the tail suspension test.
- Changes in appetite or weight gain: Depression is often associated with changes in appetite or weight gain, which is easily measured in rodents.
- Neuroanatomy: Depressed subjects display decreased hippocampal volume and rodents exposed to chronic stress or excess glucocorticoids exhibit similar signs of hippocampal loss of neurons and dendritic atrophy.
- Neuroendocrine disturbances: Disturbances of the hypothalamic–pituitary–adrenal axis (HPA) are one of the most consistent symptoms in major depression. The functionality of the HPA can be assessed by dexamethasone suppression test.
- Alterations in sleep architecture: Disturbances in the circadian rhythm and especially in the sleep architecture are often observed in depressed. In rodents, it is accessible via electroencephalography (EEG).
- Anxiety-related behavior: Anxiety is a symptom with high prevalence in depression. Therefore, animal models of depression often display altered anxiety-related behavior.

Criteria for Valid Animal Models of Depression

An appropriate animal model of human depression should fulfill the following criteria as much as possible: strong phenomenological similarities and similar pathophysiology (face validity), comparable etiology (construct validity), and common treatment (predictive validity) (Willner and Mitchell 2002; Anisman and Matheson 2005; Vollmayr et al. 2007). Again, depression is a heterogeneous disorder and its many symptoms are hard to be produced in laboratory animals. The question therefore remains whether we can know the animal is "depressed." Actually, few models of depression fully fit these validating criteria, and most models currently used rely on either actions of known antidepressants or responses to stress. It is not necessary for an "ideal" animal model of depression to exhibit all the abnormalities of depression-relevant behaviors, just like that the patients do not manifest every possible symptom of depression.

Antidepressant Screening Tests

Antidepressant screening tests, not like the models which can be defined as an [organism] or a particular state of an organism that reproduces aspects of human pathology, provide only an end-point behavioral or physiological measure designed to assess the effect of the genetic, pharmacological, or environmental manipulation.

Despair Based

Tail Suspension Test (TST)

This test was performed as described by Steru et al. (1985). A short piece of paper adhesive tape (about 6 cm) was attached along half the length of the tail (about 3 cm). The free end of the tape was attached to a 30 cm long rigid tape which hung from a horizontal bar clamped to a heavy laboratory support stand. Suspended rats were surrounded by a white wooden enclosure (45 cm high, 40 cm wide, and 40 cm deep) such that the rat's head was about 20 cm above the floor (Fig. 11.2).

For testing, each rat was suspended by its tail and observed for 6 min. An observer scored the total duration of a passive, "deadweight" hanging (immobility), between the periods of wriggling of the animal to avoid aversive situation.

A major advantage of the TST is that it is simple and inexpensive. A major disadvantage of the TST is that it is restricted to mice and limited to strains that do not tend to climb their tail. Besides, like FST, TST is sensitive to acute

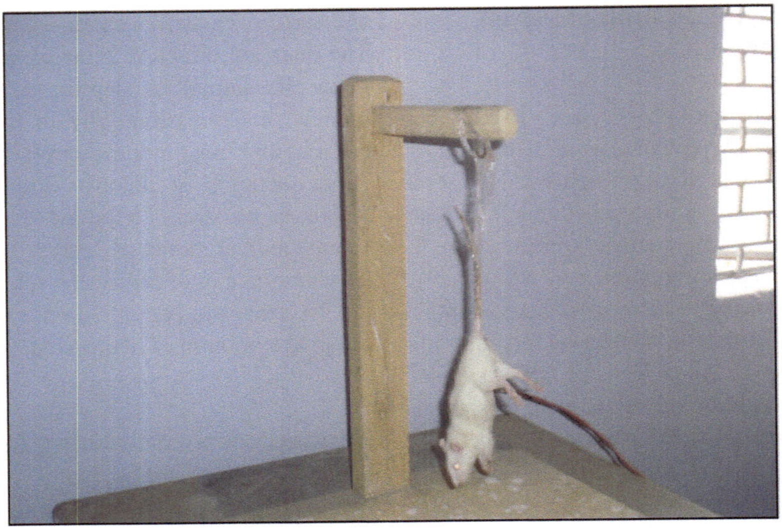

Fig. 11.2 Tail suspension test

treatment only, and its validity for non-monoamine antidepressants is uncertain.

Forced Swimming Test (FST)

The forced swimming test was performed according to the method of Porsolt et al. (1977). A vertical glass cylinder (25 cm high, 14 cm in diameter) was filled with water having a temperature of 30 °C to a depth of 20 cm. The water depth was adjusted so that the rat must swim or float without their hind limbs or tail touching the bottom. For testing, each rat was placed in the cylinder for 6 min., and the latency to float and the duration of floating (i.e., the time during which rat made only the small movements necessary to keep their heads above water) was scored. As suggested by Porsolt et al. (1977), only the data scored during the last 4 min were analyzed and presented (Fig. 11.3).

The advantages of FST are that it is low cost, a fast and reliable tool, easy to handle, and has proven its reliability across laboratories for testing potential antidepressants activities with a strong predictive validity. Besides, it allows rapid screening of large numbers of drugs. The major disadvantages of FST are that it has poor face and construct validities. The test is sensitive to acute treatment only, and its validity for non-monoamine antidepressants is uncertain.

Righting Reflex Test (RRT)

The *righting reflex* is a reflex that occurs when the body of an animal becomes inverted. It causes the body to turn around so that the animal is on its feet. The reflex may occur in the cerebral cortex, via visual cues, or in the midbrain via the labyrinthine or proprioception cues (Crabbe and Harris 1991). The reflex can be affected by alcohol consumption (Sukul and Sukul 2004).

If the animal is placed on its back, the animal quickly rights itself and assumes a normal posture. Neurological deficit is indicated by the inability to regain a normal posture within 5 min. Control and treated group rats were held upside down by the neck and lower back and dropped from a height of approximately 30 cm onto a cotton pad. The righting reflex was considered present if the rat landed on all four legs, receiving a score of “1.” Otherwise it is scored as “0” (Fig. 11.4).

Fig. 11.3 Forced swimming test (Source: English Wikipedia)

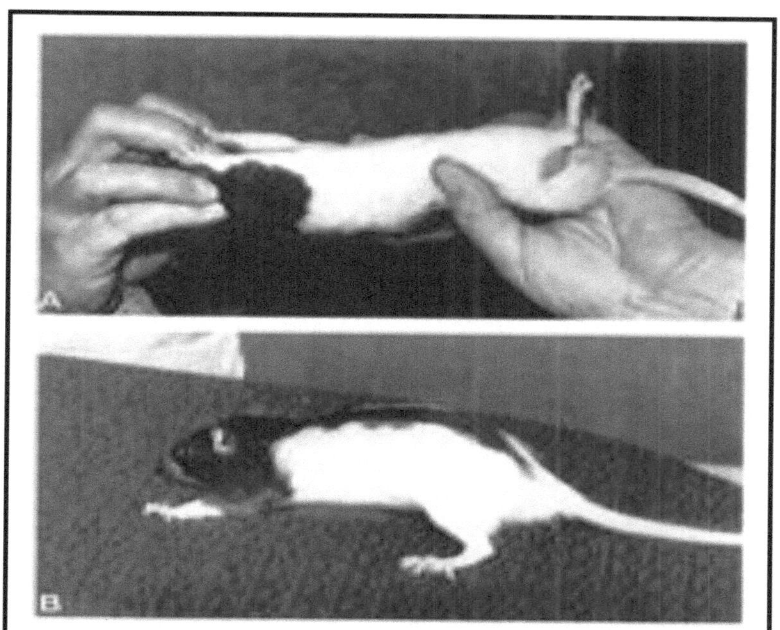

Fig. 11.4 Righting reflex illustrated (Source: English Wikipedia)

Anxiety Based

Open-Field Test

One of the most traditional and widely used methods for the assessment of the emotional state in rodents is the open-field test Tobach 1969, of which many varieties exist. Because this is a relatively simple technique and gives quantitative information on a broad range of responses, it has been used frequently in teratological studies (Di Giovanni et al. 1993; Spyker et al. 1972; Winneke et al. 1977; Cagiano et al. 1990). Locomotor activity was quantified for 5 min in an open field, a white Plexiglas box 59×59 cm with its floor divided into 16 squares. Four squares were defined as the center and the 12 squares along the walls as the periphery. Each rat was gently placed in the very center of the box and activity was scored as a line crossing when a rat removed all four paws from one square and entered another. Line crossings in the central four squares and in the peripheral 12 squares of the open field were counted separately. The time spent in the center was also scored.

Statistical Evaluation

Statistical analysis of data was done by one-factor analysis of variance (ANOVA) and Scheffe's F. Difference between groups was determined using post hoc Tukey's test followed by Student's *t*-test as described by Bennet and Franklin (1967), using software SPSS 11.0 for MS windows. In all the tests, the criterion for statistical significance was $p<0.05$.

References

Anisman H, Matheson K (2005) Stress, depression, and anhedonia: caveats concerning animal models. Neurosci Biobehav Rev 29:525–546

Bennet CA, Franklin NL (1967) Statistical analysis in chemistry and chemical industry. Wiley, New York

Cagiano R, De Salvia MA, Persichella M, Renna G, Tattoli M, Cuomo V (1990) Behavioral changes in the offspring of rats exposed to diazepam during gestation. Eur J Pharmacol 177:67–74

Carobrez AP, Bertoglio LJ (2005) Ethological and temporal analyses of anxiety-like behavior: the elevated plus-maze model 20 years on. Neurosci Biobehav Rev 29:1193–1205

Crabbe JC, Harris RA (1991) The genetic basis of alcohol and drug actions. Springer, New York. ISBN 0-306-43868-2

Di Giovanni V, Cagiano R, De Salvia MA, Giustino A, Lacombo C, Renna G, Cuomo V (1993) Neurobehavioral changes produced in rats by prenatal exposure to carbon monoxide. Brain Res 616:126–131

Engin E, Treit D (2008) The effects of intra-cerebral drug infusions on animals' unconditioned fear reactions: A systematic review. Prog Neuropsychopharmacol Biol Psychiatry 32(6):1399–1419

Fava M, Kendler KS (2000) Major depressive disorder. Neuron 28:335–341

Hasler G, Drevets WC, Manji HK, Charney DS (2004) Discovering endophenotypes for major depression. Neuropsychopharmacology 29:1765–1781

Hogg S (1996) A review of the validity and variability of the elevated plus-maze as an animal model of anxiety. Pharmacol Biochem Behav 54:21–30

Hua-Cheng Yan, Xiong Cao, Manas Das, Xin-Hong Zhu, Tian-Ming Gao (2010) Behavioral animal models of depression. Neurosci Bull 26(4):327–337

Kessler RC, Berglund P, Demler O, Jin R, Merikangas KR, Walters EE (2005) Lifetime prevalence and age-of-onset distributions of DSM-IV disorders in the National Comorbidity Survey Replication. Arch Gen Psychiatry 62(6):593–602

Lalonde R, Strazielle C (2008) Relations between open-field, elevated plus-maze, and emergence tests as displayed by C57/BL6J and BALB/c mice. J Neurosci Methods 171(1):48–52

Murray CJ, Lopez AD (1997) Alternative projections of mortality and disability by cause 1990–2020: global burden of disease study. Lancet 349:1498–1504

Pellow S, Chopin P, File SE, Briley M (1985) Validation of open-closed arm entries in an elevated plus-maze as a measure of anxiety in rat. J Neurosci Methods 14:149–167

Porsolt RD, Le Pichon M, Jalfre M (1977) Depression: a new animal model sensitive to antidepressant treatments. Nature 266:730–732

Rodgers RJ (1997) Animal models of anxiety: where next? Behav Pharmacol 8:477–496

Spyker JM, Sparber SB, Goldberg AM (1972) Subtle consequences of methylmercury exposure: behavioral deviations in offspring of treated mothers. Science 177:621–623

Steru L, Chermat R, Thierry B, Simon P (1985) The tail suspension test: a new method for screening antidepressants in mice. Psychopharmacology 85:367–370

Sukul NC, Sukul A (2004) High dilution effects: Physical and biochemical basis. Springer, New York. ISBN 1-4020-2155-0

Tobach E (1969) Experimental approaches to the study of emotional behavior. Ann N Y Acad Sci 159:621–1121

Treit D, Menard J, Royan C (1993) Anxiogenic stimuli in the elevated plus-maze. Pharmacol Biochem Behav 44:463–469

Vollmayr B, Mahlstedt MM, Henn FA (2007) Neurogenesis and depression: what animal models tell us about the link. Eur Arch Psychiatry Clin Neurosci 257:300–303

Willner P, Mitchell PJ (2002) The validity of animal models of predisposition to depression. Behav Pharmacol 13:169–188

Winneke G, Brockhaus A, Baltissen R (1977) Neurobehavioral and systemic effects of long term blood-lead elevation in rats. I: discrimination learning and open field behavior. Arch Toxicol 37:247–263

Part IV

Free Radical Stress

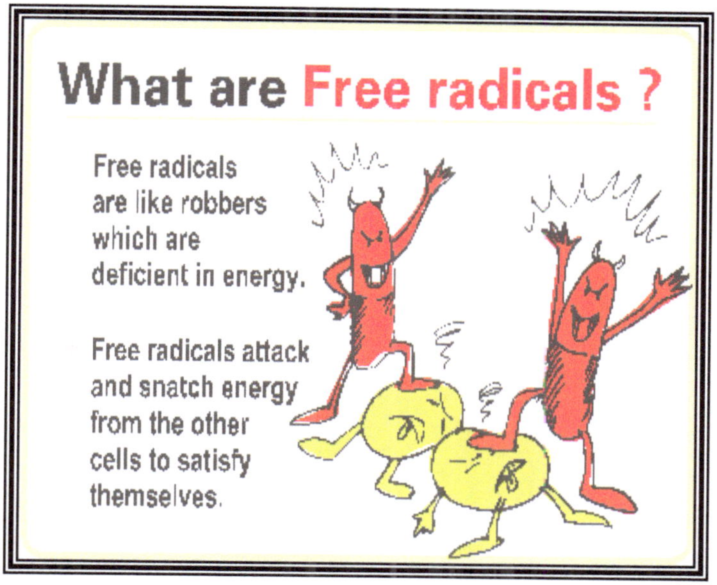

12 Mercury as a Source of Reactive Oxygen Species

Oxidation is a natural process of aging: the browning of an apple core exposed to air, or rust on metal. It is literally the dark side of oxygen.

Reactive oxygen species (ROS) include such chemical moieties as superoxide anion (O_2^-), hydrogen peroxide (H_2O_2), hydroxyl radical (HO^-), and peroxynitrite ($ONOO^-$) Boelsterli 2003. ROS are formed as a natural by-product of the normal metabolism of oxygen and has important roles in cell signaling. However, during times of environmental stress, ROS levels can increase dramatically, which can result in significant damage to cell structures. This cumulates into a situation known as *oxidative stress*. They are also generated by exogenous sources such as ionizing radiations. In humans, oxidative stress is involved in many diseases, such as atherosclerosis, Parkinson's disease, and Alzheimer's disease, but it may also be important in prevention of aging by induction of a process named mitohormesis. Reactive oxygen species can be beneficial, as they are used by the immune system as a way to attack and kill pathogens. Though ROS are produced at a basal level by the mitochondrial respiratory chain and in cytochrome p450 (CYP)-mediated oxidation processes (Boelsterli 2003), cells have endogenous antioxidants (e.g., cytochrome oxidase, superoxide dismutases, catalases, peroxidases, and glutathione, to name a few) as a means to control these substances (Buonocore et al. 2001). Certain disease states as well as pathologies resulting from toxic substances, including methylmercury, produce ROS in numbers that can overwhelm the cellular defense mechanisms and result in oxidative stress. The ROS can damage vital cellular molecules such as DNA, proteins, and lipids. Since ROS can also play a role via redox signaling in the regulation of signal transduction (Suzuki et al. 1997), excess production of these substances may prove lethal to the cell. Methylmercury-induced production of ROS will be addressed in more depth subsequently.

The actual measurement of ROS can be done by electron paramagnetic resonance spectroscopy which measures free radicals directly (Shi et al. 1998). However, this technique has some limitations, particularly for in vivo studies (Hwang and Kim 2007). Because the elusive nature of free radicals makes direct measurement of ROS impractical, these moieties are measured indirectly in many studies. Biomarkers of oxidative stress usually derive from the damage resulting when a radical interacts with biochemical (Fig. 12.1).

For example, ROS-induced DNA damage may be measured via assessment of levels of 8-hydroxydeoxyguanosine (Pilger and Rudiger 2006) and ROS-induced lipid peroxidation, via measurement of thiobarbituric acid reactive substance (Ohkawa et al. 1979). Many in vitro studies utilize tracking dyes to detect ROS (Setsukinai et al. 2003 and Uggeri et al. 2004) and calcium ions (Minta et al. 1989).

S. Nabi, *Toxic Effects of Mercury*, DOI 10.1007/978-81-322-1922-4_12,
© Springer India 2014

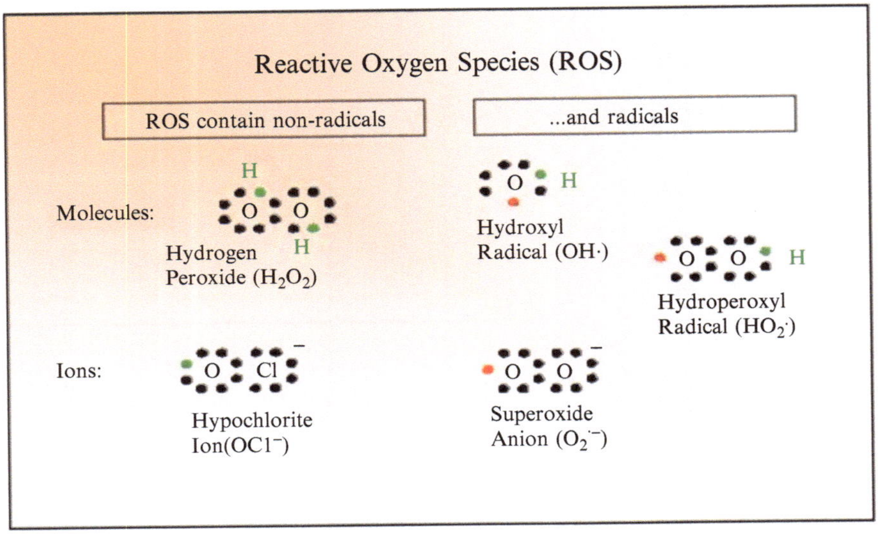

Fig. 12.1 Selected members of reactive oxygen species (ROS). ROS include both non-radicals and radicals. Most biologic molecules are non-radicals containing two electrons per orbital, which is a stable configuration in a molecule. A free radical is a molecule that can exist independently and contains one or more unpaired electrons. An unpaired electron means that there is only one electron in an orbital (shown in *red*), which is an unstable configuration and makes free radicals highly reactive (Nindl 2004)

Signaling and Damaging Effects

Normally, cells defend themselves against ROS damage with enzymes such as alpha-1-microglobulin, superoxide dismutases, catalases, lactoperoxidases, glutathione peroxidases, and peroxiredoxins. Small-molecule antioxidants such as ascorbic acid (vitamin C), tocopherol (vitamin E), uric acid, and glutathione also play important roles as cellular antioxidants. In similar manner, polyphenol antioxidants assist in preventing ROS damage by scavenging free radicals. In contrast, the antioxidant ability of the extracellular space is less – e.g., the most important plasma antioxidant in humans is uric acid.

Effects of ROS on cell metabolism are well documented in a variety of species. These include not only roles in apoptosis (programmed cell death) but also positive effects such as the induction of host defense (Rada and Leto 2008; Conner et al. 2002) genes and mobilization of ion transport systems. This implicates them in redox signaling, also known as "oxidative signaling." In particular, platelets involved in wound repair and blood homeostasis release ROS to recruit additional platelets to sites of injury. These also provide a link to the adaptive immune system via the recruitment of leukocytes.

Reactive oxygen species are implicated in cellular activity to a variety of inflammatory responses including cardiovascular disease. They may also be involved in hearing impairment via cochlear damage induced by elevated sound levels, in ototoxicity of drugs such as cisplatin, and in congenital deafness in both animals and humans. Redox signaling is also implicated in mediation of apoptosis or programmed cell death and ischemic injury. Specific examples include stroke and heart attack.

In general, harmful effects of reactive oxygen species on the cell are most often (Brooker 2011):

1. Damage of DNA
2. Oxidations of polyunsaturated fatty acids in lipids (lipid peroxidation)
3. Oxidations of amino acids in proteins
4. Oxidatively inactivate specific enzymes by oxidation of cofactors

Oxidative Damage

In aerobic organisms, the energy needed to fuel biological functions is produced in the mitochondria via the electron transport chain. In addition to energy, reactive oxygen species (ROS) with the potential to cause cellular damage are produced. ROS can damage DNA, RNA, and proteins, which, in theory, contributes to the physiology of aging.

ROS are produced as a normal product of cellular metabolism. In particular, one major contributor to oxidative damage is hydrogen peroxide (H_2O_2), which is converted from superoxide that leaks from the mitochondria. Catalase and superoxide dismutase ameliorate the damaging effects of hydrogen peroxide and superoxide by converting these compounds into oxygen and water, benign molecules. However, this conversion is not 100 % efficient, and residual peroxides persist in the cell. While ROS are produced as a product of normal cellular functioning, excessive amounts can cause deleterious effects (Patel et al. 1999). Memory capabilities decline with age, evident in human degenerative diseases such as Alzheimer's disease, which is accompanied by an accumulation of oxidative damage. Current studies demonstrate that the accumulation of ROS can decrease an organism's fitness because oxidative damage is a contributor to senescence. In particular, the accumulation of oxidative damage may lead to cognitive dysfunction, as demonstrated in a study in which old rats were given mitochondrial metabolites and then given cognitive tests. Results showed that the rats performed better after receiving the metabolites, suggesting that the metabolites reduced oxidative damage and improved mitochondrial function (Liu et al. 2002). Accumulating oxidative damage can then affect the efficiency of mitochondria and further increase the rate of ROS production (Stadtman 1992). The accumulation of oxidative damage and its implications for aging depends on the particular tissue type where the damage is occurring. Additional experimental results suggest that oxidative damage is responsible for age-related decline in brain functioning. Older gerbils were found to have higher levels of oxidized protein in comparison to younger gerbils. Treatment of old and young mice with a spin trapping compound caused a decrease in the level of oxidized proteins in older gerbils but did not have an effect on younger gerbils. In addition, older gerbils performed cognitive tasks better during treatment but ceased functional capacity when treatment was discontinued, causing oxidized protein levels to increase. This led researchers to conclude that oxidation of cellular proteins is potentially important for brain function (Carney et al. 1991).

Internal Production

Free radicals are mainly produced inside organelles, such as the mitochondrion, and also released toward the cytosol (Muller 2000; Han et al. 2001). Mitochondria convert energy for the cell into a usable form, adenosine triphosphate (ATP). The process in which ATP is produced, called oxidative phosphorylation, involves the transport of protons (hydrogen ions) across the inner mitochondrial membrane by means of the electron transport chain. In the electron transport chain, electrons are passed through a series of proteins via oxidation–reduction reactions, with each acceptor protein along the chain having a greater reduction potential than the previous. The last destination for an electron along this chain is an oxygen molecule. In normal conditions, the oxygen is reduced to produce water; however, in about 0.1–2 % of electrons passing through the chain (this number derives from studies in isolated mitochondria, though the exact rate in live organisms is yet to be fully agreed upon), oxygen is instead prematurely and incompletely reduced to give the superoxide radical (O_2^-), most well documented for complex I and complex III. Superoxide is not particularly reactive by itself, but can inactivate specific enzymes or initiate lipid peroxidation in its protonated form, hydroperoxyl HO_2. The pKa of hydroperoxyl is 4.8. Thus, at physiological pH, the majority will exist as superoxide.

If too much damage is caused to its mitochondria, a cell undergoes apoptosis or programmed cell death. Bcl-2 proteins are layered on the surface of the mitochondria, detect damage, and activate a class of proteins called Bax, which punch holes in the mitochondrial membrane, causing cytochrome C to leak out. This cytochrome C binds to Apaf-1, or apoptotic protease activating factor-1, which is free-floating in the cell's cytoplasm. Using energy from the ATPs in the mitochondrion, the Apaf-1 and cytochrome C bind together to form apoptosomes. The apoptosomes bind to and activate caspase-9, another free-floating protein. The caspase-9 then cleaves the proteins of the mitochondrial membrane, causing it to break down and start a chain reaction of protein denaturation and eventually phagocytosis of the cell.

Mitochondria are sometimes described as "cellular power plants" because they generate most of the cell's supply of adenosine triphosphate (ATP), a source of chemical energy. Reactive oxygen species (ROS) have been regarded as unwanted by-products of oxidative phosphorylation in mitochondria by the proponents of the free radical theory of aging promoted by Denham Harman. The free radical theory suggests that the use of compounds which inactivate ROS, such as so-called antioxidants, would lead to a reduction of oxidative stress and thereby produce an increase in life span. However, in clinical trials, the expected health-promoting effects of antioxidants were not found and results suggested that antioxidants may promote cancer in humans. Other studies have now shown that antioxidant supplements may be disease promoting and increase mortality in humans (Ristow and Zarse 2010). ROS may perform an essential and potentially life span-promoting role as redox signaling molecules which transduce signals from the mitochondrial compartment to other compartments of the cell. Increased formation of reactive oxygen species (ROS) within the mitochondria may cause an adaptive reaction which produces increased stress resistance and a long-term reduction of oxidative stress. This kind of reverse effect of the response to ROS stress has been named mitochondrial hormesis or mitohormesis and is hypothesized to be responsible for the respective life span-extending and health-promoting capabilities of glucose restriction and physical exercise (Ristow and Zarse 2010).

Hormesis may also be induced by endogenously produced, potentially toxic agents. For example, mitochondria consume oxygen which generates free radicals (reactive oxygen species) as an inevitable by-product. It was previously proposed on a hypothetical basis that such free radicals may induce an endogenous response culminating in increased defense capacity against exogenous radicals (and possibly other toxic compounds) (Tapia 2006). Recent experimental evidence strongly suggests that this is indeed the case and that such induction of endogenous free radical production extends life span of a model organism. Most importantly, this extension of life span is prevented by antioxidants, providing direct evidence that toxic radicals may mitohormetically exert life-extending and health-promoting effects (Schulz et al. 2007). Since mitochondrial activity was found to be increased in the before-mentioned studies, this effect cannot be explained by an excess of free radicals that might mark mitochondria for destruction by lysosomes, with the free radicals acting as a signal within the cell to indicate which mitochondria are ready for destruction, as proposed by Nick Lane (2006).

Another study on yeasts explains the underlying mechanism of the process that prevents cellular damage by reactive oxygen species (Kelley and Ideker 2009). Whether this concept applies to humans remains to be shown, although recent epidemiological findings support the process of mitohormesis and even suggest that some antioxidant supplements may increase disease prevalence in humans (Bjelakovic et al. 2007).

References

Bjelakovic G, Nikolova D, Gluud LL, Simonetti RG, Gluud C (2007) Mortality in randomized trials of antioxidant supplements for primary and secondary prevention: systematic review and meta-analysis. JAMA 297(8):842–857

Boelsterli UA (2003) The molecular basis of how chemicals disrupt biological targets. In: Mechanistic toxicology, 2nd edn. Taylor and Francis, London/New York, p 314

Brooker RJ (2011) Genetics: analysis and principles, 4th edn. McGraw-Hill Science, Boston. ISBN 978-0-07-352528-0

Buonocore G, Perrone S, Bracci R (2001) Free radicals and brain damage in the newborn. Biol Neonate 79:180–186

Carney JM, Starke-Reed PE, Oliver CN, Landum RW, Cheng MS, Wu JF, Floyd RA (1991) Reversal of age-related increase in brain protein oxidation, decrease in enzyme activity, and loss in temporal and spatial memory by chronic administration of the spin-trapping compound N-tert-butyl-alpha-phenylnitrone. Proc Natl Acad Sci 88(9):3633–3636

Conner GE, Salathe M, Forteza R (2002) Lactoperoxidase and hydrogen peroxide metabolism in the airway. Am J Respir Crit Care Med 166(12):S57–S61

Han D, Williams E, Cadenas E (2001) Mitochondrial respiratory chain-dependent generation of superoxide anion and its release into the intermembrane space. Biochem J 353(Pt 2):411–416

Hwang ES, Kim GH (2007) Biomarkers for oxidative stress status of DNA, lipids, and proteins in vitro and in vivo cancer research. Toxicology 229:1–10

Kelley R, Ideker T, Gibson G (2009) Genome-wide fitness and expression profiling implicate Mga2 in adaptation to hydrogen peroxide. PLoS Genet 5(5):e1000488

Lane N (2006) Power, sex, suicide: mitochondria and the meaning of life. Oxford University Press, Oxford. ISBN 0-19-920564-7

Liu J, Head E, Gharib AM, Yuan W, Ingersoll RT, Hagen TM, Cotman CW, Ames BN (2002) Memory loss in old rats is associated with brain mitochondrial decay and RNA/DNA oxidation: partial reversal by feeding acetyl-L-carnitine and/or R-alpha-lipoic acid. Proc Natl Acad Sci U S A 99(4):2356–2361

Minta A, Kao JPY, Tsien RY (1989) Fluorescent indicators for cytosolic calcium based on rhodamine and fluorescein chromophores. J Biol Chem 264: 8171–8178

Muller F (2000) The nature and mechanism of superoxide production by the electron transport chain: its relevance to aging. Age 23(4):227–253

Nindl G (2004) Hydrogen peroxide-from oxidative stressor to redox regulator. Cell Sci 1:1–12

Ohkawa H, Ohishi N, Yagi K (1979) Assay for lipid peroxides in animal tissue by thiobarbituric reaction. Anal Biochem 95:351–358

Patel RP, Cornwell T, Darley-Usmar VM (1999) The biochemistry of nitric oxide and peroxynitrite: implications for mitochondrial function. In: Packer L, Cadenas E (eds) Understanding the process of aging: the roles of mitochondria, free radicals, and antioxidants. Marcel Dekker, New York, pp 39–56. ISBN 0-8247-1723-6

Pilger A, Rudiger HW (2006) 8-Hydroxy-20-deoxyguanosine as a marker of oxidative DNA damage related to occupational and environmental exposure. Int Arch Occup Environ Health 80:1–15

Rada B, Leto TL (2008) Oxidative innate immune defenses by Nox/Duox family NADPH oxidases. Contrib Microbiol 15:164–187

Ristow M, Zarse K (2010) How increased oxidative stress promotes longevity and metabolic health: the concept of mitochondrial hormesis (mitohormesis). Exp Gerontol 45(6):410–418

Schulz TJ, Zarse K, Voigt A, Urban N, Birringer M, Ristow M (2007) Glucose restriction extends Caenorhabditis elegans life span by inducing mitochondrial respiration and increasing oxidative stress. Cell Metab 6(4):280–293

Setsukinai KI, Urano Y, Kakinuma K, Majima H, Nagano T (2003) Development of novel fluorescence probes that can reliably detect oxygen species and distinguish specific species. J Biol Chem 278:3170–3175

Shi X, Castranova V, Halliwell B, Vallyathan V (1998) Reactive oxygen species and silica-induced carcinogenesis. J Toxicol Environ Health B Crit Rev 1:181–197

Stadtman ER (1992) Protein oxidation and aging. Science 257(5074):1220–1224

Suzuki YJ, Forman HJ, Sevanian A (1997) Oxidants as stimulators of signal transduction. Free Radic Biol Med 22:269–285

Tapia PC (2006) Sublethal mitochondrial stress with an attendant stoichiometric augmentation of reactive oxygen species may precipitate many of the beneficial alterations in cellular physiology produced by caloric restriction, intermittent fasting, exercise and dietary phytonutrients: "Mitohormesis" for health and vitality. Med Hypotheses 66(4):832–843

Uggeri J, Gatti R, Belletti S, Scandroglio R, Corradini R, Rotoli BM, Orlandini G (2004) Calcein-AM is a detector of intracellular oxidative activity. Histochem Cell Biol 122:499–505

13 Promotion of Peroxidation and Hydroperoxidation of Lipids

Lipid peroxidation can be defined as the oxidative deterioration of lipids containing any number of carbon-carbon double bonds.

Polyunsaturated fatty acids (PUFAs) are abundant in cellular membranes and in low-density lipoproteins (LDL) (Dekkers et al. 1996). The PUFAs allow for fluidity of cellular membranes. A free radical prefers to steel electrons from the lipid membrane of a cell, initiating a free radical attack on the cell known as *lipid peroxidation*. Lipid peroxidation, a well-established mechanism of cellular injury in plants and animals, is used as an indicator of oxidative stress in cells and tissues. Lipid peroxides are unstable and decompose to form a complex series of compounds including reactive carbonyl compounds. Polyunsaturated fatty acid peroxides generate malondialdehyde (MDA) and 4-hydroxyalkenals (HAE) upon decomposition, and the measurement of MDA and HAE has been used as an indicator of lipid peroxidation. Reactive oxygen species target the carbon–carbon double bond of polyunsaturated fatty acids. The double bond on the carbon weakens the carbon–hydrogen bond allowing for easy dissociation of the hydrogen by a free radical. A free radical will steal the single electron from the hydrogen associated with the carbon at the double bond. In turn this leaves the carbon with an unpaired electron and hence becomes a free radical. In an effort to stabilize the carbon-centered free radical, molecular rearrangement occurs. The newly arranged molecule is called a conjugated diene (CD). The CD then very easily reacts with oxygen to form a peroxyl radical. The peroxyl radical steals an electron from another lipid molecule in a process called propagation. This process then continues in a chain reaction (Halliwell and Gutteridge 1985).

Oxidative stress that occurs in the cells, as a consequence of an inequity between the prooxidant/antioxidant systems, causes injury to biomolecules such as nucleic acids, proteins, structural carbohydrates, and lipids (Sies and Cadenas 1985). Among these targets, the peroxidation of lipids is basically damaging because the formation of lipid peroxidation products leads to spread of free radical reactions. The general process of lipid peroxidation consists of three stages: initiation, propagation, and termination (Catala 2006). The initiation phase of lipid peroxidation includes hydrogen atom abstraction. Several species can abstract the first hydrogen atom and include the radicals: hydroxyl (•OH), alkoxyl (RO•), peroxyl (ROO•), and possibly HO_2• but not H_2O_2 or O_2^{-}• (Gutteridge 1988). The membrane lipids, mainly phospholipids, containing polyunsaturated fatty acids are predominantly susceptible to peroxidation because of a hydrogen atom abstraction from a methylene (–CH_2–) group, which contains only one electron and leaves at the back an unpaired electron on the carbon, -•CH–. The presence of a double bond in the fatty acid weakens the C–H

S. Nabi, *Toxic Effects of Mercury*, DOI 10.1007/978-81-322-1922-4_13,
© Springer India 2014

bonds on the carbon atom nearby to the double bond and thus facilitates H• subtraction. The initial reaction of •OH with polyunsaturated fatty acids produces a lipid radical (L•), which in turn reacts with molecular oxygen to form a lipid peroxyl radical (LOO•). The LOO• can abstract hydrogen from a neighboring fatty acid to produce a lipid hydroperoxide (LOOH) and a second lipid radical (Catala 2006). The LOOH formed can suffer reductive cleavage by reduced metals, such as Fe^{++}, producing lipid alkoxyl radical (LO•). Both alkoxyl and peroxyl radicals stimulate the chain reaction of lipid peroxidation by abstracting additional hydrogen atoms (Buettner 1993). Peroxidation of lipids can disturb the assembly of the membrane, causing changes in fluidity and permeability, alterations of ion transport, and inhibition of metabolic processes (Nigam and Schewe 2000). Injury to mitochondria induced by lipid peroxidation can direct to further ROS generation (Green and Reed 1998). In addition, LOOH can break down, frequently in the presence of reduced metals or ascorbate, to reactive aldehyde products, including malondialdehyde (MDA), 4-hydroxy-2-nonenal (HNE), 4-hydroxy-2-hexenal (4-HHE), and acrolein (Esterbauer et al. 1991; Uchida et al. 1999; Kehrer and Biswal 2000; Lee et al. 2001). A great variety of compounds are formed during lipid peroxidation of membrane phospholipids. Lipid peroxidation is one of the major outcomes of free radical-mediated injury to tissue. Peroxidation of fatty acyl groups occurs mostly in membrane phospholipids. Peroxidation of lipids can greatly alter the physicochemical properties of membrane lipid bilayers, resulting in severe cellular dysfunction. In addition, a variety of lipid by-products are produced as a consequence of lipid peroxidation, some of which can exert adverse and/or beneficial biological effects (Fig. 13.1).

Several mechanisms by which Hg exerts its adverse effects have been hypothesized in acute exposure to high levels, among them impairment of intracellular calcium homeostasis (Aschner et al. 2007; Dietrich et al. 2005; Sirois and Atchison 2000) and alteration of glutamate homeostasis (Aschner et al. 2000) were reported. However, it is believed that the major process responsible for triggering the toxicity is the oxidative stress. Hg has a great affinity for –SH groups, attaching thiol-containing proteins and small molecular weight thiols, such as glutathione (GSH) (Clarkson 1997), decreasing the main nonprotein thiol involved in the antioxidant cellular defense against reactive oxygen species (ROS) (Sarafian 1999; Grotto et al. 2009). Moreover, Hg is related to changes in the activities of antioxidant enzymes, such as superoxide dismutase (Ariza et al. 1998) and catalase (Hussain et al. 1999), besides inducing ROS production, especially of H_2O_2 and O_2^-(Lund et al. 1991), and promoting oxidative stress and lipid peroxidation (Carvalho et al. 2007; Jina et al. 2007; Stringari et al. 2008). Malondialdehyde (MDA) is one of the most known and important secondary products of lipid peroxidation, and it has been used as a biomarker of cell membrane injury (Esterbauer et al. 1991) in many diseases, including cardiovascular pathogenesis (Heinecke 1998). Acute exposure to MeHg also induced concentration-dependent cytotoxicity in endothelial cells and a reduction in NO synthase (NOS) activity (Kishimoto et al. 1995).

In animal tissue, largest amount of lipids occur in the brain followed by the liver, pancreas, heart muscles, diaphragm, and neck muscles (Bloor 1944). In the brain, the lipids account for half of the dry weight and most of the structural architecture of membranes in the brain (Ordy and Kaack 1975).

The present experiment was designed to study the promotion of peroxidation and hydroperoxidation of lipids by low dose of methylmercury intoxication, and its connection to protein concentration of tissues was also evaluated. In addition, antioxidant effect of vitamin E and acetyl-L-carnitine against methylmercury toxicity was also assessed.

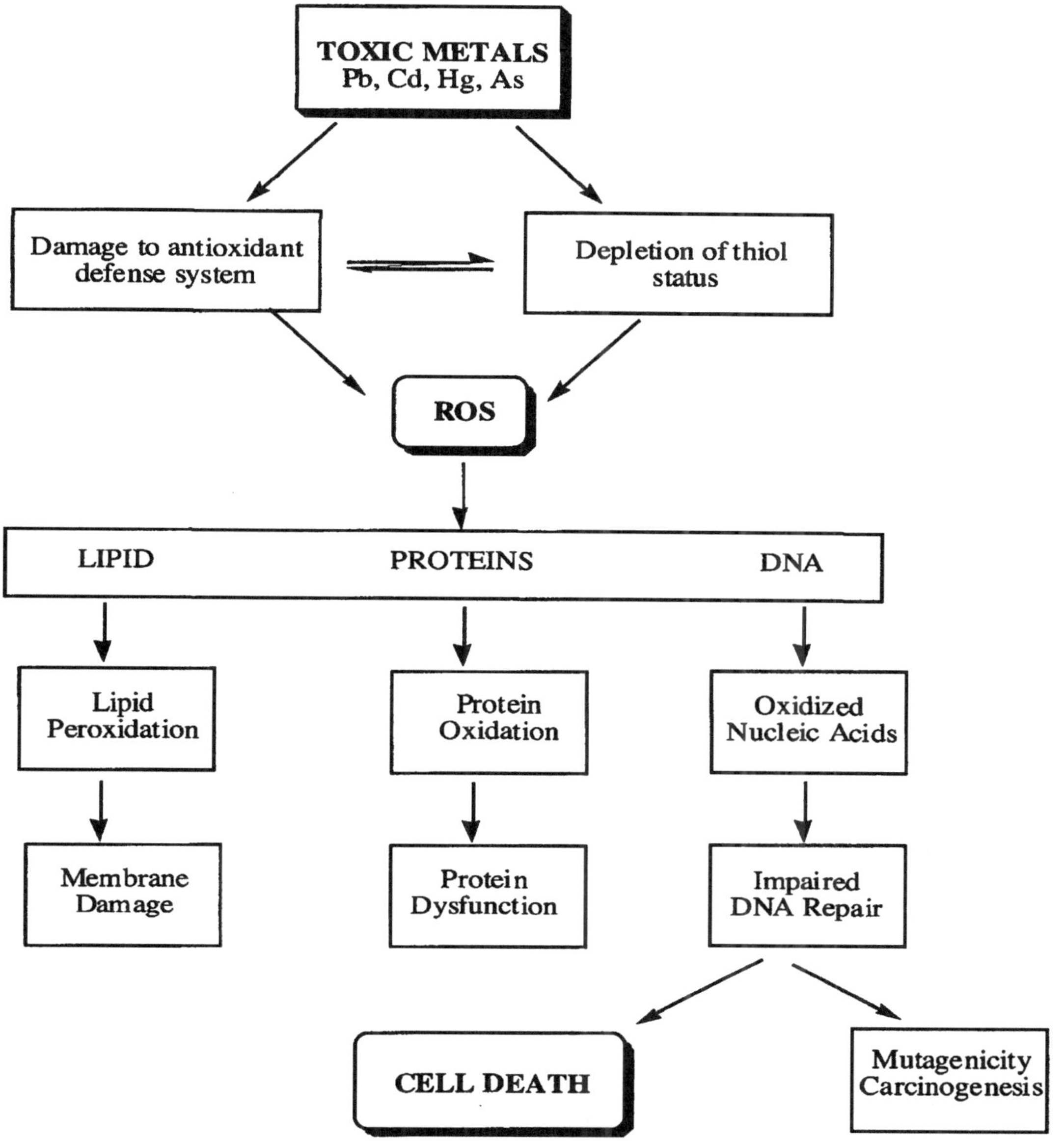

Fig. 13.1 Possible mechanisms for metal-induced oxidative stress (Ercal et al. 2001)

References

Ariza ME, Bijur GN, Williams MV (1998) Lead and mercury mutagenesis: role of H_2O_2, superoxide dismutase, and xanthine oxidase. Environ Mol Mutagen 31:352–361

Aschner M, Yao CP, Allen JW, Tan KH (2000) Methylmercury alters glutamate transport in astrocytes. Neurochem Int 37:188–206

Aschner M, Syversen T, Souza DO, Rocha JBT, Farina M (2007) Involvement of glutamate and reactive oxygen species in methylmercury neurotoxicity. Braz J Med Biol Res 40:285–291

Bloor WR (1944) Biochemistry of the fatty acids and their compounds, the lipids. Am J Med Sci 207:278

Buettner GR (1993) The pecking order of free radicals and antioxidants: lipid peroxidation, α-tocopherol, and ascorbate. Arch Biochem Biophys 300:535–543

Carvalho MC, Franco JL, Ghizonia H, Kobusa K, Nazaria EM, Rocha JBT, Nogueira CW, Dafred AL, Mullera YMR, Farina M (2007) Effects of 2, 3-dimercapto-1-propanesulfonic acid (DMPS) on methylmercury-induced locomotor deficits and cerebellar toxicity in mice. Toxicology 239:195–203

Catala A (2006) An overview of lipid peroxidation with emphasis in outer segments of photoreceptors and the chemiluminescence assay. Int J Biochem Cell Biol 38:1482–1495

Clarkson TW (1997) The toxicity of mercury. Crit Rev Clin Lab Sci 34:369–403

Dekkers JC, Von Doormen LJP, Han CGK (1996) The role of antioxidant vitamins and enzymes in the prevention of exercise-induced muscle damage. Sports Med 21:213–238

Dietrich MO, Mantese CE, Anjos GD, Souza DO, Farina M (2005) Motor impairment induced by oral exposure to methylmercury in adult mice. Environ Toxicol Pharmacol 19:169–175

Ercal N, Gurer-Orhan H, Aykin-Burns N (2001) Toxic metals and oxidative stress part I: mechanisms involved in metal induced oxidative damage. Curr Top Med Chem 1:529–539

Esterbauer H, Schaur RJ, Zollner H (1991) Chemistry and biochemistry of 4-hydroxynoneal malondialdehyde and related aldehydes. Free Radic Biol Med 11:81–128

Green DR, Reed JC (1998) Mitochondria and apoptosis. Science 281:1309–1312

Grotto D, Barcelos GRM, Valentini J, Antunes MGA, Angeli JPA, Garcia SC, Barbosa F (2009) Low levels of methylmercury induce DNA damage in rats: protective effects of selenium. Arch Toxicol 83:249–254

Gutteridge JMC (1988) Lipid peroxidation: some problems and concepts. In: Halliwell B (ed) Oxygen radicals and tissue injury. Upjohn Co. and FASEB, Bethesda, pp 9–19

Halliwell B, Gutteridge JMC (1985) The chemistry of oxygen radicals and other oxygen-derived species. In: Free radicals in biology and medicine. Oxford University Press, New York, pp 20–64

Heinecke JW (1998) Oxidants and antioxidants in the pathogenesis of atherosclerosis: implications for the oxidized low density lipoprotein hypothesis. Atherosclerosis 141:1–15

Hussain S, Atkinson A, Thompson SJ, Khan AT (1999) Accumulation of mercury and its effect on antioxidant enzymes in brain, liver, and kidneys of mice. J Environ Sci Health B 34:645–660

Jina X, Loka E, Bondya G, Caldwella D, Muellera R, Kapala K, Cheryl A, Taylora M, Kubowc S, Mehtaa R, Chan HM (2007) Modulating effects of dietary fats on methylmercury toxicity and distribution in rats. Toxicology 230:22–44

Kehrer JP, Biswal SS (2000) The molecular effects of acrolein. Toxicol Sci 57:6–15

Kishimoto T, Oguri T, Tada M (1995) Effect of methylmercury (CH_3HgCl) injury on nitric oxide synthase (NOS) activity in cultured human umbilical vascular endothelial cells. Toxicology 103:1–7

Lee YW, Ha MS, Kim YK (2001) Role of reactive oxygen species and glutathione in inorganic mercury-induced injury in human glioma cells. Neurochem Res 26:1187–1193

Lund BO, Miler DM, Woods JS (1991) Mercury-induced H_2O_2 formation and lipid peroxidation in vitro in rat kidney mitochondria. Biochem Pharmacol 42:181–187

Nigam S, Schewe T (2000) Phospholipase A_2s and lipid peroxidation. Biochim Biophys Acta 1488:167–181

Ordy JM, Kaack B (1975) Neurochemical changes in composition, metabolism and neurotransmitters in human brain with age. In: Ordy JM, Brizzee KR (eds) Neurobiology of aging. Plenum Press, London, p 264

Sarafian TA (1999) Methylmercury-induced generation of free radical: biological implications. Met Ions Biol Syst 36:415–444

Sies H, Cadenas E (1985) Oxidative stress: damage to intact cells and organs. Philos Trans R Soc Lond B Biol Sci 311:617–631

Sirois JE, Atchison WD (2000) Methylmercury affects multiple subtypes of calcium channels in rat cerebellar granule cells. Toxicol Appl Pharmacol 167:1–11

Stringari J, Nunes AKC, Franco JL, Bohrer D, Garcia SC, Dafre AL, Milatovic D, Souza DO, Rocha JBT, Aschner M, Farina M (2008) Prenatal methylmercury exposure hampers glutathione antioxidant system ontogenesis and causes long-lasting oxidative stress in the mouse brain. Toxicol Appl Pharmacol 227:147–154

Uchida K, Shiraishi M, Naito Y, Torii Y, Nakamura Y, Osawa T (1999) Activation of stress signaling pathways by the end product of lipid peroxidation: 4-hydroxy-2-nonenal is a potential inducer of intracellular peroxide production. J Biol Chem 274:2234–2242

Results and Conclusions 14

Protocol

For various sets of biochemical studies, different groups comprising ten rats each were used. Rats from group I served as control, while rats of groups II, III, IV, and V were used as experimental sets. Group II rats were given 2 mg/kg body weight of methylmercury chloride for 14 days, and for the next 14 days, they were kept untreated. Animals of group III received MeHgCl (2 mg/kg body weight) for 14 days and vitamin E (100 mg/kg body weight) for the next 14 days. Group IV animals were given MeHgCl (2 mg/kg body weight) for 14 days, and for the next 14 days, they were treated with acetyl-L-carnitine (100 mg/kg body weight). Group V animals were treated with MeHgCl (2 mg/kg body weight) for 14 days, and for the next 14 days, they were given vitamin E (100 mg/kg body weight) and acetyl-L-carnitine (100 mg/kg body weight) in combination. In combined therapy, acetyl-L-carnitine was always administered at a gap of 30 min after vitamin E as per Sood et al. (1997). Methylmercury chloride and acetyl-L-carnitine were diluted in physiological saline, while vitamin E was given as such. All groups were treated once a day orally through intragastric intubation. The intake of drinking water and food by rats was examined daily, and rats were weighed every other day for weight change assessment due to toxic metal. The animals were sacrificed later on the scheduled day by cervical dislocation, and immediately brains, spinal cords, hearts, lungs, and pancreases were taken out and kept on ice. Tissues were weighed, both in control and treated animals, to observe the weight changes. Brains were separated into the cerebrum, cerebellum, and brain stem. The tissues were later processed for the assay of lipid peroxidation, lipid hydroperoxidation, and protein concentration by standard methods described in detail in Methodology chapter.

Results

Lipid Peroxide

In methylmercury chloride-treated rats, the contents of lipid peroxide in nervous tissues were lowest in the brain stem and highest in the spinal cord. In non-nervous tissues, the pancreas was more affected by methylmercury chloride intoxication than the lung and heart. The lowest change was analyzed in the lung. The levels of lipid peroxide in the above-mentioned tissues showed values increasing in the following order: nervous tissues (spinal cord–cerebellum–cerebrum–brain stem) and non-nervous tissues (pancreas–heart–lung). The alterations in lipid peroxide levels were more acute in nervous tissues as compared to non-nervous tissues.

S. Nabi, *Toxic Effects of Mercury*, DOI 10.1007/978-81-322-1922-4_14,
© Springer India 2014

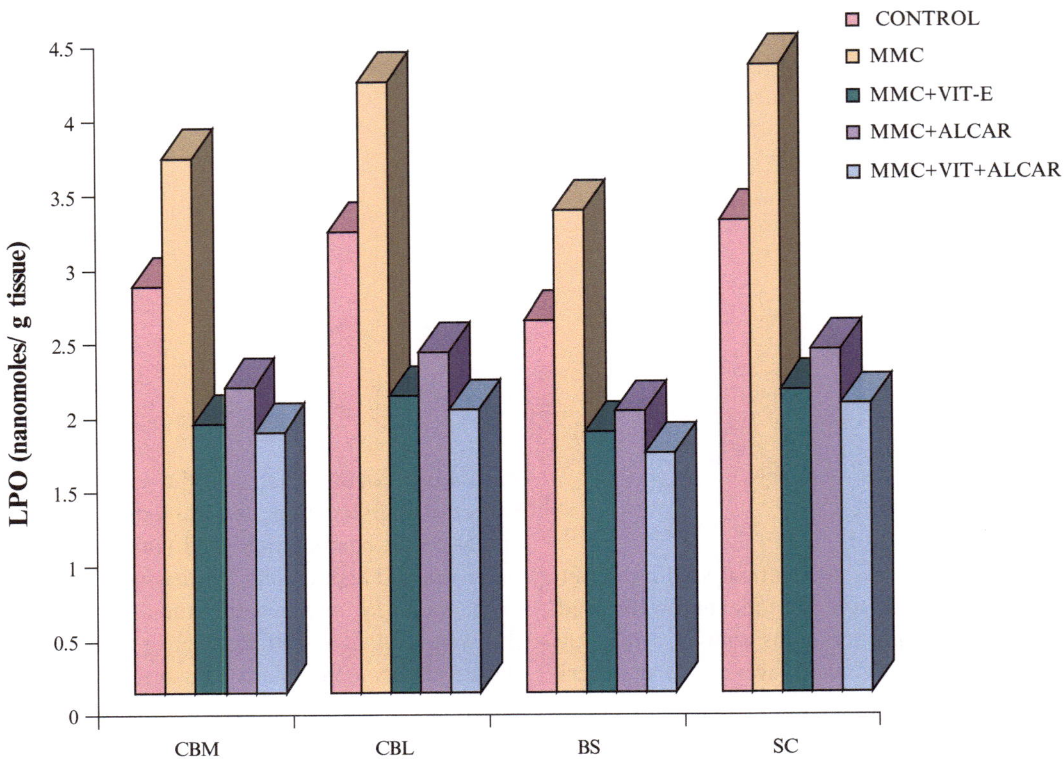

Fig. 14.1 Effect of methylmercury chloride on lipid peroxide (LPO) level in nervous tissues of rats. Values represent mean ± SEM. *CBM* cerebrum, *CBL* cerebellum, *BS* brain stem, *SC* spinal cord

The above-mentioned changes are diagrammatically represented in Figs. 14.1 and 14.2.

Effect of Methylmercury Chloride on Lipid Peroxide Level

After continuous 14-day treatment with methylmercury chloride, the lipid peroxide levels were significantly elevated in various regions of CNS and other tissues. The increase in lipid peroxide contents in the spinal cord was +32.38 %, in cerebellum +32.15 %, in cerebrum +31.38 %, in brain stem +30.00 %, in pancreas +29.31 %, in heart +28.82 %, and in lung +28.31 % when experimental rats were compared with the control ones. ANOVA revealed that the difference between the control and treated animals was significant at $p<0.001$ and $p<0.01$ in nervous tissues and at $p<0.05$ in non-nervous tissues.

Effect of Vitamin E and Acetyl-L-Carnitine on Lipid Peroxide Level

A significant depletion in the lipid peroxide level was observed in various regions of the brain, spinal cord, and other tissues of rats when they were administered vitamin E and acetyl-L-carnitine after methylmercury chloride treatment, separately or in combination orally for 14 days. Interestingly when methylmercury chloride-intoxicated rats were treated with vitamins, a remarkable recovery of tissues from free radical damage was observed. However, the recovery was more significant in rats treated with vitamins in combined way. Among the various tissues, the spinal cord showed the best results during vitamin therapy causing approximately −51.78 % and −45.36 % recovery with vitamin E and acetyl-L-carnitine, respectively, when given separately. But when

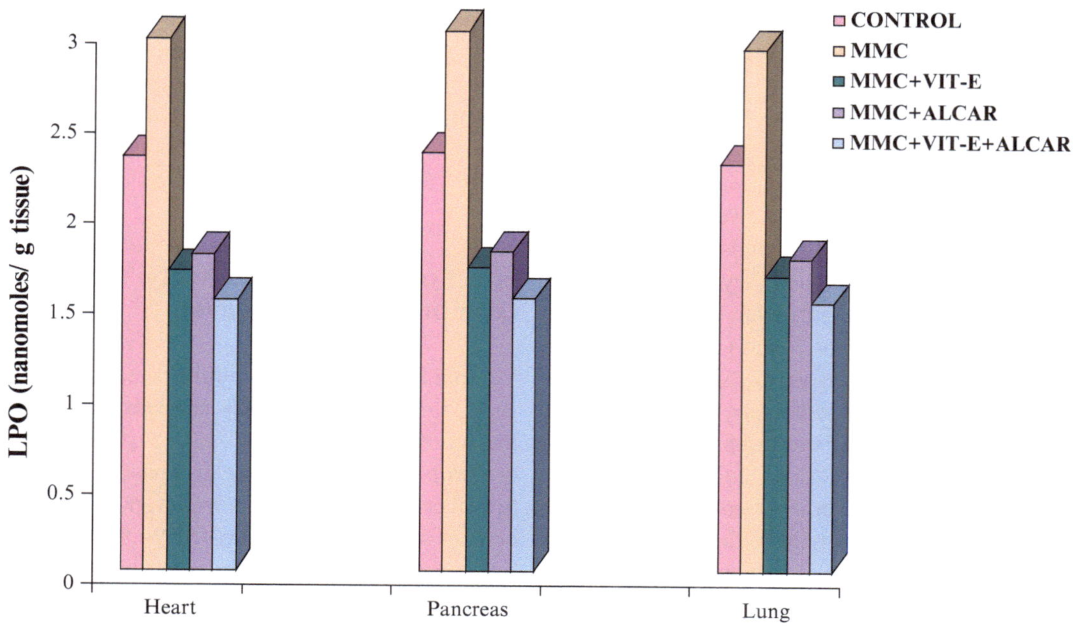

Fig. 14.2 Effect of methylmercury chloride on lipid peroxide (LPO) level in non-nervous tissues of rats. Values represent mean ± SEM

antioxidants were given in combination, the spinal cord showed −53.91 % recoveries. In the cerebrum, cerebellum, and brain stem, the recovery was −49.72 %, −51.33 %, and −46.15 % with vitamin E, respectively, and −43.05 %, −44.28 %, and −41.84 % with acetyl-L-carnitine, respectively, and −51.38 %, −53.77 %, and −50.46 % with vitamin E + acetyl-L-carnitine, respectively. In the pancreas, vitamin E showed much more recovery than acetyl-L-carnitine, but the protection conferred by their combined therapy was much better than separate treatment. In the lung, the recovery was lowest.

Lipid Hydroperoxidation

The amount of lipid hydroperoxide found in various regions of CNS and other tissues of rats treated with methylmercury chloride is given below. The lipid hydroperoxide levels were highest in the spinal cord followed by cerebellum, cerebrum, brain stem, pancreas, heart, and lung in decreasing order.

Effect of Methylmercury Chloride on Lipid Hydroperoxide Level

Methylmercury chloride-induced lipid hydroperoxide elevation was observed in brain parts, spinal cord, and other non-nervous tissues. The levels of lipid hydroperoxide were elevated progressively after 14-day methylmercury chloride treatment in the spinal cord (+24.24 %), cerebellum (+23.84 %), cerebrum (+23.66 %), brain stem (+23.60 %), pancreas (+22.61 %), heart (+20.78 %), and lung (+18.86 %) when compared with control rats. ANOVA analysis showed high significant elevation of LHPO in the spinal cord, cerebellum, and cerebrum ($p<0.001$) and moderately significant in the brain stem and pancreas ($p<0.01$) and less significant in the heart and lung ($p<0.05$).

Effect of Vitamin E and Acetyl-L-Carnitine on Lipid Hydroperoxide Level

When vitamin E and acetyl-L-carnitine were administered orally, separately, or in combination for 14 days to pre-toxicated animals, a significant change in the values of lipid hydroperoxide was noticed in different nervous and non-nervous tissues mentioned above when they were compared with their respective control values. The recovery power of vitamin E was found to be higher than acetyl-L-carnitine even at the same dose level when they were given separately, but in combined treatment they reflected the best protection against methylmercury chloride-induced oxidative stress. After vitamin E treatment, cerebrum showed −38.65 % declines in lipid hydroperoxide level. Accordingly cerebellum showed −39.82 % decreases, brain stem showed −38.36 %, and spinal cord, heart, pancreas, and lung displayed −40.24 %, −36.00 %, −37.84 %, and −38.73 %, respectively. In case of acetyl-L-carnitine, highest decrement was shown by the spinal cord with −39.43 % decreases, while heart showed minimum decrease by −33.84 %. Combined vitamin therapy displayed highest and lowest decline in the spinal cord and heart, respectively, having −40.85 % and −36.92 % recovery, respectively.

Protein

The levels of protein in various parts of CNS and other tissue homogenate are depicted below. The amount of total protein in various tissues of rats was found to vary in the following increasing order: spinal cord < brain stem < cerebrum = cerebellum < heart < pancreas < lung.

Effect of Methylmercury Chloride on Protein Level

Protein levels were significantly decreased in various tissues of rats after 2 mg/kg body weight of methylmercury treatment for 14 days orally via gavage ($p<0.001, 0.01, 0.05$) when compared with control animals. ANOVA showed significant decrease of protein level in the spinal cord by −20.12 % decline, brain stem displayed −19.81 % decrease, cerebrum showed −19.80 %, cerebellum showed −19.80 %, while lung, heart, and pancreas displayed −18.73 %, −19.28 %, and −19.16 % decrement, respectively.

Effect of Vitamin E and Acetyl-L-Carnitine on Protein Level

Vitamin E and acetyl-L-carnitine protected the tissues against methylmercury-mediated toxicity in the following order: spinal cord > brain stem > cerebrum > cerebellum > heart > pancreas > lung. Spinal cord showed +57.07 % elevation in protein content by vitamin E, while +53.94 % recoveries by acetyl-L-carnitine. Similarly, the minimum recovery shown by lung was +45.47 % and +45.45 % increase by vitamin E and acetyl-L-carnitine, respectively. Maximum recovery was seen in combined therapy rather than in separated ones. In combined treatment, spinal cord showed +59.52 % increases in protein level while lung displayed +48.40 % recoveries. In separated cases, vitamin E showed stronger antioxidant properties than acetyl-L-carnitine at the same dose level against metal toxicity (Figs. 14.3 and 14.4).

Effect of Methylmercury, Vitamin E, and Acetyl-L-Carnitine on Tissue Weight of Rats

Tissue weight differences in rats treated with methylmercury chloride and antioxidants (vitamin E and acetyl-L-carnitine) are diagrammatically illustrated in Fig. 14.5. The decrease in weight was in the order of brain > heart > lung > pancreas.

Effect of Methylmercury Chloride on Tissue Weight

In methylmercury-treated group, the sequence of change in tissue weight was −20.20 % decline in brain, −16.76 % in heart, lung with −16.38 % decline, and pancreas having −16.13 % decline.

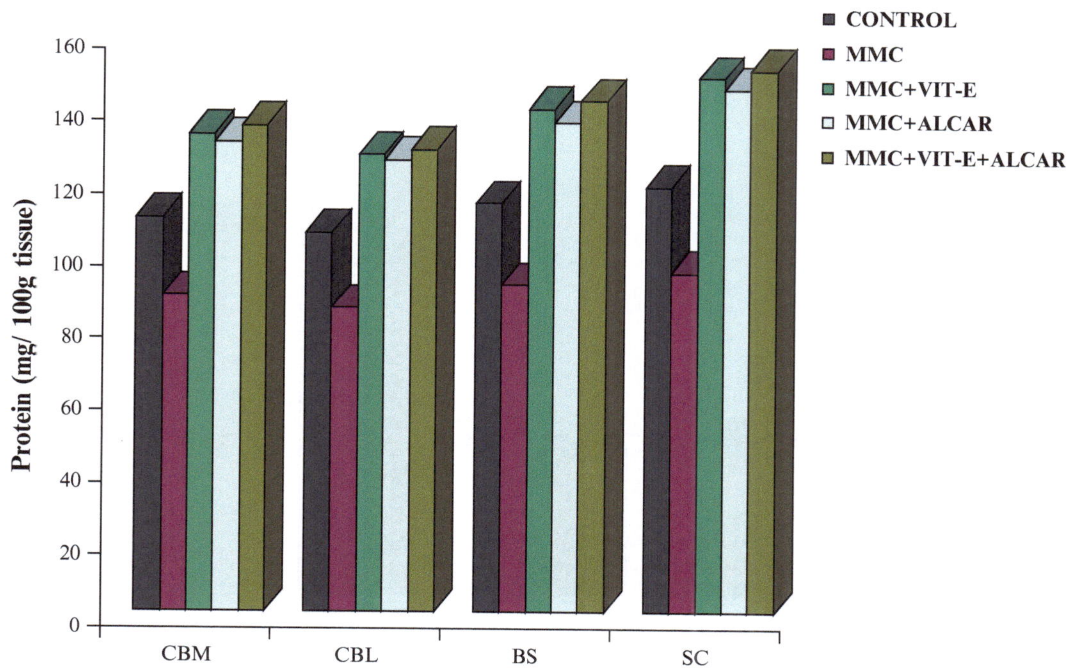

Fig. 14.3 Effect of methylmercury chloride on protein level in nervous tissues of rats. Values represent mean ± SEM. *CBM* cerebrum, *CBL* cerebellum, *BS* brain stem, *SC* spinal cord

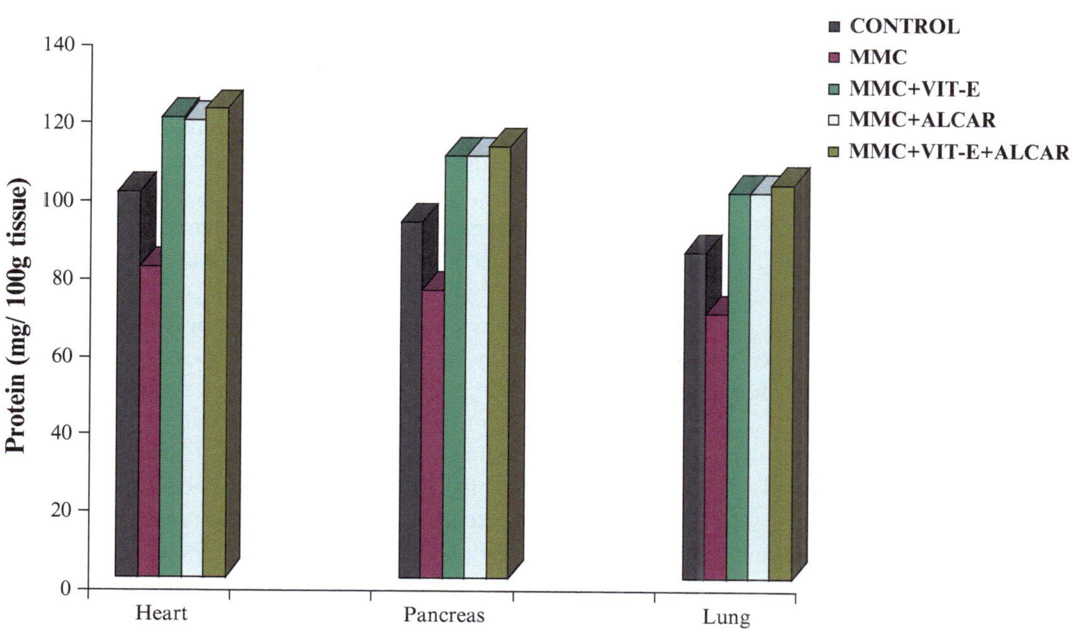

Fig. 14.4 Effect of methylmercury chloride on protein level in non-nervous tissues of rats. Values represent mean ± SEM

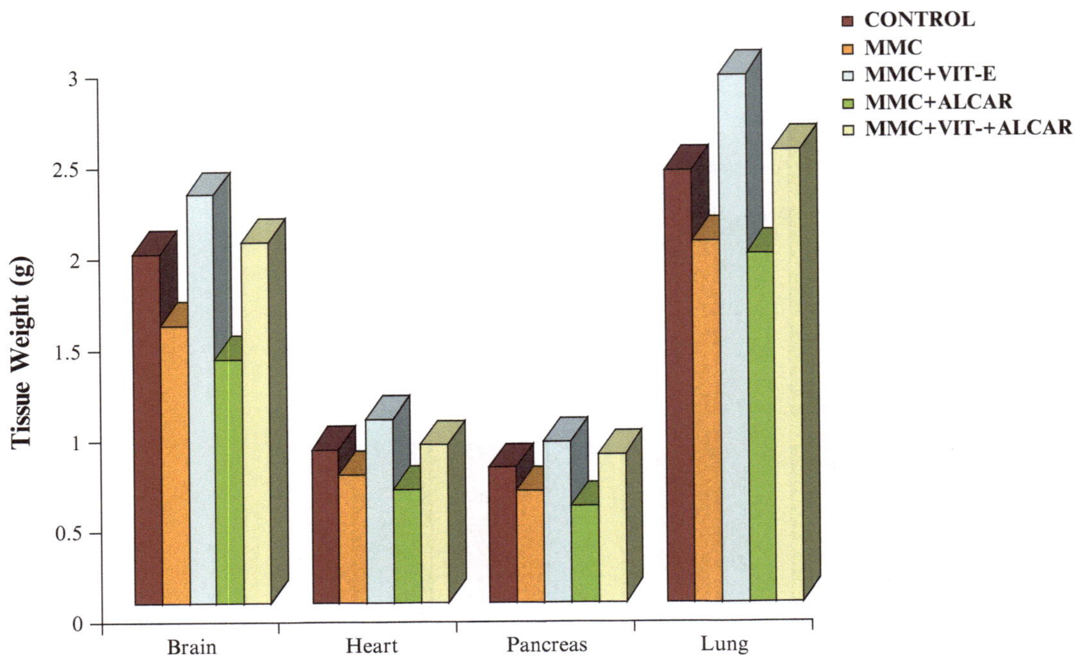

Fig. 14.5 Effect of methylmercury chloride on tissue weight of rats. Values represent mean ± SEM

Effect of Vitamin E and Acetyl-L-Carnitine on Tissue Weight

Significant changes occurred in tissue weight of rats treated with antioxidants separately or in combination. The tissue weight changes in vitamin E-treated rats are as follows: lung with +45.72 % increase, heart showing +43.66 % recovery, pancreas with +43.54 % elevation, and brain +46.10 %. In the case of acetyl-L-carnitine, the sequence follows the order as pancreas −12.90 % > brain −12.33 % > heart −11.26 % > lung −3.51 %. The rats treated with antioxidants in combined form do not show any significant change in tissue weight.

Discussion

Oxidative stress occurs when the normal balance between the oxidative events and the antioxidant defenses is disrupted either by loss of reducing agents/antioxidant enzymes or by increased amount of oxidizing species. Mercury compounds are known to induce the formation of ROS and there are a number of reports concerning oxidative stress and antioxidant enzyme activities on mercury exposure (Berntssen et al. 2003; Sanfeliu et al. 2003; Shanker and Aschner 2003; Yee and Choi 1994). The final effect depends on several factors, such as the nature of the metal, mode, and mechanism of administration, species, gender, tissue, and age. The necessity of developing sensitive procedures to detect toxic effects of heavy metals is given by the fact that human food can be highly contaminated by heavy metals, for example, up to 10 ppm mercury has been found in European predatory fish. Berglund and Berlin (1969) reported that fish from southern Sweden contained 0.02–10 ppm mercury. Up to 3.3 ppm MMC have been found in pike (Diehl and Schelenz 1971) and 0.55 ppm mercury in perch (Schelenz and Diehl 1973). Knoppler and Dorn (1977, personal communication) detected up to 0.27 ppm mercury in fish from Lake Constance. Mercury in fish exists largely as a methylmercury compound; methylmercury in total mercury averaged 70 % in a study of Bache et al. (1971) and 91 % in investigations of Diehl and Schelenz (1971).

One notable factor of environmental and occupational pollution is an accumulation of such elements including heavy metals like cadmium, mercury, and lead which are resistant to biological degradation. Both occupational and environmental exposures to heavy metals remain a serious problem (Lumb 1995; Evans 1998; Korenekova et al. 2002). Organic and inorganic mercury compounds have been reported to be potent toxic and/or carcinogenic agents in humans and animals, although the exact mode of action is still not certain. Numerous studies indicated that mercuric ions can interact with glutathione (GSH) in the presence of hydrogen peroxide, leading to the generation of reactive oxygen species. However, the mechanism of the generation of radical species including the redox cycle of the metal and the formation of oxidative DNA modification by mercury compounds has not been studied. The reactive oxygen species subsequently induced lipid peroxidation measured by the thiobarbituric acid reaction for malondialdehyde in liver, kidney, lung, testes, and serum. Increased MDA was reduced in these target tissues after pretreatment with antioxidants and chelators in methylmercury-treated rats.

The results of this study indicated that the lipid peroxidation is one of the molecular mechanisms for cell injury in acute methylmercury chloride poisoning and is associated with a decrease of cellular protein contents. Several in vivo and in vitro studies have demonstrated that both inorganic mercury and methylmercury induce oxidative stress (Tran et al. 2007; Chen et al. 2005; Girardi and Elias 1995) by increasing the intracellular levels of ROS and modifying enzyme activities (Girardi and Elias 1995). Farhana et al. (2006) reported high concentration of TBARS and decreased protein contents in rat brain of methylmercury chloride-treated rats. Lipid peroxidation was also increased in plasma of rats treated with mercury (Hijova et al. 2005). In our study, methylmercury chloride increased lipid peroxides and hydroperoxides significantly in cerebrum, cerebellum, brain stem, spinal cord, heart, lung, and pancreas while decreased protein concentration in the above-mentioned tissues. Oxidative stress may contribute to the development of neurodegenerative disorders caused by mercury intoxication (Husain et al. 1997).

Lund et al. (1993) have also demonstrated that the administration of mercury as Hg^{2+} (1.5 or 2.25 mg $HgCl_2$/kg) to rats resulted in increased hydrogen peroxide formation, glutathione depletion, and lipid peroxidation in kidney mitochondria. The exposure of mice to mercury, chromium, or silver results in enhanced production of MDA in liver and kidneys (Rungby and Ernst 1992). Lipid peroxidation was also stimulated in liver, kidney, and brain of mice administered methylmercury chloride via drinking water in a study reported by Helle and Ole (1993).

Numerous earlier reports have shown evidence of ROS involvement in methylmercury chloride toxicity such as those of Helle and Ole (1993) in brain parts, liver, and kidney. However, the present study is probably the first to suggest the role of ROS in heart, lung, and pancreas of rats treated with methylmercury. The increase in LPO and LHPO in cerebrum, cerebellum, brain stem, spinal cord, heart, lung, and pancreas after a low dose of methylmercury chloride indicates a condition of oxidative stress. No direct relation between the organ deposition of Hg and the level of lipid peroxidation and hydroperoxidation in these organs was found. This might be due to organ differences in biotransformation of CH_3Hg^+ to inorganic mercury and in the capacity of the antioxidative defense system.

In the present study, it was noted that along with increase in LPO and LHPO, methylmercury intoxication leads to a significant decrease in protein concentration as well. This suggests that besides directly affecting proteins through oxyradical stress, methylmercury also indirectly inhibits protein synthesis in nervous and non-nervous tissues on short-term low dose exposure. There are various theories which attribute methylmercury inhibition of protein synthesis to direct interaction of protein synthetic machinery (Sarafian et al. 1984).

Antioxidants have previously been demonstrated to modify methylmercury-induced toxicity. Increase of the dietary vitamin E level from 50 to 500 ppm protected against methylmercury chloride toxicity in rats (Welsh 1979) and Japanese quails (Welsh and Soares 1976) by decreasing the clinical toxicity and increasing the survival rate of the animals. In golden hamsters, no toxic symptoms were observed after coadministration of methylmercury chloride and vitamin E. In animals, given MMC alone, severe neuronal necrosis was demonstrated by light and electron microscopy (Chang et al. 1978). Vitamin E crosses through the blood–brain barrier and is effective in various central nervous system diseases, i.e., Alzheimer's disease (Sano et al. 1997; Grundman 2000; Patra et al. 2001). The present study demonstrated that rats treated with 100 mg/kg body weight of vitamin E after MMC intoxication significantly decreased free radicals in nervous and non-nervous tissues. Vitamin E also increased the protein contents in the above-mentioned tissues. Taken the above facts and our results together, it is suggested that oxidative injury, especially lipid peroxidation and hydroperoxidation via a powerful oxidant (e.g., hydroxyl radicals), may play an important role in tissue degeneration whether nervous or non-nervous and that vitamin E may be one of the most useful protective antioxidants against the cytotoxicity of methylmercury in rat tissues.

Acetyl-L-carnitine alone administered to rats intoxicated with MMC also mitigated the adverse effects of this heavy metal. Acetyl-L-carnitine treatment reduced the tissue TBARS levels. This may be due to the enhancement of the fatty acid transport by carnitine into mitochondria for energy production, thereby lowering the availability of lipids for peroxidation. Acetyl-L-carnitine inhibits microsomal peroxidation (Sushamakumari et al. 1989), and carnitine and its esters have been shown to partially inhibit iron-induced lipid peroxidation in liposome by chelating free iron (Arduini 1992). However, our study is perhaps the first one to report the role of acetyl-L-carnitine in the treatment of methylmercury chloride-induced oxidative stress. ALCAR also increased the concentration of protein in tissues showing that it is a useful antioxidant for MMC toxicity. But the antioxidant capacity of ALCAR was little less than vitamin E. When both antioxidants were given in combination, they showed wonderful results revealing that the incorporation of dietary antioxidants like vitamin E and ALCAR in routine diet from early age may help combat the risk of developing degenerative diseases due to methylmercury chloride toxicity in ensuing years.

Our results also showed changes in body weight, food intake, water intake, and tissue weight after methylmercury treatment. Furthermore, the results of these parameters were antagonistic when treated with antioxidants. The adverse effects on these parameters may be due to hormonal imbalance inside the body caused by this metal. Thus, it is clear that metal-induced damage depends on many factors such as the animal species, tissues, and cell types, as well as the level of metal inside the cells, and the antioxidants are the agents ameliorating these adverse effects.

Conclusion

In summary, oxidative stress is present during methylmercury intoxication. The concentrations of MDA in tissues were observed to increase significantly in rats treated with this metal. Treatment with vitamin E and acetyl-L-carnitine reduced sensitivity to oxidative stress. On the other hand, vitamin E plus acetyl-L-carnitine could cause complex alterations in the antioxidant system and could minimize the oxidative stress induced by methylmercury chloride.

References

Arduini A (1992) Carnitine and its acyl esters as secondary antioxidants? Am Heart J 123:1726–1727

Bache CA, Gutenmann WH, Lisk DJ (1971) Residues of total mercury and methylmercuric salts in lake trout as a function of age. Science 172:951–952

Berglund F, Berlin M (1969) Human risk evaluation for various populations in Sweden due to methylmercury in fish. In: Miller MW, Berg GG (eds) Chemical Fallout, current research on persistent pesticides. Chas C Thomas, Springfield, pp 423–432

Berntssen MHG, Aatland A, Handy RD (2003) Chronic dietary mercury exposure causes oxidative stress, brain lesions, and altered behavior in Atlantic salmon (Salmo Salar) parr. Aquat Toxicol 65:55–72

Chang LW, Gilbert M, Sprecher J (1978) Modification of methylmercury neurotoxicity by vitamin E. Environ Res 17:356–366

Chen C, Qu L, Li B, Xing L, Jia G, Wang T, Gao Y, Zhang P, Li M, Chen W, Chai Z (2005) Increased oxidative DNA damage, as assessed by urinary 8-hydroxy-2-deoxyguanosine concentrations, and serum redox status in persons exposed to mercury. Clin Chem 51:759–767

Diehl JF, Schelenz R (1971) Quecksilber in Lebensmitteln. Med Ernahrung 12:241–249

Evans HL (1998) Mercury. In: Rom WN (ed) Environmental and occupational medicine. Lippincott-Raven, Philadelphia, pp 993–999

Farhana Z, Shameem JR, Soghra KH, Rizwan HK (2006) Effect of methylmercury induced free radical stress on nucleic acids and protein: implications on cognitive and motor functions. Indian J Clin Biochem 21:149–152

Girardi G, Elias MM (1995) Mercury chloride effects on rat renal redox enzymes activities: SOD protection. Free Radic Biol Med 18:61–66

Grundman M (2000) Vitamin E and Alzheimer disease: the basis for additional clinical trials. Am J Clin Nutr 71:630S–636S

Helle RA, Ole A (1993) Effects of dietary α-Tocopherol and β-Carotene on lipid peroxidation induced by methyl mercuric chloride in mice. Pharmacol Toxicol 73:192–201

Hijova E, Nistiar F, Sipulova A (2005) Changes in ascorbic acid and malondialdehyde in rats after exposure to mercury. Bratisl Lek Listy 106:248–251

Husain S, Rodgers DA, Duhart HM, Ali SF (1997) Mercuric chloride-induced reactive oxygen species and its effect on antioxidant enzymes in different regions of rat brain. J Environ Sci Health B 32:359–409

Korenekova B, Skalicka M, Nad P (2002) Cadmium exposure of cattle after long-term emission from polluted area. Trace Elem Electrolytes 19:97–99

Lumb G (1995) Metal toxicity. In: Craighead JE (ed) Pathology of environmental and occupational disease. Mosby Year Book, St. Louis, pp 41–56

Lund BO, Miller DM, Woods JS (1993) Studies on Hg-II induced H_2O_2 formation and oxidative stress in vivo and in vitro in rat kidney mitochondria. Biochem Pharmacol 45:2017–2024

Patra RC, Swarup D, Dwivedi SK (2001) Antioxidant effects of alpha tocopherol, ascorbic acid and L-methionine on lead induced oxidative stress to the liver, kidney and brain in rats. Toxicology 162:81–88

Rungby J, Ernst E (1992) Experimentally induced lipid peroxidation after exposure to chromium, mercury or silver: interactions with carbon tetrachloride. Pharmacol Toxicol 70:205–207

Sanfeliu C, Sebastia J, Cristofol R, Rodriguez-Farre E (2003) Neurotoxicity of organomercurial compounds. Neurotox Res 5:283–306

Sano M, Ernesto C, Thomas RG, Klauber MR, Schafer K, Grundman M, Woodbury P, Growdon J, Cotman CW, Pfeiffer E, Schneider LS, Thal LJ (1997) A controlled trial of selegiline, alpha-tocopherol, or both as treatment for Alzheimer's disease: the Alzheimer's Disease Cooperative Study. N Engl J Med 336:1216–1222

Sarafian TA, Cheung MK, Verity MA (1984) In vitro methylmercury inhibition of protein synthesis in neonatal cerebellar perikarya. Neuropathol Appl Neurobiol 10:85–100

Schelenz R, Diehl JF (1973) Quecksilbergehalte in lebensmitteln des deutschen Marktes. Z Lebensmitt Untersuch Forschg 151:369–375

Shanker G, Aschner A (2003) Methylmercury-induced reactive oxygen species formation in neonatal cerebral astrocytic cultures is attenuated by antioxidants. Mol Brain Res 110:85–91

Sood PP, Bapu C, Sinha N, Rao AP (1997) Cholesterol and triglyceride fluctuations in mice tissues during methylmercury intoxication and monothiols and vitamin therapy. J Nutr Environ Med 7:155–162

Sushamakumari S, Jaydeep A, Suresh, Kumar JS, Venogopal PM (1989) Effect of carnitine on malondialdehyde, taurine and glutathione levels in the heart of rats subjected to myocardial stress by isoproterenol. Indian J Exp Biol 27:134–137

Tran D, Moody AJ, Fisher AS, Foulkes ME, Jha AN (2007) Protective effects of selenium on mercury-induced DNA damage in mussel haemocytes. Aquat Toxicol 84:11–18

Welsh SO (1979) The protective effect of vitamin E and N, N′-diphenyl-p-phenylenediamine (DPPD) against methylmercury toxicity in the rat. J Nutr 109:1673–1681

Welsh SO, Soares JH (1976) The protective effect of vitamin E and selenium against methylmercury toxicity in the Japanese quail. Nutr Rep Int 13:43

Yee S, Choi BH (1994) Methylmercury poisoning induces oxidative stress in the mouse brain. Exp Mol Pathol 60:188–196

Part V

Toxic Responses of Antioxidant Defense System

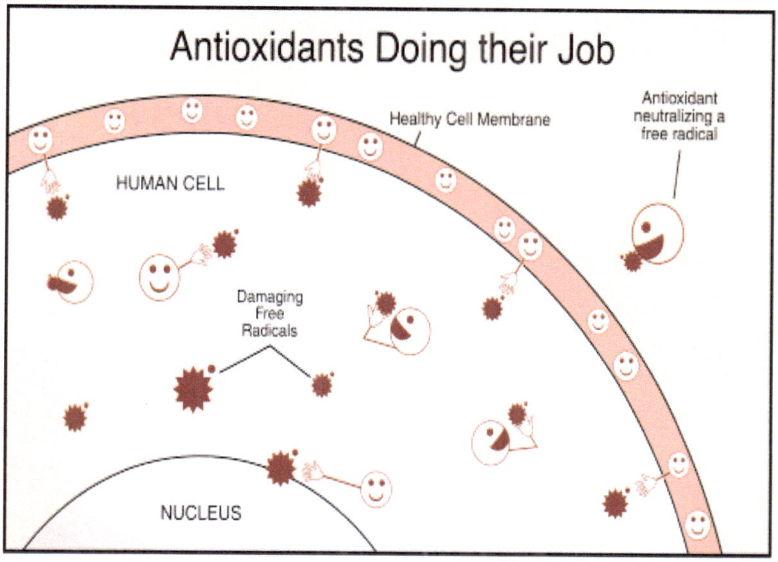

15 Endogenous Antioxidants

Antioxidants are compounds already in your body, but you need more than what the body produces

(Joe Vinson)

An antioxidant is a molecule that inhibits the oxidation of other molecules. Oxidation is a chemical reaction that transfers electrons or hydrogen from a substance to an oxidizing agent. Oxidation reactions can produce free radicals. In turn, these radicals can start chain reactions. When the chain reaction occurs in a cell, it can cause damage or death to the cell. Antioxidants terminate these chain reactions by removing free radical intermediates and inhibit other oxidation reactions. They do this by being oxidized themselves, so antioxidants are often reducing agents such as thiols, ascorbic acid, or polyphenols (Sies Helmut 1997). Antioxidants are vitamins, minerals, enzymes, or plant-derived nutrients called phytonutrients, found in food. They do what their name implies: antioxidation. Antioxidant means "against oxidation." Antioxidants work to protect lipids from peroxidation by radicals. Antioxidants are effective because they are willing to give up their own electrons to free radicals. When a free radical gains the electron from an antioxidant, it no longer needs to attack the cell, and the chain reaction of oxidation is broken (Dekkers et al. 1996). After donating an electron, an antioxidant becomes a free radical by definition. Antioxidants in this state are not harmful because they have the ability to accommodate the change in electrons without becoming reactive. The human body has an elaborate antioxidant defense system. Antioxidants are manufactured within the body and can also be extracted from the food humans eat such as fruits, vegetables, seeds, nuts, meats, and oil. There are two lines of antioxidant defense within the cell. The first line, found in the fat-soluble cellular membrane, consists of vitamin E, beta-carotene, and coenzyme Q (Kaczmarski et al. 1999). Of these, vitamin E is considered the most potent chain-breaking antioxidant within the membrane of the cell. Inside the cell, water-soluble antioxidant scavengers are present. These include vitamin C, glutathione peroxidase, superoxide dismutase (SOD), and catalase (Dekkers et al. 1996).

Antioxidants are important additives in gasoline. These antioxidants prevent the formation of gums that interfere with the operation of internal combustion engines (Werner Dabelstein et al. 2007). Although oxidation reactions are crucial for life, they can also be damaging; plants and animals maintain complex systems of multiple types of antioxidants, such as glutathione, vitamin C, and vitamin E as well as enzymes such as catalase, superoxide dismutase, and various peroxidases. Low levels of antioxidants, or inhibition of the antioxidant enzymes, cause oxidative stress and may damage or kill cells.

As oxidative stress appears to be an important part of many human diseases, the use of

S. Nabi, *Toxic Effects of Mercury*, DOI 10.1007/978-81-322-1922-4_15,
© Springer India 2014

antioxidants in pharmacology is intensively studied, particularly as treatments for stroke and neurodegenerative diseases. Moreover, oxidative stress is both the cause and the consequence of disease.

Antioxidants are widely used in dietary supplements and have been investigated for the prevention of diseases such as cancer, coronary heart disease, and even altitude sickness. Although initial studies suggested that antioxidant supplements might promote health, later large clinical trials with a limited number of antioxidants detect no benefit and even suggested that excess supplementation with certain putative antioxidants may be harmful (Jha et al. 1995; Baillie et al. 2009; Bjelakovic et al. 2007). Antioxidants also have many industrial uses, such as preservatives in food and cosmetics and to prevent the degradation of rubber and gasoline.

Superoxide Dismutase (SOD) is a metalloenzyme whose active center is occupied by copper and zinc, sometimes manganese or iron. SOD plays an extremely important role in the protection of all aerobic life systems, including man, against oxygen toxicity (and the free radicals derived from oxygen).

The enzyme superoxide dismutase (SOD) catalyzes the dismutation of superoxide into oxygen and hydrogen peroxide. SOD is an endogenously produced intracellular enzyme present essentially in every cell inside the body. There are at least three forms of superoxide dismutase in nature. Human erythrocytes contain an SOD enzyme with divalent copper and divalent zinc. Chicken liver mitochondria and *E. coli* contain a form with trivalent manganese. *E. coli* also contains a form of the enzyme with trivalent iron. The Cu-Zn enzyme is a dimer of molecular weight 32,500. The two subunits are joined by a disulfide bond. Superoxide dismutases are enzymes that play major roles in the protection of cells against oxidative damage. The two major forms of superoxide dismutase (SOD) in humans are the mitochondrial manganese SOD and the cytosolic copper/zinc SOD. MnSOD is transcribed in the nucleus and has a mitochondrial targeting sequence, thereby localizing it to the mitochondrial matrix. Copper/zinc SOD is present in the cytoplasm of the cell. It is isolated from beef liver and has been used intra-articularly for degenerative joint disorders as an anti-inflammatory agent. SOD is also marketed as a nutritional supplement. The genes that control the formation of SOD are located on chromosomes 21, 6, and 4. When superoxide dismutase comes in contact with superoxide, it reacts with it and forms hydrogen peroxide. The stoichiometry of this reaction is that for each two superoxide radicals encountered by SOD, one H_2O_2 is formed. This hydrogen peroxide is dangerous in the cell because it can easily transform into a hydroxyl radical (via reaction with Fe^{2+}: Fenton chemistry), one of the most destructive free radicals. As an enzyme, SOD has particular value as an antioxidant that can help to protect against cell destruction. It has the distinct ability to neutralize superoxide, one of the most damaging free radical substances in nature. Like so many other protective compounds which naturally occur in the body, it decreases with age, making cells much more vulnerable to the oxidants which cause aging and disease. It occurs naturally in broccoli, Brussels sprouts, wheat grass, and in the majority of green plants.

Catalase is a common enzyme found in nearly all living organisms which are exposed to oxygen, where it functions to catalyze the decomposition of hydrogen peroxide to water and oxygen (Chelikani et al. 2004). Catalase has one of the highest turnover numbers of all enzymes; one molecule of catalase can convert millions of molecules of hydrogen peroxide to water and oxygen per second (Goodsell 2004). Catalase is a tetramer of four polypeptide chains, each over 500 amino acids long (Boon et al. 2007). It contains four porphyrin heme (iron) groups that allow the enzyme to react with the hydrogen peroxide. The optimum pH for human catalase is approximately 7 (Maehly and Chance 1954) and has a fairly broad maximum (the rate of reaction does not change appreciably at pHs between 6.8 and 7.5) (Aebi 1984). The pH optimum for other catalases varies between 4 and 11 depending on the species (BRENDA 2009). The optimum temperature also varies by species (Toner et al. 2007). Catalase is concentrated in peroxisomes located next to the mitochondria but formed in the rough endoplasmic reticulum and located everywhere in the cell. The reaction of catalase in the decomposition of hydrogen peroxide is:

$$2H_2O_2 \rightarrow 2H_2O + O_2$$

Hydrogen peroxide is a harmful by-product of many normal metabolic processes: to prevent damage, it must be quickly converted into other, less dangerous substances. To this end, catalase is frequently used by cells to rapidly catalyze the decomposition of hydrogen peroxide into less reactive gaseous oxygen and water molecules (Gaetani et al. 1996). The true biological significance of catalase is not always straightforward to assess: Mice genetically engineered to lack catalase are phenotypically normal, indicating that this enzyme is dispensable in animals under some conditions (Ho et al. 2004). Some human beings have very low levels of catalase (acatalasia) yet show few ill effects. It is likely that the predominant scavengers of H_2O_2 in normal mammalian cells are peroxiredoxins rather than catalase.

Glutathione (GSH) is a tripeptide. It contains an unusual peptide linkage between the amine group of cysteine and the carboxyl group of the glutamate side chain. Glutathione, an antioxidant, helps to protect cells from reactive oxygen species such as free radicals and peroxides (Pompella et al. 2003). Glutathione is nucleophilic at sulfur and attacks poisonous electrophilic conjugate acceptors.

Thiol groups are kept in a reduced state at a concentration of approximately ~5 mM in animal cells. In effect, glutathione reduces any disulfide bond formed within cytoplasmic proteins to cysteines by acting as an electron donor. In the process, glutathione is converted to its oxidized form glutathione disulfide (GSSG). Glutathione is found almost exclusively in its reduced form, since the enzyme that reverts it from its oxidized form, glutathione reductase, is constitutively active and inducible upon oxidative stress. In fact, the ratio of reduced glutathione to oxidized glutathione within cells is often used scientifically as a measure of cellular toxicity (Pastore et al. 2003).

In healthy cells and tissues, more than 90 % of the total glutathione pool is in the reduced form (GSH) and less than 10 % exists in the disulfide form (GSSG). An increased GSSG-to-GSH ratio is considered indicative of oxidative stress.

Glutathione has multiple functions:

1. It is the major endogenous antioxidant produced by the cells, participating directly in the neutralization of free radicals and reactive oxygen compounds, as well as maintaining exogenous antioxidants such as vitamins C and E in their reduced (active) forms.
2. Through direct conjugation, it detoxifies many xenobiotics (foreign compounds) and carcinogens, both organic and inorganic.
3. It is essential for the immune system to exert its full potential, e.g., (1) modulating antigen presentation to lymphocytes, thereby influencing cytokine production and type of response (cellular or humoral) that develops; (2) enhancing proliferation of lymphocytes thereby increasing magnitude of response; (3) enhancing killing activity of cytotoxic T cells and NK cells; and (4) regulating apoptosis, thereby maintaining control of the immune response.
4. It plays a fundamental role in numerous metabolic and biochemical reactions such as DNA synthesis and repair, protein synthesis, prostaglandin synthesis, amino acid transport, and enzyme activation. Thus, every system in the body can be affected by the state of the glutathione system, especially the immune system, the nervous system, the gastrointestinal system, and the lungs.

Hg has a great affinity for –SH groups, attaching thiol-containing proteins and small molecular-weight thiols, such as glutathione (GSH) (Clarkson 1997), decreasing the main nonprotein thiol involved in the antioxidant cellular defense against reactive oxygen species (ROS) (Sarafian 1999; Grotto et al. 2009). Moreover, Hg is related to changes in the activities of antioxidant enzymes, such as superoxide dismutase (Ariza et al. 1998) and catalase (Hussain et al. 1999), besides inducing ROS production, especially of H_2O_2 and O_2^- (Lund et al. 1991), and promoting oxidative stress and lipid peroxidation (Carvalho et al. 2007; Jina et al. 2007; Stringari et al. 2008).

References

Aebi H (1984) Catalase in vitro. Methods Enzymol 105:121–126

Ariza ME, Bijur GN, Williams MV (1998) Lead and mercury metagenesis: role of H_2O_2, superoxide dismutase, and xanthine oxidase. Environ Mol Mutagen 31:352–361

Baillie JK, Thompson AAR, Irving JB, Bates MGD, Sutherland AI, MacNee W, Maxwell SRJ, Webb DJ (2009) Oral antioxidant supplementation does not prevent acute mountain sickness: double blind randomized placebo-controlled trial. QJM 102(5):341–348

Bjelakovic G, Nikolova D, Gluud LL, Simonetti RG, Gluud C (2007) Mortality in randomized trials of antioxidant supplements for primary and secondary prevention: systematic review and meta-analysis. JAMA 297(8):842–857

Boon EM, Downs A, Marcey D (2007) Catalase: H_2O_2: H_2O_2 oxidoreductase. Catalase structural tutorial text

BRENDA (2009) The comprehensive enzyme information system. Department of Bioinformatics and Biochemistry, Technical University of Braunschweig, Germany

Carvalho MC, Franco JL, Ghizonia H, Kobusa K, Nazaria EM, Rocha JBT, Nogueira CW, Dafred AL, Mullera YMR, Farina M (2007) Effects of 2, 3-dimercapto-1-propanesulfonic acid (DMPS) on methylmercury-induced locomotor deficits and cerebellar toxicity in mice. Toxicology 239:195–203

Chelikani P, Fita I, Loewen PC (2004) Diversity of structures and properties among catalases. Cell Mol Life Sci 61:192–208

Clarkson TW (1997) The toxicity of mercury. Crit Rev Clin Lab Sci 34:369–403

Dabelstein W, Reglitzky A, Schütze A, Reders K. (2007) Automotive fuels. In: Ullmann's encyclopedia of industrial chemistry, Wiley-VCH, Weinheim.

Dekkers JC, Von Doormen LJP, Han CGK (1996) The role of antioxidant vitamins and enzymes in the prevention of exercise-induced muscle damage. Sports Med 21:213–238

Gaetani G, Ferraris A, Rolfo M, Mangerini R, Arena S, Kirkman H (1996) Predominant role of catalase in the disposal of hydrogen peroxide within human erythrocytes. Blood 87:1595–1599

Goodsell DS (2004) Catalase. Molecule of the month. RCSB Protein Data Bank

Grotto D, Barcelos GRM, Valentini J, Antunes MGA, Angeli JPA, Garcia SC, Barbosa F (2009) Low levels of methylmercury induce DNA damage in rats: protective effects of selenium. Arch Toxicol 83:249–254

Ho YS, Xiong Y, Ma W, Spector A, Ho D (2004) Mice lacking catalase develop normally but show differential sensitivity to oxidant tissue injury. J Biol Chem 279:32804–32812

Hussain S, Atkinson A, Thompson SJ, Khan AT (1999) Accumulation of mercury and its effect on antioxidant enzymes in brain, liver, and kidneys of mice. J Environ Sci Health B 34:645–660

Jha P, Flather M, Lonn E, Farkouh M, Yusuf S (1995) The antioxidant vitamins and cardiovascular disease: a critical review of epidemiologic and clinical trial data. Ann Intern Med 123(11):860–872

Jina X, Loka E, Bondya G, Caldwella D, Muellera R, Kapala K, Cheryl A, Taylora M, Kubowc S, Mehtaa R, Chan HM (2007) Modulating effects of dietary fats on methylmercury toxicity and distribution in rats. Toxicology 230:22–44

Kaczmarski M, Wojicicki J, Samochowiee L, Dutkiewicz T, Sych Z (1999) The influence of exogenous antioxidants and physical exercise on some parameters associated with production and removal of free radicals. Pharmazie 54:303–306

Lund BO, Miler DM, Woods JS (1991) Mercury-induced H_2O_2 formation and lipid peroxidation in vitro in rat kidney mitochondria. Biochem Pharmacol 42:181–187

Maehly A, Chance B (1954) The assay of catalases and peroxidases. Methods Biochem Anal 1:357–424

Pastore A, Piemonte F, Locatelli M, Lo Russo A, Gaeta LM, Tozzi G, Federici G (2003) Determination of blood total, reduced, and oxidized glutathione in pediatric subjects. Clin Chem 47:1467–1469

Pompella A, Visvikis A, Paolicchi A, De Tata V, Casini AF (2003) The changing faces of glutathione, a cellular protagonist. Biochem Pharmacol 66:1499–1503

Sarafian TA (1999) Methylmercury-induced generation of free radical: biological implications. Met Ions Biol Syst 36:415–444

Sies H (1997) Oxidative stress: oxidants and antioxidants. Exp Physiol 82(2):291–295

Stringari J, Nunes AKC, Franco JL, Bohrer D, Garcia SC, Dafre AL, Milatovic D, Souza DO, Rocha JBT, Aschner M, Farina M (2008) Prenatal methylmercury exposure hampers glutathione antioxidant system ontogenesis and causes long-lasting oxidative stress in the mouse brain. Toxicol Appl Pharmacol 227:147–154

Toner K, Sojka G, Ellis R (2007) A quantitative enzyme study; CATALASE. bucknell.edu

16 Deregulation of Antioxidant Activities

Methylmercury is one of the most toxic forms of Hg exerting its major toxic effects on the central nervous system (Clarkson et al. 2003). Several mechanisms by which Hg exerts its adverse effects have been hypothesized in acute exposure to high levels; among them, impairment of intracellular calcium homeostasis (Aschner et al. 2007; Dietrich et al. 2005; Sirois and Atchison 2000) and alteration of glutamate homeostasis (Aschner et al. 2000) were reported. However, it is believed that the major process responsible for triggering the toxicity is the oxidative stress.

The aim of the present study was to evaluate the low-dose methylmercury-induced deregulation of the activities of the antioxidant enzymes superoxide dismutase (SOD) and catalase (CAT). We also analyzed its effects on TSH, GSH, and depression-like behavior in male albino rats because methylmercury is known to be an environmental neurotoxicant potentially causing neuropsychological disorders in humans (Gilbert and Grant-Webster 1995). Furthermore, epidemiological and experimental studies have clearly shown that the developing nervous system is particularly vulnerable to MeHg toxicity. Severe neurotoxic effects of prenatal exposure to high doses of MeHg were established in humans after MeHg disasters in Japan and Iraq (Amin-Zaki et al. 1979; Harada 1995) and confirmed in animal studies (Burbacher et al. 1990). Later, developmental exposure to low doses of methylmercury contained in seafood was found to be a risk factor for cognitive disorders (e.g., memory, attention, and language problems) in children and adolescents in the fish-eating population of the Faroe Islands (Debes et al. 2006; Grandjean et al. 1997). This fact raised researchers' interest in studying the effects of prolonged low-dose exposure in animal models, representing a chronic pattern of exposure in humans.

Protocol

For various sets of biochemical studies, different groups comprising ten rats each were used. Rats from Group-I served as control, while rats of Group-II, -III, -IV, and -V were used as experimental sets. Group-II rats were given 2 mg/kg body weight of methylmercury chloride for 14 days, and for the next 14 days, they were kept untreated. Animals of Group-III received MeHgCl (2 mg/kg body weight) for 14 days and vitamin E (100 mg/kg body weight) for the next 14 days. Group-IV animals were given MeHgCl (2 mg/kg body weight) for 14 days, and for the next 14 days, they were treated with acetyl-L-carnitine (100 mg/kg body weight). Group-V animals were treated with MeHgCl (2 mg/kg body weight) for 14 days, and for the next 14 days, they were given vitamin E (100 mg/kg body weight) and acetyl-L-carnitine (100 mg/kg body weight) in combination. In combined therapy, acetyl-L-carnitine was always administered at a gap of 30 min after vitamin E as per Sood et al. (1997).

S. Nabi, *Toxic Effects of Mercury*, DOI 10.1007/978-81-322-1922-4_16,
© Springer India 2014

Methylmercury chloride and acetyl-L-carnitine were diluted in physiological saline, while vitamin E was given as such. All groups were treated once a day orally through intragastric intubation. The intake of drinking water and food by rats was examined daily, and rats were weighed every other day for weight change assessment due to toxic metal. On the 29th day, the rats were tested for tail suspension test and forced swim test. The rats were sacrificed later on the scheduled day by cervical dislocation, and immediately, the brains, spinal cords, hearts, lungs, and pancreases were taken out and kept on ice. Tissues were weighed, both in control and treated rats to observe the weight changes. Brains were separated into cerebrum, cerebellum, and brain stem. The tissues were later processed for the assay of superoxide dismutase (SOD), catalase (CAT), total sulfhydryl (TSH), and glutathione (GSH) by standard methods described in detail in Methodology chapter.

Results

Effect of Methylmercury Chloride, Vitamin E, and Acetyl-L-Carnitine on Antioxidant System

The results presented below show the methylmercury associated alterations of superoxide dismutase (SOD), catalase (CAT), total sulfhydryl (TSH), and free or reduced glutathione (GSH) in different nervous and nonnervous tissues following 14-day continuous treatment. Exogenous antioxidant effect was also depicted in the results.

Superoxide Dismutase (SOD)

The activity of SOD observed in various regions of the CNS, heart, lung, and pancreas is given below. The highest decrease in SOD activity was found in the cerebellum, followed by the cerebrum, brain stem, spinal cord, heart, and pancreas, while the lung showed the lowest decrease.

Effect of Methylmercury Chloride on SOD Activity

Methylmercury exposure decreased SOD concentration significantly in comparison to controls in studied nervous and nonnervous tissues. The highest inhibition of SOD activity was found in cerebellum at −26.00 %. The observed order for SOD decrement is cerebellum −26.00 % > cerebrum −25.75 % > brain stem −24.39 % > spinal cord −24.13 % > heart −23.37 % > pancreas −23.05 % > lung −22.80 %. ANOVA revealed that the difference between the treated and control rats was significant at $p<0.001$, $p<0.01$, and $p<0.05$.

Effect of Vitamin E and Acetyl-L-Carnitine on SOD Activity

A significant increase in SOD level was observed in various regions of the brain, spinal cord, and other tissues of rats when they were administered vitamin E and acetyl-L-carnitine after methylmercury chloride treatment, separately or in combination orally for 14 days. Interestingly, when methylmercury chloride-intoxicated rats were treated with vitamins, a remarkable rise of SOD was observed. However, the recovery was more significant in animals treated with vitamins in a combined way. Among the various tissues, the cerebellum showed the best results during vitamin therapy causing approximately +72.47 % and +71.11 % recovery with vitamin E and acetyl-L-carnitine, respectively, when given separately. But when antioxidants were given in combination, cerebellum showed +74.65 % recoveries. In the cerebrum, brain stem, and spinal cord, the recovery was +70.55 %, +67.44 %, and +63.93 % with vitamin E, respectively; +68.80 %, +64.22 %, and +62.42 % with acetyl-L-carnitine, respectively; and +72.59 %, +68.91 %, and +65.75 % with vitamin E + acetyl-L-carnitine, respectively. In the heart, vitamin E showed much more recovery of +62.03 % than acetyl-L-carnitine with +59.66 %, but the protection conferred by their combined therapy was much better than separate treatment. In the lung, the recovery was lowest.

Catalase (CAT)

The specific activity of catalase in different regions of the brain, spinal cord, heart, lung, and pancreas of rats is diagrammatically represented in Figs. 16.1 and 16.2. The highest decline in catalase activity was found in cerebrum followed by cerebellum, spinal cord, brain stem, pancreas, lung, and heart.

Effect of Methylmercury Chloride on CAT Activity

Significant inhibition of catalase activity in various parts of the CNS, heart, lung, and pancreas following 14-day treatment was observed. ANOVA revealed that the difference between the control and experimental rats was highly significant in cerebrum with −27.03 % decrease followed by cerebellum showing −26.09 % change compared to control ($p<0.001$). Decrease was moderately significant in brain stem with −25.89 % decrement and spinal cord by −25.92 % ($p<0.01$) and less significant in heart by −22.85 %, pancreas with −23.93 % decrease, and lung with −23.69 %, respectively ($p<0.05$).

Effect of Vitamin E and Acetyl-L-Carnitine on CAT Activity

Administration of vitamin E after pretreatment with methylmercury chloride to male rats when compared with their control counterparts showed elevation in the activity of catalase in all the regions of the CNS, heart, lung, and pancreas with maximum and minimum values in the cerebrum +75.49 % and heart +59.72 %, respectively. When rats were treated with acetyl-L-carnitine alone orally for 14 days after methylmercury chloride pretreatment, recovery was assessed to be less than vitamin E. But vitamins in combined treatment showed amazing results. The maximum recovery was seen in cerebrum with +77.45 %

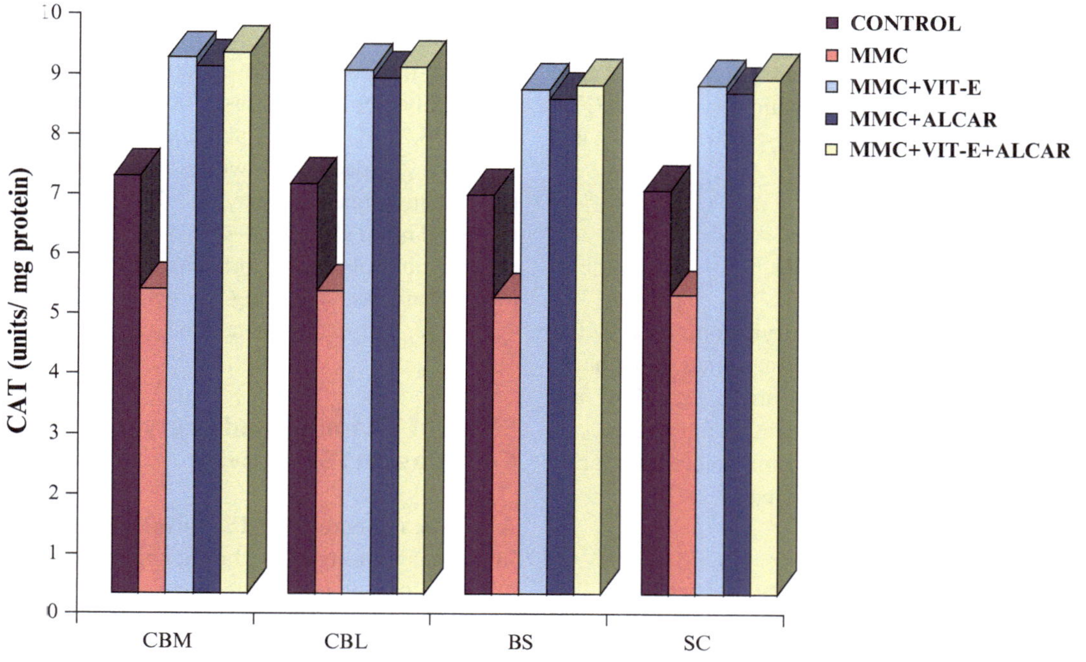

Fig. 16.1 Effect of methylmercury chloride on catalase (CAT) level in nervous tissues of rats. Values represent mean ± SEM. *CBM* cerebrum, *CBL* cerebellum, *BS* brain stem, *SC* spinal cord

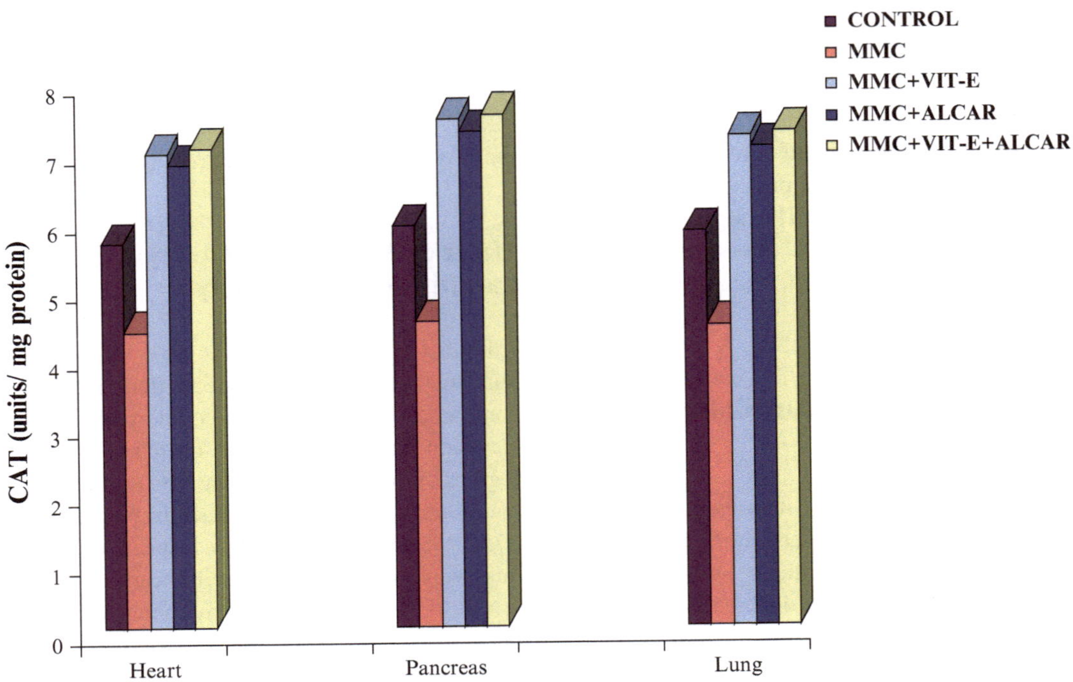

Fig. 16.2 Effect of methylmercury chloride on catalase (CAT) level in nonnervous tissues of rats. Values represent mean ± SEM

increases. The order of the recovery in catalase activity in different nervous and nonnervous tissues was cerebrum > cerebellum > spinal cord > brain stem > pancreas > lung > heart. ANOVA showed the significance at $p<0.001$, $p<0.01$, and $p<0.005$.

Total Sulfhydryl (TSH)

The levels of total sulfhydryl group observed in various regions of the brain, spinal cord, heart, lung, and pancreas of rats are given below. The highest decrease of TSH was found in the cerebellum followed by cerebrum, spinal cord, brain stem, heart, lung, and pancreas.

Effect of Methylmercury Chloride on TSH Activity

Significant inhibition of total sulfhydryl group levels in various parts of the CNS, heart, lung, and pancreas following 14-day treatment was observed. ANOVA revealed that the difference between the control and experimental animals was highly significant at $p<0.001$ in cerebellum with −28.27 % changes and cerebrum showing −28.18 % declines, while moderately at $p<0.01$ in spinal cord having −27.95 % decrement and brain stem with −27.55 % change compared to controls, less significant at $p<0.05$ in the heart, pancreas, and lung showing −27.08 %, −26.38 %, and −26.67 % decline, respectively.

Effect of Vitamin E and Acetyl-L-Carnitine on TSH Activity

A significant increase in TSH level was observed in various regions of the brain, spinal cord, and other tissues of rats when they were administered vitamin E and acetyl-L-carnitine after methylmercury chloride treatment, separately or in combination orally for 14 days. Interestingly, when methylmercury chloride-intoxicated rats were treated with vitamins, a remarkable rise of TSH level was observed. However, the recovery was more significant in rats treated with vitamins in a combined way. Among the various tissues, the cerebellum showed the best results during vitamin therapy causing approximately +83.25 % and +82.66 % recovery with vitamin E and acetyl-L-carnitine, respectively, when given separately. But when antioxidants were given in combination, cerebellum showed +84.25 % recoveries. In the cerebrum, brain stem, and spinal cord, the recovery was +83.01 %, +79.28 %, and +81.12 % with vitamin E, respectively; +82.42 %, +78.61 %, and +80.53 % with acetyl-L-carnitine, respectively; and +83.17 %, +80.05 %, and +81.21 % with vitamin E + acetyl-L-carnitine, respectively. In the heart, vitamin E showed much more recovery of +77.48 % than acetyl-L-carnitine having +76.70 % increases, but the protection conferred by their combined therapy was much better than separate treatment. In the pancreas, the recovery was lowest.

Free Sulfhydryl Group (GSH)

In rats treated with methylmercury chloride, the amounts of GSH in different parts of the brain, spinal cord, heart, lung, and pancreas were found to vary in increasing rank as brain stem < spinal cord < cerebrum < cerebellum < pancreas < lung < heart.

Effect of Methylmercury on GSH Activity

After 14-day methylmercury chloride oral treatment, the contents of GSH were inhibited significantly in all the regions of the brain, spinal cord, heart, lung, and pancreas. The maximal reduction of GSH was in brain stem showing −31.56 % decreases and the lowest in heart with −24.31 % declines. ANOVA of all the regions of the CNS and heart, lung, and pancreas showed the difference between controls and experimental animals at $p<0.001$, 0.01, and 0.05.

Effect of Vitamin E and Acetyl-L-Carnitine on GSH Activity

Vitamin E and acetyl-L-carnitine administration orally for 14 days after methylmercury pretreatment was found to induce significant increment of GSH contents in different regions of the brain, spinal cord, heart, lung, and pancreas. However, the recovery was better when vitamins were administered in combined manner. The maximum GSH elevation was seen in the brain stem with +97.41 % change followed by spinal cord of +95.00 % increase, cerebrum +94.20 %, cerebellum +91.32 %, pancreas +85.95 %, lung +73.57 %, and heart with +71.78 % increase in vitamin E treatment, while in acetyl-L-carnitine administration, the percent change was as follows: brain stem +95.94 %, spinal cord +93.53 %, cerebrum +92.71 %, cerebellum +90.00 %, pancreas +84.62 %, lung +72.43 %, and heart +70.44 % increase. In combined vitamin therapy, the increment was in the order of brain stem +99.63 %, spinal cord +96.67 %, cerebrum +95.51 %, cerebellum +92.83 %, pancreas +87.47 %, lung +74.33 %, and heart +75.04 % elevation.

Discussion

The brain is particularly susceptible to the harmful effects of oxidative stress, a condition caused by an imbalance between the generation of free radicals and the cell's protective antioxidant system (Duchen 2000). Free radicals have one or more unpaired electrons and thus have the ability to react with crucial cell components such as proteins, DNA, and lipids. Enzymes like catalase, peroxides, and superoxide dismutase (SOD) and nonenzymatic compounds, including GSH and TSH, are involved in the cellular defense against free radicals. In addition to inducing lipid peroxidation (by means of MDA levels) and altering GSH concentrations, $HgCl_2$ was also reported to affect other antioxidant enzyme activities of cells. Ariza et al. (1998) showed that Hg^{+2} induces H_2O_2 formation and stimulates the activities of copper–zinc SOD and xanthine oxidase in AS52 cells.

The present study evaluated the effects of low levels of methylmercury on antioxidant enzyme activities in different tissues of rats, and its effect on depression-like behavior in rats was also assessed. Our findings demonstrated reduced levels of total (TSH) and free (GSH) sulfhydryl in methylmercury-treated rats when compared with the control group. GSH is the main nonprotein thiol involved in the antioxidant cellular defense against the toxic effects of reactive oxygen species produced naturally in the organism or from the metabolism of xenobiotics (Meister 1983). It is known that the most important mechanism for Hg-induced oxidative damage is its strong reactivity with –SH components (Rabenstein and Fairhurst 1975; Li et al. 2007), which can lead to diminishing the antioxidant reserves, contributing to the increase in reactive species production, damaging lipid membranes, proteins, and DNA (Halliwell and Chirico 1993). Besides, biliary excretion is the major pathway for MeHg elimination, and this mechanism appears to be dependent on GSH bioavailability (Refsvik 1978; Dutczak and Ballatori 1994), which is bound in MeHg by a direct chemical interaction, forming GS–$HgCH_3$ complexes (Dutczak and Ballatori 1994). In a study carried out by Thompson, it was observed that exposure to high levels of MeHg throughout the gestation resulted in GSH depletion in the fetus (Thompson et al. 2000). In agreement, Stringari et al. (2006) also observed reduced GSH levels in the cerebellum of mice treated with high doses of MeHg (7 mg/kg per day). Besides GSH as an antioxidant, free radicals are detoxified by various antioxidant enzymes, such as CAT and SOD. We observed decreased CAT activity in our rats treated with low doses of MeHg, which could be explained by a direct inhibition of the enzyme by the MeHg (Abdel- Hamid et al. 2001). Contrary to our findings, Hussain et al. (1999) had observed increased CAT activity in mice exposed to inorganic mercury. SOD is involved in dismutation of the superoxide anion (O_2^-) to hydrogen peroxide (H_2O_2) and oxygen, and the results of the present study demonstrate a decrease in the SOD enzyme activity in rats treated with methylmercury chloride. This decrease in the activity of SOD, observed in the present study, indicates either reduced synthesis of the enzyme or elevated degradation or inactivation of the enzyme during the toxic metal treatment. Catalase is inhibited by O_2^- (Kono and Fridovich 1982). Superoxide anion served to convert catalase to the ferroxy and ferryl states, which are inactive forms of the enzyme (Freeman and Crapo 1982). Hence, the lower activity of CAT observed in methylmercury chloride-treated rats may be due to lower levels of SOD, which was similar to the reported results of the present study or may be due to inactivation of catalase owing to excess production of ROS. CAT and SOD are considered to be indispensable for the survival of the cell against deleterious effects of hydroperoxides. The combined action of SOD and CAT provides an efficient mechanism for removal of free radicals from the cell (Husain et al. 1996). Thus, their decline in the present study proves that their absence may cause cell death due to free radical damage.

Under physiological conditions, intracellular antioxidant enzymes, such as SOD, CAT, and GSH-Px, eliminate ROS, thereby playing an integral role in the antioxidative stress defenses of the cell (Bukowska 2004). We found that when rats were post-treated with vitamins, methylmercury chloride–induced decreases in the levels of SOD, CAT, TSH, and GSH activities were prevented, thus showing that vitamin E and acetyl-L-carnitine proved to have protective effect against methylmercury toxicity.

Conclusion

In summary, the present study revealed that the peroxide-removing antioxidative system of nervous and nonnervous tissues of rats was adversely affected by low levels of methylmercury, and this effect was prevented by posttreatment with combination of vitamin E and acetyl-L-carnitine. Thus, it is concluded that dietary supplementation of these two vitamins might be useful in populations that are occupationally exposed to methylmercury chloride.

References

Abdel-Hamid HA, Fahmy FC, Sharaf IA (2001) Influence of free radicals on cardiovascular risk due to occupational exposure to mercury. J Egypt Public Health Assoc 76:53–69

Amin-Zaki L, Majeed MA, Elhassani SB, Clarkson TW, Greenwood MR, Doherty RA (1979) Prenatal methylmercury poisoning: clinical observations over five years. Am J Dis Child 133:172–177

Ariza ME, Bijur GN, Williams MV (1998) Lead and mercury metagenesis: role of H_2O_2, superoxide dismutase, and xanthine oxidase. Environ Mol Mutagen 31:352–361

Aschner M, Yao CP, Allen JW, Tan KH (2000) Methylmercury alters glutamate transport in astrocytes. Neurochem Int 37:188–206

Aschner M, Syversen T, Souza DO, Rocha JBT, Farina M (2007) Involvement of glutamate and reactive oxygen species in methylmercury neurotoxicity. Braz J Med Biol Res 40:285–291

Bukowska B (2004) 2, 4, 5-T and 2, 4, 5-TCP induces oxidative damage in human erythrocytes: the role of glutathione. Cell Biol Int 28:557–563

Burbacher TM, Rodier PM, Weiss B (1990) Methylmercury developmental neurotoxicity: a comparison of effects in humans and animals. Neurotoxicol Teratol 12:191–202

Clarkson TW, Magos L, Myers GJ (2003) The toxicology of mercury-current exposures and clinical manifestations. N Engl J Med 349:1731–1737

Debes F, Budtz-Jorgensen E, Weihe P, White RF, Grandjean P (2006) Impact of prenatal methylmercury exposure on neurobehavioral function at age 14 years. Neurotoxicol Teratol 28:363–375

Dietrich MO, Mantese CE, Anjos GD, Souza DO, Farina M (2005) Motor impairment induced by oral exposure to methylmercury in adult mice. Environ Toxicol Pharmacol 19:169–175

Duchen MR (2000) Mitochondria and Ca (2^+) in cell physiology and pathophysiology. Cell Calcium 28:339–348

Dutczak WJ, Ballatori N (1994) Transport of the glutathione–methylmercury complex across liver canalicular membranes on reduced glutathione carriers. J Biol Chem 269:9746–9751

Freeman BA, Crapo JD (1982) Biology of disease-free radicals and tissue injury. Lab Investig 47:412–426

Gilbert SG, Grant-Webster KS (1995) Neurobehavioral effects of developmental methylmercury exposure. Environ Health Perspect 103:135–142

Grandjean P, Weihe P, White RF, Debes F, Araki S, Yokoyama K, Murata K, Sorensen N, Dahl R, Jorgensen PL (1997) Cognitive deficit in 7-year-old children with prenatal exposure to methylmercury. Neurotoxicol Teratol 19:417–428

Halliwell B, Chirico S (1993) Lipid peroxidation: its mechanism, measurement, and significance. Am J Clin Nutr 57:715–725

Harada M (1995) Minamata disease: methylmercury poisoning in Japan caused by environmental pollution. Crit Rev Toxicol 25:1–24

Husain K, Dube SN, Sugenolran K, Singh R, Das Gupta S, Somani SM (1996) Effect of topically applied sulphur mustard on antioxidant enzymes in blood cells and body tissues of rat. J Appl Toxicol 16:245–248

Hussain S, Atkinson A, Thompson SJ, Khan AT (1999) Accumulation of mercury and its effect on antioxidant enzymes in brain, liver, and kidneys of mice. J Environ Sci Health B 34:645–660

Kono Y, Fridovich I (1982) Superoxide radical inhibits catalase. J Biol Chem 257:5751–5754

Li Y, Yan XP, Chen C, Xia YL, Jiang Y (2007) Human serum albumin–mercurial species interactions. J Proteome Res 6:2277–2286

Meister A (1983) Selective modification of glutathione metabolism. Science 220:472–477

Rabenstein DL, Fairhurst MT (1975) Nuclear magnetic resonance studies of the solution chemistry of metal complexes XI: binding of methylmercury by sulfhydryl-containing amino acids and by glutathione. J Am Chem Soc 97:2086–2092

Refsvik T (1978) Excretion of methyl mercury in rat bile: the effect of diethylmaleate, cyclohexene oxide and acrylamide. Acta Pharmacol Toxicol 42:135–141

Sirois JE, Atchison WD (2000) Methylmercury affects multiple subtypes of calcium channels in rat cerebellar granule cells. Toxicol Appl Pharmacol 167:1–11

Sood PP, Bapu C, Sinha N, Rao AP (1997) Cholesterol and triglyceride fluctuations in mice tissues during methylmercury intoxication and monothiols and vitamin therapy. J Nutr Environ Med 7:155–162

Stringari J, Meotti FC, Souza DO, Santos ARS, Farina M (2006) Postnatal methylmercury exposure induces hyperlocomotor activity and cerebellar oxidative stress in mice: dependence on the neurodevelopmental period. Neurochem Res 31:563–569

Thompson SA, White CC, Krejsa CM, Eaton DL, Kavanagh TJ (2000) Modulation of glutathione and glutamate-L-cysteine ligase by methylmercury during mouse development. Toxicol Sci 57:141–146

Part VI

Systemic Toxicology

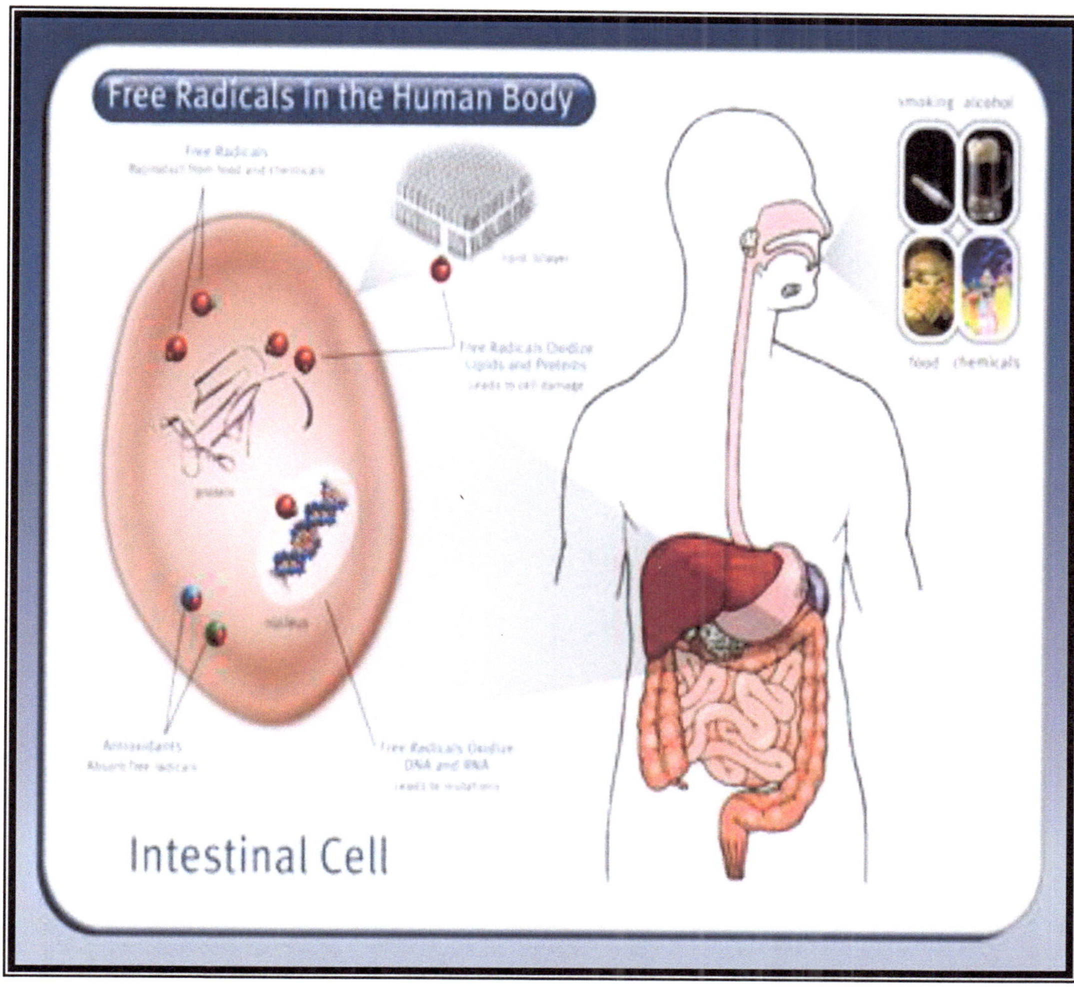

Toxic Responses of the Plasma Total Antioxidant Power (FRAP)

17

Production of free radicals and other reactive species, including free radicals, is an integral part of human metabolism. Potentially harmful reactive oxygen species (ROS) are produced as a consequence of normal aerobic metabolism (Gutteridge 1994; Halliwell 1994). These "free radicals" are usually removed or inactivated in vivo by a team of antioxidants (Gutteridge 1994; Halliwell 1994; Halliwell and Gutteridge 1990; Frei et al. 1988, 1989). Individual members of the antioxidant defense team are deployed to prevent generation of ROS, to destroy potential oxidants, and to scavenge ROS. Thus, oxidative stress-induced tissue damage is minimized. However, an absolute or relative deficiency of antioxidant defenses may lead to a situation of increased oxidative stress, and this may be associated with both the causes and consequences of a variety of disorders, including coronary heart disease and cancer (Aruoma 1994; Knight 1995; Cheeseman and Slater 1993; Frei 1995; Schwartz et al. 1993; Cerutti 1994; Emerit 1994; Gutteridge 1995). Because of the high potential to damage vital biological systems, reactive species have now been incriminated in aging and in more than 100 disease states (Ames et al. 1993; Halliwell et al. 1992).

Tests which measure the combined antioxidant effect of the nonenzymatic defenses in biological fluids may be useful in providing an index of ability to resist oxidative damage (Gutteridge 1995; Ghiselli et al. 1995; Miller et al. 1993; Popov and Lewin 1994; Wayner et al. 1987; Whitehead et al. 1992; Lissi et al. 1995). Most tests of "total antioxidant power" used to date have measured the ability of plasma to withstand the oxidative effects of reactive species purposefully generated in the reaction mixture. Depletion of antioxidants is denoted by a change in signal, such as rate of oxygen utilization (Wayner et al. 1987) or chemiluminescence (Popov and Lewin 1994; Wayner et al. 1987; Whitehead et al. 1992; Lissi et al. 1995). Measurement of these signals requires specialized equipment, and such tests can be time-consuming as well as technically demanding. This limits the availability of such tests and, furthermore, makes clinical evaluative studies difficult.

A biological antioxidant has been defined as "any substance that when present at low concentrations compared to those of an oxidisable substrate significantly delays or prevents oxidation of that substrate" (Halliwell and Gutteridge 1995). This definition is clear and covers every member of the antioxidant defense team. However, unless an antioxidant prevents the generation of ROS, for example, by metal chelation or enzyme-catalyzed removal of a potential oxidant (Gutteridge 1994; Halliwell 1994), a redox reaction still occurs. The difference is that the oxidizing species reacts with the antioxidant instead of the "substrate," i.e., the antioxidant reduces the oxidant. In simple terms then, nonenzymatic antioxidants such as ascorbic acid can be described as reductants, and inactivation of oxidants by reductants can be described as redox reactions in which one reactive species is reduced at the expense of the oxidation of another.

S. Nabi, *Toxic Effects of Mercury*, DOI 10.1007/978-81-322-1922-4_17,
© Springer India 2014

Living organisms have developed complex antioxidant systems to counteract reactive species and to reduce their damage. These antioxidant systems include enzymes such as superoxide dismutase, catalase, and glutathione peroxidase; macromolecules such as albumin, ceruloplasmin, and ferritin; and an array of small molecules, including ascorbic acid, a-tocopherol, b-carotene, ubiquinol-10, reduced glutathione (GSH), 3 methionine, uric acid, and bilirubin (Yu 1994). Oxidative stress can be reduced with the provision of additional antioxidants. Antioxidants are closely related with the prevention of degenerative illness, such as cardiovascular and neurological diseases, cancer, and oxidative stress dysfunctions (Bolck 1992; Diplock 1995; Halliwell 1996).

Several methods (Wayner et al. 1985; Glazer 1990; Ghiselli et al. 1995; Miller et al. 1993; Whitehead et al. 1992; Cao et al. 1993, 1995; Benzie and Strain 1996) have been developed to assess the total antioxidant capacity of human serum or plasma because of the difficulty in measuring each antioxidant component separately and the interactions among different antioxidant components in the serum or plasma. Various methods are known to measure the total antioxidant capacity of biological samples, but we tried the FRAP assay, which depends upon the reduction of ferric tripyridyltriazine (Fe (III)-TPTZ) complex to the ferrous tripyridyltriazine (Fe (II)-TPTZ) by a reductant at low pH. Fe (II)-TPTZ has an intensive blue color and can be monitored at 593 nm (Benzie and Strain 1996). The FRAP assay measures the ferric-to-ferrous iron reduction in the presence of antioxidants and is very simple and convenient in terms of its operation.

Protocol

For various sets of biochemical studies, different groups comprising ten rats each were used. Rats from Group-I served as control, while rats of Group-II, -III, -IV, and -V were used as experimental sets. Group-II rats were given 2 mg/kg body weight of methylmercury chloride for 14 days, and for the next 14 days, they were kept untreated. Animals of Group-III received MeHgCl (2 mg/kg body weight) for 14 days and vitamin E (100 mg/kg body weight) for the next 14 days. Group-IV animals were given MeHgCl (2 mg/kg body weight) for 14 days, and for the next 14 days, they were treated with acetyl-L-carnitine (100 mg/kg body weight). Group-V animals were treated with MeHgCl (2 mg/kg body weight) for 14 days, and for the next 14 days, they were given vitamin E (100 mg/kg body weight) and acetyl-L-carnitine (100 mg/kg body weight) in combination. In combined therapy, acetyl-L-carnitine was always administered at a gap of 30 min after vitamin E as per Sood et al. (1997). Methylmercury chloride and acetyl-L-carnitine were diluted in physiological saline, while vitamin E was given as such. All groups were treated once a day orally through intragastric intubation. The intake of drinking water and food by rats was examined daily, and rats were weighed every other day for weight change assessment due to toxic metal. Animals of all the groups were tested for elevated plus maze and righting reflex test on the 29th day. At the end of the treatment, overnight-fasted rats were sacrificed by cervical dislocation and immediately blood drawn by cardiac puncture. The blood from each rat in a given group was collected using heparin as anticoagulant. Blood was mixed gently by inversion two to three times and immediately cooled to 4 °C in a refrigerator. The samples were centrifuged at 2,500 rpm for 30 min. Plasma was aliquoted and either stored at 4 °C or frozen at–20 °C for future use.

Results

Effect of Methylmercury Chloride on "Total Antioxidant Power" (FRAP)

The FRAP assay gives fast, reproducible results with plasma, with a single antioxidant in pure solution and with mixtures of antioxidants in aqueous solution and added to plasma. The data summarized demonstrate the impact of methylmercury on plasma concentrations of total antioxidants in rats. Fourteen-day treatment of methylmercury showed that a total plasma antioxidant level was reduced from a control value of –26.81 %.

Effect of Vitamin E and Acetyl-L-Carnitine on FRAP Assay

Treatment of rats with vitamin E and acetyl-L-carnitine for 14 days resulted in an increase of total antioxidant levels by +82.47 % and +71.37 %, respectively, which is significantly higher than the values in treated group. The combined vitamin therapy showed better results displaying an increase of +97.84 % compared to treated group.

Discussion

The FRAP assay is quick and simple to perform, and the reaction is reproducible and linearly related to the molar concentration of the antioxidant(s) present. As in other tests of oxidative stress and antioxidant defense, FRAP reaction conditions are far from physiological, and it must be noted that in vitro testing of plasma may not reflect in vivo hierarchies or activities (Halliwell and Gutteridge 1995; Buettner 1993). The clinical utility of measuring antioxidant, or reducing, power in biological fluids remains to be established. Nevertheless, the FRAP assay appears to be an attractive and potentially useful test Reagents are inexpensive, the procedure is speedy and straightforward, results are highly reproducible over a wide concentration range, and the equipment required is of a type commonly found in biochemical laboratories. Clinical studies, comparing different population groups, measuring changes in FRAP values associated with particular pathological states, and monitoring FRAP values during various treatment strategies, are now needed. Nonenzymatic antioxidants are important regulators of reactive oxygen species (ROS) produced in extracellular milieu and represent the first line of defense against them. In the present study, extracellular total antioxidant capacity as plasma ferric reducing power (FRAP) decreased in rats subjected to low dose of methylmercury chloride treatment, since blood plasma is a rich source of antioxidants and plays a central role in redistribution of antioxidants to various organs of the body (Serafini et al. 2000). Methylmercury-induced increase in oxidant and depletion of antioxidant activity in our study can lead to overall oxidant–antioxidant imbalance in the animal body.

Conclusion

To conclude, therefore, the FRAP assay offers a putative index of antioxidant defense of potential use to and within the technological reach of every laboratory and researcher interested in oxidative stress and its effects. The aim of this study was to measure the antioxidant defense system all in one (total antioxidant capacity) of plasma of rats treated with low dose of methylmercury. Combining all, results suggest that vitamin E and ALCAR provide protection against methylmercury chloride-induced oxidative impairments, via their antioxidative property.

References

Ames BN, Shigenaga MK, Hagen TM (1993) Oxidants, antioxidants, and the degenerative diseases of aging. Proc Natl Acad Sci 90:7915–7922

Aruoma OI (1994) Nutrition and health effects of free radicals and antioxidants. Food Chem Toxicol 32:671–683

Benzie IFF, Strain JJ (1996) The ferric reducing ability of plasma (FRAP) as a measure of "antioxidant power": the FRAP assay. Anal Biochem 239:70–76

Bolck G (1992) The data support a role of antioxidants in reducing cancer risks. Nutr Rev 50:207–213

Buettner GR (1993) The pecking order of free radicals and antioxidants: lipid peroxidation, α-tocopherol, and ascorbate. Arch Biochem Biophys 300:535–543

Cao G, Alessio HM, Cutler RG (1993) Oxygen-radical absorbance capacity assay for antioxidants. Free Radic Biol Med 14:303–311

Cao G, Verdon CP, Wu AHB, Wang H, Prior RL (1995) Automated oxygen radical absorbance capacity assay using the COBAS FARA II. Clin Chem 41: 1738–1744

Cerutti PA (1994) Oxy-radicals and cancer. Lancet 344:862–863

Cheeseman KH, Slater TF (1993) Free radicals in medicine. Br Med Bull 49:479–724

Diplock A (1995) Antioxidant nutrients-efficacy in disease prevention and safety. Biochemist 17:16–18

Emerit I (1994) Reactive oxygen species, chromosome mutation and cancer. Free Radic Biol Med 16:99–109

Frei B (1995) Cardiovascular disease and nutrient antioxidants: role of low-density lipoprotein oxidation. Crit Rev Food Sci Nutr 35:83–98

Frei B, Stocker R, Ames BN (1988) Antioxidant defenses and lipid peroxidation in human blood plasma. Proc Natl Acad Sci 85:9748–9752

Frei B, England L, Ames BN (1989) Ascorbate is an outstanding antioxidant in human blood plasma. Proc Natl Acad Sci 86:6377–6381

Ghiselli A, Serafini M, Maiani G, Azzini E, Ferro-Luzzi A (1995) A fluorescence-based method for measuring total plasma antioxidant capability. Free Radic Biol Med 18:29–36

Glazer AN (1990) Phycoerythrin fluorescence-based assay for reactive oxygen species. Methods Enzymol 186:161–168

Gutteridge JMC (1994) Biological origin of free radicals, and mechanisms of antioxidant protection. Chem Biol Interact 91:133–140

Gutteridge JMC (1995) Lipid peroxidation and antioxidants as biomarkers of tissue damage. Clin Chem 41:1819–1828

Halliwell B (1994) Free radicals and antioxidants: a personal view. Nutr Rev 53:253–265

Halliwell B (1996) Antioxidants. In: Ziegler EE, Filer LJ (eds) Present knowledge in nutrition. ILSI Press, Washington, DC, pp 596–603

Halliwell B, Gutteridge JMC (1990) The antioxidants of human extracellular fluids. Arch Biochem Biophys 280:1–8

Halliwell B, Gutteridge JMC (1995) The definition and measurement of antioxidants in biological systems. Free Radic Biol Med 18:125–126

Halliwell B, Gutteridge JMC, Cross CE (1992) Free radicals, antioxidants, and human disease: where are we now? J Clin Lab Med 119:598–620

Knight JA (1995) The process and theories of aging. Ann Clin Lab Sci 25:1–12

Lissi E, Salim-Hanna M, Pascual C, del Castillo MD (1995) Free Radic Biol Med 18:153–158

Miller NJ, Rice-Evans C, Davies MJ, Gopinathan V, Milner A (1993) A novel method for measuring antioxidant capacity and its application to monitoring the antioxidant status in premature neonates. Clin Sci 84:407–412

Popov IN, Lewin G (1994) Photochemiluminescent detection of antiradical activity: II. Testing of nonenzymic water-soluble antioxidants. Free Radic Biol Med 17:267–271

Schwartz JL, Antoniades DZ, Zhao S (1993) Molecular and biochemical reprogramming of oncogenesis through the activity of prooxidants and antioxidants. Ann N Y Acad Sci 686:262–278

Serafini M, Natella F, Scaccni N (2000) Total antioxidant capacity of assess redox status: critical view and experimental data. Free Radic Biol Med 29:1106–1114

Sood PP, Bapu C, Sinha N, Rao AP (1997) Cholesterol and triglyceride fluctuations in mice tissues during methylmercury intoxication and monothiols and vitamin therapy. J Nutr Environ Med 7:155–162

Wayner DDM, Burton GW, Ingold KU, Locke S (1985) Quantitative measurement of the total, peroxyl radical-trapping antioxidant capacity of human blood plasma by controlled peroxidation. FEBS Lett 187:33–37

Wayner DDM, Burton GW, Ingold KU, Barclay LRC, Locke SJ (1987) The relative contributions of vitamin E, urate, ascorbate and proteins to the total peroxyl radical-trapping antioxidant activity of human blood plasma. Biochim Biophys Acta 924:408–419

Whitehead TP, Thorpe GHG, Maxwell SRJ (1992) Enhanced chemiluminescent assay for antioxidant capacity in biological fluids. Anal Chim Acta 266: 265–277

Yu BP (1994) Cellular defenses against damage from reactive oxygen species. Physiol Rev 74:139–162

18 Toxic Responses of the Plasma Xanthine Oxidase (XO)

Xanthine oxidase is an important source of free oxygen radicals, especially superoxide anions, because it is able to use molecular oxygen instead of NAD+ as electron acceptor in the two-step conversion of hypoxanthine to uric acid. This reaction mechanism has been confirmed in both animal and human studies (Connors et al. 1981; Brown et al. 1988; Ferreira et al. 1988; Pietz et al. 1988) and is probably the primary source of reactive oxygen species after reperfusion of ischemic tissues (Saugstad and Aasen 1980).

Free radical production and oxidative stress is one of the mechanisms of tissue damage in heavy metal toxicity. Xanthine oxidase is an enzyme involved in catabolism of purine bases, adenine, and guanine. An enzyme exists in two interconvertible forms, as xanthine oxidase and xanthine dehydrogenase. Higher level of xanthine oxidase activity against xanthine dehydrogenase increases production of free radicals and lipid peroxidation level.

Many recent reports implicate oxygen-free radicals as a mediator of ischemia reperfusion injury, in which xanthine oxidase (XO), which generates toxic O_2 as a metabolite, could be responsible for tissue injuries in biological systems (McCord 1985; Granger et al. 1986). In this pathological system, enhancement of cellular adenosine catabolism (after ATP degradation) yields high levels of hypoxanthine and xanthine, resulting coincidentally in effective supply of a substrate for XO. On the other hand, conversion of xanthine dehydrogenase (XD) to XO is also accelerated. Thus, these two simultaneous processes seem to accelerate the generation of free radicals, and remarkably elevated production of O_2 becomes possible (McCord 1985). Furthermore, there is now evidence that the level of XO in plasma is elevated in adult respiratory distress syndrome (ARDS) (Grum et al. 1987) and that XO mediates lung injury by neutrophil elastase and hyperoxia (Rodell et al. 1987).

Protocol

For various sets of biochemical studies, different groups comprising ten rats each were used. Rats from Group-I served as control, while rats of Group-II, -III, -IV, and -V were used as experimental sets. Group-II rats were given 2 mg/kg body weight of methylmercury chloride for 14 days, and for the next 14 days, they were kept untreated. Animals of Group-III received MeHgCl (2 mg/kg body weight) for 14 days and vitamin E (100 mg/kg body weight) for the next 14 days. Group-IV animals were given MeHgCl (2 mg/kg body weight) for 14 days, and for the next 14 days, they were treated with acetyl-L-carnitine (100 mg/kg body weight). Group-V animals were treated with MeHgCl (2 mg/kg body weight) for 14 days, and for the next 14 days, they were given vitamin E (100 mg/kg body weight) and acetyl-L-carnitine (100 mg/kg body weight) in combination. In the combined therapy, acetyl-L-carnitine was always administered at a gap of 30 min after vitamin E as per

S. Nabi, *Toxic Effects of Mercury*, DOI 10.1007/978-81-322-1922-4_18,
© Springer India 2014

Sood et al. (1997). Methylmercury chloride and acetyl-L-carnitine were diluted in physiological saline, while vitamin E was given as such. All groups were treated once a day orally through intragastric intubation. The intake of drinking water and food by rats was examined daily, and rats were weighed every other day for weight change assessment due to toxic metal. Animals of all the groups were tested for elevated plus maze and righting reflex test on the 29th day. At the end of the treatment, overnight-fasted rats were sacrificed by cervical dislocation and immediately blood drawn by cardiac puncture. The blood from each rat in a given group was collected using heparin as anticoagulant. Blood was mixed gently by inversion two to three times and immediately cooled to 4 °C in a refrigerator. The samples were centrifuged at 2,500 rpm for 30 min. Plasma was aliquoted and either stored at 4 °C or frozen at−20 °C for future use.

Results

Effect of Methylmercury Chloride, Vitamin E, and Acetyl-L-Carnitine on Xanthine Oxidase

Activity

Xanthine oxidase is known to be an important biological source of free radicals. There is overwhelming evidence to suggest that serum xanthine oxidase activity is significantly increased in various pathological disorders, like hepatitis, inflammatory disease, hypercholesterolemia, atherosclerosis, ischemic-reperfusion, carcinogenesis, aging, and diabetes and that free radicals generated in the enzymatic processes are involved in oxidative damage. Thus, it may be possible that the inhibition of this enzymatic pathway by the compounds that have both antiradical and xanthine oxidase inhibitory properties may have additional therapeutic importance in the treatment of the above diseases. Therefore, we have investigated the possible inhibitory effect of vitamin E and acetyl-L-carnitine on plasma xanthine oxidase activity in methylmercury-treated rats. Plasma xanthine oxidase activity was significantly increased by +69.34 % in treated rats in comparison to control rats. Vitamin E and acetyl-L-carnitine showed significant effect in preventing this increase in plasma xanthine oxidase activity and was decreased by−24.33 % and −22.40 %, respectively, when compared with the corresponding values in methylmercury chloride-treated animals. These results demonstrate that both the above antioxidants and their potent antioxidant activity also significantly inhibit the elevated levels of xanthine oxidase in plasma, indicating a dual therapeutic benefit in the treatment of the above disorders. The results are depicted diagrammatically in Fig. 19.1, shown in the next chapter.

Discussion

Xanthine oxidase has a pivotal role in ischemia and reperfusion injuries in some animal models (Granger et al. 1986). Xanthine oxidase activity is relatively low under normal circumstances. In adult rats, plasma xanthine oxidase activity amounts to approximately 8–10 mU/mL (Anderson et al. 1991; Terada et al. 1992), and in healthy adults, activity ranges from 0 to 1.20 mU/mL (Majkic-Singh et al. 1987).

Xanthine dehydrogenase (XDH) can be converted reversibly to xanthine oxidase (XO) by oxidation of cysteine residues or irreversibly by limited proteolysis. XO has high reactivity toward O_2 but negligible reactivity toward NAD+. As XO can reduce molecular oxygen to superoxide and hydrogen peroxide, XO is thought to be one of the key enzymes producing reactive oxygen species. Increased activity of xanthine oxidase to dehydrogenase form increases production of free radicals as lipid peroxidation can lead to gout and mutagenesis. Superoxide anion, hydrogen peroxide, and hydroxyl radical are formed as by-products in reaction. Inhibitors of xanthine oxidase reaction have therapeutic application in treating of gout and hepatic injury (Sokol et al. 1998; Nishino et al. 2005).

In this study, our data on xanthine oxidase activity of plasma revealed an increase due to methylmercury chloride treatment. As xanthine oxidase is an important source of free oxygen radicals, its elevated level in this case suggests a role for oxidant injury in animals. Therefore, it is reasonable to mention that elevated xanthine oxidase activity in the plasma would facilitate tissue injuries by the generation of O_2^-, due to its distribution to various organs where it must be then converted to more toxic oxygen metabolites (i.e., OH). This work thus clarifies that the oxygen-free radical generated by xanthine oxidase is a pathological principle in methylmercury chloride toxicity in rats and that a therapeutic approach by elimination of the radical oxygen seems possible.

Conclusion

It is concluded that exposure to low dose of methylmercury induces differential alteration of oxidative stress markers in male albino rats, and these changes are consistently reversed by antioxidant treatment indicating the possible involvement of free radicals in such differential metal-induced stress responses.

The results demonstrate that methylmercury disrupts the redox status of AS52 cells by enhancing the activities of XO. Furthermore, these results also demonstrate that there is a causal relationship between the induction of H_2O_2 by this heavy toxic metal and mutagenesis.

References

Anderson BO, Moore EE, Moore FA, Leff JA, Terada LS, Harken AH, Repine JE (1991) Hypovolemic shock promotes neutrophil sequestration in lungs by a xanthine oxidase-related mechanism. J Appl Physiol 71:1862–1865

Brown JM, Terada LS, Grosso MA, Whitmann DJ, Velasco SE, Patt A, Harken AH, Repine JE (1988) Xanthine oxidase produces hydrogen peroxide which contributes to reperfusion injury of ischemic, isolated, perfused rat hearts. J Clin Invest 81:1297–1301

Connors MC, Harkness RA, Simmonds RJ, Hytten FE (1981) The measurement of hypoxanthine, xanthine, inosine and uridine in umbilical cord and fetal blood samples as a measure of fetal asphyxia. Br J Obstet Gynaecol 88:381–390

Ferreira R, Llesuy S, Milei J, Scordo D, Hourquebil H, Motteni L, de Palma C, Boveris A (1988) Assessment of myocardial oxidative stress in patients after myocardial revascularization. Am Heart J 115:307–312

Granger DN, Hollwarth ME, Parks DA (1986) Ischemia-reperfusion injury: role of oxygen-derived free radicals. Acta Physiol Scand Suppl 548:47–63

Grum CM, Ragsdale RA, Ketai LH, Simon RH (1987) Plasma xanthine oxidase activity in patients with adult respiratory distress syndrome. J Crit Care 2:22–26

Majkic-Singh N, Bogavac I, Kalimanovska V, Jelie Z, Spasic S (1987) Spectrophotometric assay of xanthine oxidase with 2:2′-azino-di (3-ethylbenzthiazoline-6-sulphonate) (ABTS) as chromogen. Clin Chim Acta 162:29–36

McCord JM (1985) Oxygen-derived free radicals in post-ischemic tissue injury. N Engl J Med 312:159–163

Nishino T, Okamoto K, Kawaguchi Y, Matsumura HHT, Eger BT, Pai EF, Nishino T (2005) Mechanism of the conversion of xanthine dehydrogenase to xanthineoxidase. J Biol Chem 280(I26):24888–24894

Pietz J, Guttenberg N, Gluck L (1988) Hypoxanthine: marker for asphyxia. Obstet Gynecol 72:762–766

Rodell TC, Cheronis JC, Ohnemus CL, Piermattei DJ, Repine JE (1987) Xanthine oxidase mediates elastase-induced injury to isolated lungs and endothelium. J Appl Physiol 63:2159–2163

Saugstad OD, Aasen AO (1980) Plasma hypoxanthine concentrations in pigs: a prognostic aid in hypoxia. Eur Surg Res 12:123–129

Sokol RJ, Devereaux MW, Khandwala R (1998) Effect of oxypurinol, a xanthine oxidase inhibitor, on hepatic injury in the bile duct-ligated rat. Pediatr Res 44(3):397–401

Sood PP, Bapu C, Sinha N, Rao AP (1997) Cholesterol and triglyceride fluctuations in mice tissues during methylmercury intoxication and monothiols and vitamin therapy. J Nutr Environ Med 7:155–162

Terada LS, Dormish JJ, Shankey PF, Leff JA, Anderson BO, Refine JE (1992) Circulating xanthine oxidase mediates lung neutrophil sequestration after intestinal ischemia – reperfusion. Am J Physiol 263:L394–L401

Toxic Responses of the Plasma Glutathione Peroxidase (GSH-Px) 19

Selenium (Se) acts as an antagonist to the toxic effects of many heavy metals, including mercury (Yoneda and Suzuki 1997; Falnoga and Tusek-Znidaric 2007). This essential trace element (Combs and Combo 1984) is important in many biological functions, since it is a cofactor in selenium-containing enzymes, especially in antioxidant enzymes such as glutathione peroxidase (GSH-Px) (Rotruck et al. 1973). This enzyme is effective in catalyzing the decomposition of hydrogen peroxides and lipid peroxides. When GSH-Px activity is inhibited, peroxides formed during oxidative stress can propagate cell damage. Measurement of this enzyme activity is an indirect and noninvasive method that could be used to assess oxidant stress. The activity of this enzyme has been used to assess body selenium status and nutritional requirements (Levander 1991).

The present work was designed to assess the effects of low dose of methylmercury on GSH-Px activity in order to estimate the antioxidant capacity in the cellular environment.

Results

Effect of Methylmercury Chloride, Vitamin E, and Acetyl-L-Carnitine on Glutathione Peroxidase Activity (GSH-Px)

Oxidative stress may be increased due to methylmercury treatment owing to higher production of reactive oxygen species (ROS), such as superoxide radical, hydroxyl radical, hydrogen peroxide, and/or deficiency in the antioxidant defense systems. The increased production of ROS has been attributed to protein glycation and/or glucose auto-oxidation due to hyperglycemic environment. An impaired radical scavenger function has been linked to the decreased activity of enzymatic and nonenzymatic scavengers of free radicals. Therefore, the status of antioxidant enzymes, such as catalase, SOD, GSH-Px, GST, and GSH concentrations in plasma, is highly important.

The data depicted in Fig. 19.1 showed a significant decrease of −60.78 % in the activity of GSH-Px by methylmercury from normal control. Administration of vitamin E and acetyl-L-carnitine to treated rats resulted in a remarkable increase in GSH-Px activity by +37.87 % and +36.75 %, respectively, when results were compared with controls. These results indicate that the decrease in GSH-Px in plasma was prevented by vitamin E and acetyl-L-carnitine, demonstrating their potent antioxidant property.

Discussion

Several in vivo and in vitro studies have demonstrated that both inorganic mercury and methylmercury induce oxidative stress (Tran et al. 2007; Chen et al. 2005; Girardi and Elias 1995) by increasing the intracellular levels of ROS and modifying enzyme activities (Girardi and Elias 1995). In this study, the antioxidant enzyme GSH-Px, which has the ability to

S. Nabi, *Toxic Effects of Mercury*, DOI 10.1007/978-81-322-1922-4_19,
© Springer India 2014

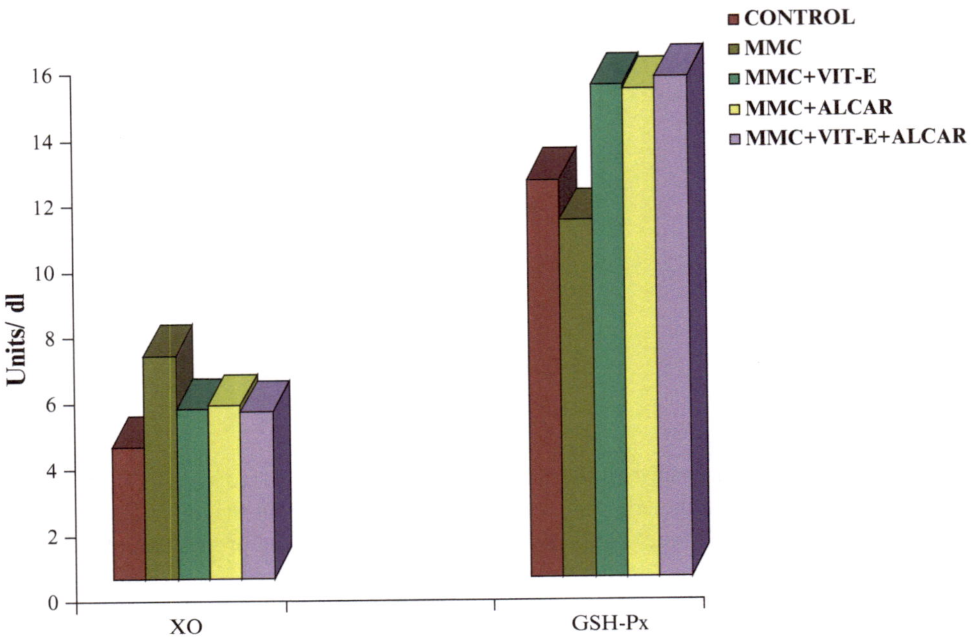

Fig. 19.1 Effect of methylmercury chloride on xanthine oxidase (*XO*) and glutathione peroxidase (*GSH-Px*) level in plasma of rats. Values represent mean ± SEM

reduce hydroperoxides in the presence of reduced GSH, had its activity significantly decreased in plasma of the group of rats treated with methylmercury when compared with the control group. However, while decreased GSH-Px activity after MeHg exposure was also observed by Bulat et al. (1998) in subjects occupationally exposed to elemental Hg. Chen et al. (2005) reported an increase in GSH-Px activity after exposure to Hg. In a study by Denise et al. (2008), it was observed that the lower the GSH-Px activity, the greater the DNA damage, showing that the decrease in GSH-Px activity may be associated with DNA damage, as this enzyme has an important role in detoxification of peroxides, thus contributing to the protection of the cell integrity.

The mechanism of action of selenium in modifying methylmercury toxicity is not known, but it does not appear to involve a decrease in tissue mercury concentration. Brain mercury levels are not decreased by selenium at the time selenium is exerting a protective effect (Ohi et al. 1976; Ganther et al. 1973). Selenite increased the uptake of methylmercury in rat brain (Prohaska and Ganther 1977). In some cases where methylmercury is fed for long periods together with high levels of selenium, the mercury levels are well above the level (about 10 ppm) normally considered to cause obvious signs of central nervous system toxicity.

Another mechanism that has been considered is that some of the toxic effects of mercury result from complexing of essential, biologically active forms of selenium (El-Begearmi et al. 1977). The only functional form of selenium thus far identified in animals is the selenoenzyme.

Glutathione peroxidase (Hoekstra 1975; Ganther 1975), a deficiency of this enzyme, is induced by feeding silver (Wagner et al. 1975). In such a case, supplementation with additional selenium permits more of the biologically active selenium compound to be synthesized (Wagner 1975), thus overcoming a conditioned nutritional deficiency of selenium induced by the metal. Although there is some evidence that methylmercury may induce signs of selenium deficiency (Froseth et al. 1974; Welsh 1974), it was not found to be particularly effective in decreasing glutathione peroxidase levels when fed in the diet of rats (Wagner 1975) or cats, in contrast to silver, and in some tissues such as liver, methylmercury tended to elevate the activity of this enzyme in rats (Wagner 1975). Injection of a subacute dose of methylmercury caused a slight decrease in rat brain glutathione peroxidase activity (Prohaska and Ganther 1977). The possibility that selenium might alter the distribution of mercury in brain and thereby protect some critical site has also been investigated (Prohaska and Ganther 1977; Chen et al. 1975), with negative results. Injection of 75Se-selenite did not alter the subcellular distribution of 203 Hg administered as 203 Hg-methylmercury, nor did 203 Hg and 75Se tend to concentrate in any particular fraction of cytosol fractionated by gel filtration. When rats labeled at 7 days of age with a physiological dose of 75Se-selenite were later injected with 203 Hg-methylmercury, there was some shift of 755en from the cytosol to the mitochondrial fraction of the brain (Prohaska and Ganther 1977); both 203 Hg and 75Se tended to follow the protein concentration in fractions of cytosol separated by gel filtration chromatography, but no evidence for chromatography of the isotopes in a particular fraction was observed, unlike the case for mercuric chloride plus equimolar selenite (Chen et al. 1974, 1975; Burk et al. 1974).

The standard of recommended intake levels of selenium is under debate (Rayman 2000). The UK reference nutrient intake (RNI) of 75 μg per day for men and 60 μg per day for women has been determined as the intake believed to be necessary to maximize the activity of the antioxidant selenoenzyme GPx in plasma (MacPherson et al. 1997). The American recommended dietary allowance (RDA), set at 55 μg per day for both men and women, is based on the investigations of the selenium intake required to achieve plateau concentrations of plasma GPx (National Academy Press 2000). The WHO/FAO/IAEA expert group recommended an intake level of only 40 μg per day for men and 30 μg per day for women, assuming only two-thirds of the full expression of GPx activity is required (WHO 1996). However, as Rayman (2000) points out, if levels of GPx activity saturation are determined using platelets rather than plasma, then the intake levels needed should be approximately 80–100 μg per day. Additionally, intake levels which saturate plasma GPx activity are insufficient to optimize the immune response and reduce cancer risk. This insufficiency is amplified at intake levels suggested by the WHO/FAO/IAEA which only accommodate two-thirds of plasma GPx activity. Currently, the UK and other European countries have intake levels of approximately half the RNI, and areas of China have intakes of less than 19 μg per day for men and less than 13 μg per day for women. Likewise, low selenium soils are prevalent in many areas of the world including New Zealand, Russia, and Africa, thus compromising the selenium status of these populations.

As with other heavy metal-induced oxidative stress, high levels of dietary ά-tocopherol protected against methylmercury chloride-induced hepatic lipid peroxidation and enhanced the activity of selenium-dependent glutathione peroxidase activity (Anderson and Anderson 1993). However, excess dietary β-carotene did not provide protection and had no effect on glutathione peroxidase activity. The differences in response to ά-tocopherol and β-carotene may be related to distribution and site-specific effects of the methylmercury chloride.

Exogenous supplementation of antioxidants has been reported to exert protective effect in various pathological states in which free radicals are involved (Vendemiale et al. 1999). In the present study, posttreatment with vitamin E and acetyl-L-carnitine was found to reverse both the oxidative stress and behavioral changes associated with exposure to methylmercury. The fact that treatment with antioxidants attenuated both metal-induced behavioral suppression and brain oxidative stress in a consistent manner clearly suggests that alteration of prooxidant–antioxidant balance might be associated to methylmercury chloride-induced neurobehavioral changes.

Conclusion

In conclusion, this study provides substantial evidence that suggests that GPx is an important target of and a major cellular defense against MeHg-induced cytotoxicity presumably because this enzyme is essential for counteracting the pro-oxidative effects of MeHg both in vitro and in vivo. Our results show that MeHg is able to reduce GPx activity in rat models and that inhibition of this antioxidant enzyme causes a significant augmentation of MeHg-induced impairment of cell viability and oxidative stress.

References

Anderson HR, Anderson O (1993) Effects of dietary α-tocopherol and β-carotene on lipid peroxidation induced by methylmercuric chloride in mice. Pharmacol Toxicol 73:192–201

Bulat P, Dujic I, Potkonjak B, Vidakovic A (1998) Activity of glutathione peroxidase and superoxide dismutase in workers occupationally exposed to mercury. Int Arch Occup Environ Health 71:S37–S39

Burk RF, Foster KA, Greenfield PM, Kiker KW, Hannon JP (1974) Binding of simultaneously administered inorganic selenium and mercury to a rat plasma protein. Proc Soc Exp Biol Med 145(3):782–785

Chen RW, Whanger PD, Fang SC (1974) Diversion of mercury binding in rat tissues by selenium: a possible mechanism of protection. Pharmacol Res Commun 6(6):571–579

Chen RW, Lacy VL, Whanger PD (1975) Effect of selenium on methylmercury binding to subcellular and soluble proteins in rat tissues. Res Commun Pathol Pharmacol 12(2):297–308

Chen C, Qu L, Li B, Xing L, Jia G, Wang T, Gao Y, Zhang P, Li M, Chen W, Chai Z (2005) Increased oxidative DNA damage, as assessed by urinary 8-hydroxy-2-deoxyguanosine concentrations, and serum redox status in persons exposed to mercury. Clin Chem 51:759–767

Combs GF, Combo SB (1984) The nutritional biochemistry of selenium. Annu Rev Nutr 4:257–280

Denise G, Gustavo RMB, Juliana V, Lusania MGA, Jose Pedro FA, Solange CG, Fernando B (2008) Low levels of methylmercury induce DNA damage in rats: protective effects of selenium. Arch Toxicol 83:249–254

El-Begearmi MM, Sunde ML, Ganther HE (1977) A mutual protective effect of mercury and selenium in Japanese quail. Poult Sci 56(1):313–322

Falnoga I, Tusek-Znidaric M (2007) Selenium-mercury interactions in man and animals. Biol Trace Elem Res 119:212–220

Froseth JA, Piper RC, Carlson JR (1974) Relationship of dietary selenium and oral methylmercury to blood and tissue selenium and mercury concentrations and deficiency-toxicity signs in swine. Fed Proc 33:660

Ganther HE (1975) Selenoproteins. Chem Scr 8A:79

Ganther HE, Wagner PA, Sunde ML (1973) Protective effects of selenium against heavy metal toxicities. In: Hemphill D (ed) Trace substances in environmental health, vol 6. University of Missouri, Columbia, pp 247–252

Girardi G, Elias MM (1995) Mercury chloride effects on rat renal redox enzymes activities: SOD protection. Free Radic Biol Med 18:61–66

Hoekstra WG (1975) Biochemical function of selenium and its relation to vitamin E. Fed Proc 34(11): 2083–2089

Levander OA (1991) Scientific rationale for the 1989 recommended dietary allowances for selenium. J Am Diet Assoc 91:1572–1576

MacPherson A, Barclay MNI, Scott R, Yates RWS (1997) Loss of Canadian wheat lowers selenium intake and status of the Scottish population. In: Fischer PWF, L'Abbe MR, Cockell KA, Gibson RS (eds) Trace elements in man and animals 9: proceedings of the ninth international symposium on trace elements in man and animals. NRC Research Press, Ottawa, pp 203–205

National Academy Press (2000) Standing Committee on the Scientific Evaluation of Dietary Reference Intakes of the Food and Nutrition Board, Institute of Medicine, the National Academics with Health Canada. Dietary reference intakes for vitamin C, vitamin E, selenium, and carotenoids. Washington DC

Ohi G, Nishigaki S, Seki H, Tamura Y, Maki T (1976) Efficacy of selenium in tuna and selenite in modifying methylmercury intoxication. Environ Res 12(1):49–58

Prohaska JR, Ganther HE (1977) Interactions between selenium and methylmercury in rat brain. Chem Biol Interact 16(2):155–167

Rayman M (2000) The importance of selenium to human health. Lancet 356:233–241

Rotruck IT, Pope AL, Ganther HE, Swanson AB, Hafeman DG, Hockstra WG (1973) Selenium: biochemical role as a component of glutathione peroxidase. Science 179:588–590

Tran D, Moody AJ, Fisher AS, Foulkes ME, Jha AN (2007) Protective effects of selenium on mercury-induced DNA damage in mussel haemocytes. Aquat Toxicol 84:11–18

Vendemiale G, Grattagliano I, Altomare E (1999) An update on the role of free radicals and antioxidant defense in human disease. Int J Clin Lab Res 29:49–55

Wagner PA (1975) Studies on the interaction of selenium with silver and methylmercury in the rat. Ph.D. thesis, University of Wisconsin, Madison

Wagner PA, Hoekstra WG, Ganther HE (1975) Alleviation of silver toxicity by selenite in the rat in relation to tissue glutathione peroxidase. Proc Soc Exp Biol Med 148(4):1106–1110

Welsh S (1974) Physiological effects of methylmercury toxicity: interaction of methylmercury with selenium, tellurium and vitamin E. Ph.D. thesis, University of Maryland, College Park, MD

WHO, Food and Agriculture Organization, International Atomic Energy Agency Expert Group (1996) Trace elements in human nutrition and health. WHO, Geneva

Yoneda S, Suzuki KT (1997) Equimolar Hg-Se complex binds to selenoprotein P. Biochem Biophys Res Commun 231:7–11

Part VII

Neurobehavioral Toxicology

20 Impairment of Emotional Behavior

Behavior is a mirror in which every one displays his own image

[Johann Wolfgang von Goethe]

Behavior is a major function whereby animals adapt to changes in the environment. Spontaneous behavior is dependent on integration of sensory input to motor output. This reveals the capability of animals to habituate to their novel environment, integrating new information with that previously attained, and is a measure of cognitive function. The cholinergic involvement in behavior is well known (Narahashi et al. 2000; Russel and Kelly 1982). The cholinergic system is implicated in regulating general brain excitability during arousal and sleep–wake cycles, and the basal forebrain complex plays a special role in learning and memory function (Bear et al. 1996). It has been suggested that this system in particular is involved in the aging processes (Bartus et al. 1982).

Changes in behavior can be evidence for chemical pollution of our natural environment. Behavior is an important endpoint for studying environmental toxicants in mammals because it can reveal effects on the nervous system.

The rat is the most frequently used mammal for behavioral experiments, and several methods can be used to assess behavioral changes that result from modifications to its nervous system. Objective neurological tests can be used to establish a "behavioral baseline" for a normal animal, and to study the effects of drugs or a lesion on behavior.

Comparisons between humans and other animals are complicated by limitations in the direct evaluation of subjective states (i.e., unlike humans, animals do not have proactive direct communication of their self-perception). Nonhuman animals have no verbal language to express the subtleties of psychological states (emotional, motivational, cognitive, etc.). As a consequence, animal studies must be sensitive to the ways in which species can communicate their affective states. Studies of animal behavior must identify well-defined descriptive categories, avoiding redundancies and overlap, and monitor frequencies, intensities, sequences, patterns, and trends (Martin and Bateson 1986). Abnormalities may emerge by disruptions in sequences or by unpredictable fluctuations in intensity. It is equally critical to measure motivational levels and relate them to behavior.

To evaluate the emotional behavior in laboratory animals, it should be possible to measure emotions directly, to classify types of emotions, and indeed to identify emotions in animals that may have relevance to human emotional states. Recognizing the difficulty in specifying emotions, due to the fact that the complexities of overt behavior must be ascribed to some underlying emotional state, the experimenter nevertheless attempts to classify emotions despite this

S. Nabi, *Toxic Effects of Mercury*, DOI 10.1007/978-81-322-1922-4_20,
© Springer India 2014

limitation. Because of the subjective nature of emotional states, animal analogues of such states as anxiety have been difficult to design. Even though several models centering on particular aspects of the emotional behavior in rodents have been developed in recent years, none has been evaluated thoroughly for its efficacy in developmental behavioral toxicity testing. Most of these testing methods have been validated behaviorally and physiologically and appear to be useful for distinguishing anxiogenic and anxiolytic effects within several classes of drugs.

One of the most traditional and widely used methods for the assessment of the emotional state in rodents is the open-field test (Tobach 1969), of which many varieties exist. Computerized open-field equipment have been recently developed like Image Motion Analyzer and Videotrack System. Because open-field test is a relatively simple technique and gives quantitative information on a broad range of responses, it has been used frequently in teratological studies (Di Giovanni et al. 1993; Spyker et al. 1972; Cagiano et al. 1990). A flat area bounded by walls is divided into squares, and several activities are scored (number of center and peripheral squares entered per unit time, latency to leave the center area, rearing, grooming, etc.).

In the open-field situation, other responses such as defecation and urination can also be measured. Open-field activity scores seem to reflect both emotional reactivity and exploratory behavior, whereas defecation primarily reflects emotional reactivity. Even though the results have not always been consistent, an inverse relationship between exploratory activity and the emotional state of the animal has been suggested, and activity has frequently been inversely correlated with defecation levels (Rodier 1978). However, according to Nortan (1989), the notion that the open-field test can be used to measure general autonomic reactivity, or emotionality, is not substantiated by the evidence. In this regard, for example, different measurements of autonomic reactivity (i.e., cardiac rate and defecation) do not show parallel changes with habituation, and activity in the open field is not correlated with corticosterone levels (Candland and Nagy 1969; Stem et al. 1973).

There is an increasing concern about neurological deficits in humans, and the environmental contaminants have been proposed as possible causes of learning and emotional disturbances at young age and neurodegenerative diseases in later life (Landrigan et al. 2005; National Research Council 2000). Methylmercury is known to be an environmental neurotoxicant potentially causing neuropsychological disorders in humans (Gilbert and Grant-Webster 1995). Furthermore, epidemiological and experimental studies have clearly shown that the developing nervous system is particularly vulnerable to methylmercury toxicity. Severe neurotoxic effects of prenatal exposure to high doses of methylmercury were established in humans after methylmercury disasters in Japan and Iraq (Amin-Zaki et al. 1979; Harada 1995) and confirmed in animal studies (Burbacher et al. 1990). Later, developmental exposure to low doses of methylmercury contained in the seafood was found to be a risk factor for cognitive disorders (e.g., memory, attention, and language problems) in children and adolescents in studies conducted on the fish-eating population of the Faeroe Islands (Debes et al. 2006; Grandjean et al. 1997). This fact raised research interest in studying effects of prolonged low-dose intoxication in animal models, representing a chronic pattern of exposure in humans (Weiss et al. 2005).

Studies using rodent animal models have proven useful in simulating neurobehavioral effects of methylmercury exposure (Spyker and Smithberg 1972; Spyker et al. 1972). However, most neurological effects observed in rodents have been reported from studies that utilized prenatal exposure to methylmercury (Goulet et al. 2003; Kakita et al. 2000; Kim et al. 2000; Newland and Rasmussen 2000; Newland et al. 2004; Rasmussen and Newland 2001; Rossi et al. 1997; Sakamoto et al. 2002; Salvaterra et al. 1973; Su and Okita 1976). Still, comparatively little is known about the behavioral effects of methylmercury exposure on young adults at low to moderate exposure levels.

We designed the present experiment as a model for the chronic exposure of rats to fairly low dose of methylmercury (2 mg/kg body weight), which allowed us to analyze the long-lasting effects in male rats with a focus on emotional behavior.

Open-Field Behavior (OFB) Apparatus

The OFB apparatus used in the present study was similar to that used by Tobach (1969). Briefly, it consisted of a wooden, circular open arena (82 cm diameter) surrounded by a wall (31 cm high). The wooden floor was marked with three centric circles which were divided into segments by lines radiating from the center. These 25 units of approximately equal size were used to score ambulation of the animals during the test. Two types of stimuli were presented to the animals: Noise (78 dB, ref. intensity 2×10^{-4} dyn/cm^2) was produced by an oscillator through four loudspeakers and light (165 FC) was shown by four lamps. A translucent glass screen enclosed the arena on all sides, the front side having a glass door through which the animals were placed in the arena.

Protocol

For various sets of biochemical studies, different groups comprising ten rats each were used. Rats from Group-I served as control, while rats of Group-II, -III, -IV, and -V were used as experimental sets. Group-II rats were given 2 mg/kg body weight of methylmercury chloride for 14 days, and for the next 14 days, they were kept untreated. Animals of Group-III received MeHgCl (2 mg/kg body weight) for 14 days and vitamin E (100 mg/kg body weight) for the next 14 days. Group-IV animals were given MeHgCl (2 mg/kg body weight) for 14 days, and for the next 14 days, they were treated with acetyl-L-carnitine (100 mg/kg body weight). Group-V animals were treated with MeHgCl (2 mg/kg body weight) for 14 days, and for the next 14 days, they were given vitamin E (100 mg/kg body weight) and acetyl-L-carnitine (100 mg/kg body weight) in combination. In the combined therapy, acetyl-L-carnitine was always administered at a gap of 30 min after vitamin E as per Sood et al. (1997). Methylmercury chloride and acetyl-L-carnitine were diluted in physiological saline, while vitamin E was given as such. All groups were treated once a day orally through intragastric intubation. The intake of drinking water and food by rats was examined daily, and rats were weighed every other day for weight change assessment due to toxic metal. All groups were tested for open-field test on the 0th, 7th, 15th, 22nd, and 29th days. The animals were sacrificed later on the scheduled day by cervical dislocation, and immediately, the brains, spinal cords, hearts, lungs, and pancreases were taken out and kept on ice. Tissues were weighed both in control and treated animals to observe the weight changes. Brains were separated into cerebrum, cerebellum, and brain stem. The tissues were later processed for the assay of biochemical parameters.

OFB Procedure

After treatment, each animal of both the treated and control groups was exposed on the 0th, 7th, 15th, 22nd, and 29th days for 5 min in the apparatus, and the ambulation, preening, rearing, and center crossing responses were recorded by a three-channeled hand-operated counter.

1. Ambulation

   It is defined as the walking score derived from the number of radical segments crossed by the animals. The placement of all the four limbs in one segment was taken as one unit of ambulation.
2. Preening

   Preening response was determined by the number of times the animal scratched its face with the forelimbs.
3. Rearing

   A rearing score of one was awarded when the rat stood on its hind limbs with the support of the wall and two for standing without support.

Fig. 20.1 Open-field equipment displaying a rat treated with low dose of MMC

4. Center Crossing

   Center crossing was scored each time the animal placed at least two paws in the 6 in. diameter circle outlined in the center of the open field (Fig. 20.1).

Results

The low dose of 2 mg/kg body weight of methylmercury chloride produced overt signs of toxicity in treated rats. The methylmercury chloride-treated rats showed signs of significantly altered behavior compared to control rats. To evaluate the rats for more subtle changes in activity levels, we used this standard behavior test named open-field activity.

Open-field behavior study was observed with MeHgCl toxicosis on the following parameters: (1) Ambulation, (2) preening, (3) rearing, and (4) center crossing.

Effect on Ambulation Score

Significant decline in ambulation score was observed in methylmercury chloride-treated group from the seventh day of toxication in comparison to control rats. On the 22nd and 29th days, the decrease noticed was maximum about −50.45 % and −54.23 %, respectively. This negative effect caused by the metal was mitigated by the vitamins. In this case, vitamin E proved to be more beneficial than acetyl-L-carnitine. Vitamin E showed +85.49 % increases in ambulation score on the 29th day while there were only +84.12 % increases in this parameter by acetyl-L-carnitine on the same day. The results were best in case of combined vitamin therapy showing about +98.26 % increment on the 29th day assessment (Fig. 20.2). ANOVA showed change on $p<0.001$, $p<0.01$, and $p<0.05$ level.

Effect on Preening Score

Preening activity also showed remarkable decrement from the seventh up to last day of toxicosis in methylmercury chloride group as compared to controls. There was about −87.03 % decrease in preening score on the 29th day in methylmercury chloride-treated group. Preening score was statistically increased when vitamins were inoculated to rats. Vitamin E and acetyl-L-carnitine displayed +89.01 % and +80.21 % increment, respectively, on the last day of treatment. Recovery was maximal in MeHgCl+vitamin E+acetyl-L-carnitine group of about +98.90 % increase on the 29th day of treatment (Fig. 20.3). Significance was shown on $p<0.001$, $p<0.01$, and $p<0.05$ level by ANOVA.

Effect on Rearing Score

The rearing score was reported to show significant depletion on $p<0.001$, $p<0.01$, and $p<0.05$ level in methylmercury chloride-treated rats from the 15th day, and the maximum change was observed on the last day of treatment showing about −49.62 % decrease. Results are depicted in Fig. 20.4. This parameter was also elevated significantly by antioxidants. Vitamin E and acetyl-L-carnitine displayed +74.18 % and +77.69 % increment, respectively, on the 29th day of treatment. Recovery was maximal in MeHgCl+vitamin E+acetyl-L-carnitine group of about +78.19 % increase on the 29th day of treatment. Results are graphically shown in Fig. 20.4.

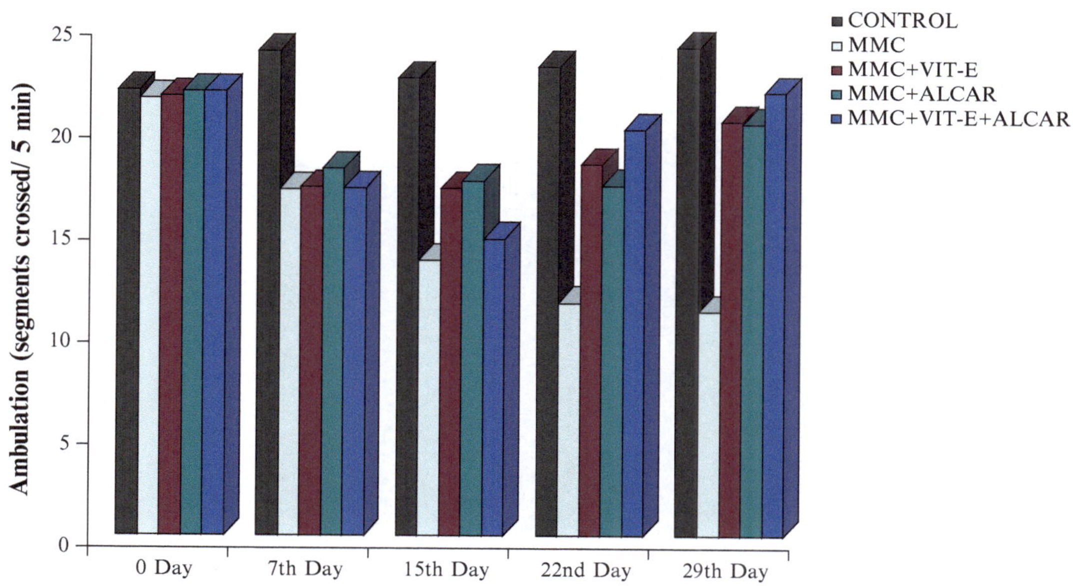

Fig. 20.2 Effect of methylmercury chloride on ambulation score of rats in an open-field chamber. Values represent mean ± SEM

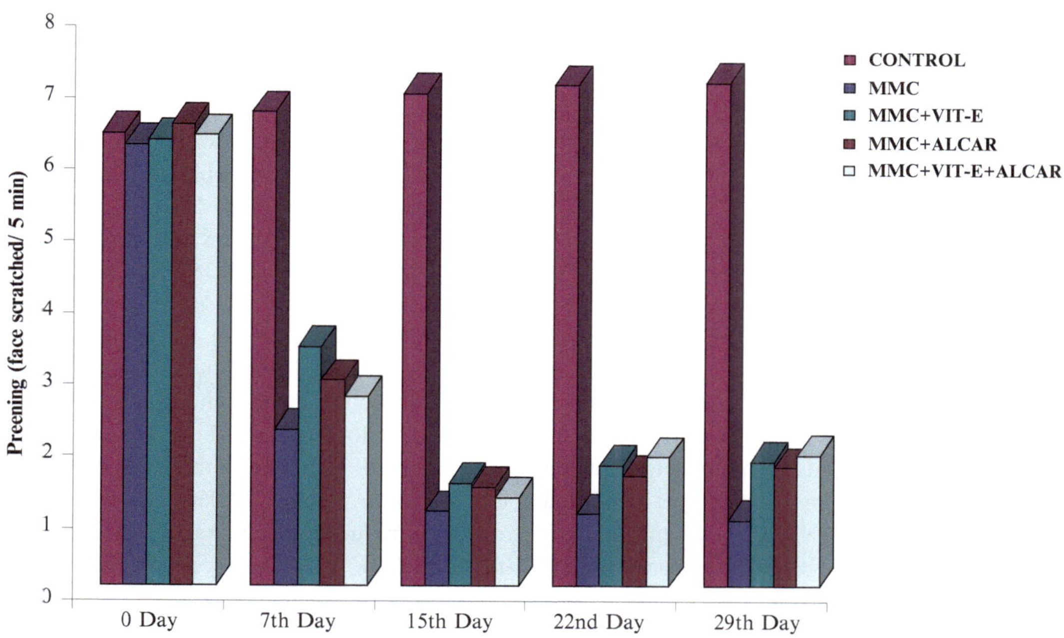

Fig. 20.3 Effects of methylmercury chloride on preening score in an open-field chamber. Values represent mean ± SEM

Effect on Center Crossing Score

Center crossing activity also showed a noticeable change from the 7th up to 29th days due to methylmercury chloride toxicosis at $p<0.001$, $p<0.01$, and $p<0.05$ level of significance by −56.90 % decline. This parameter was also ameliorated by vitamin E and acetyl-L-carnitine

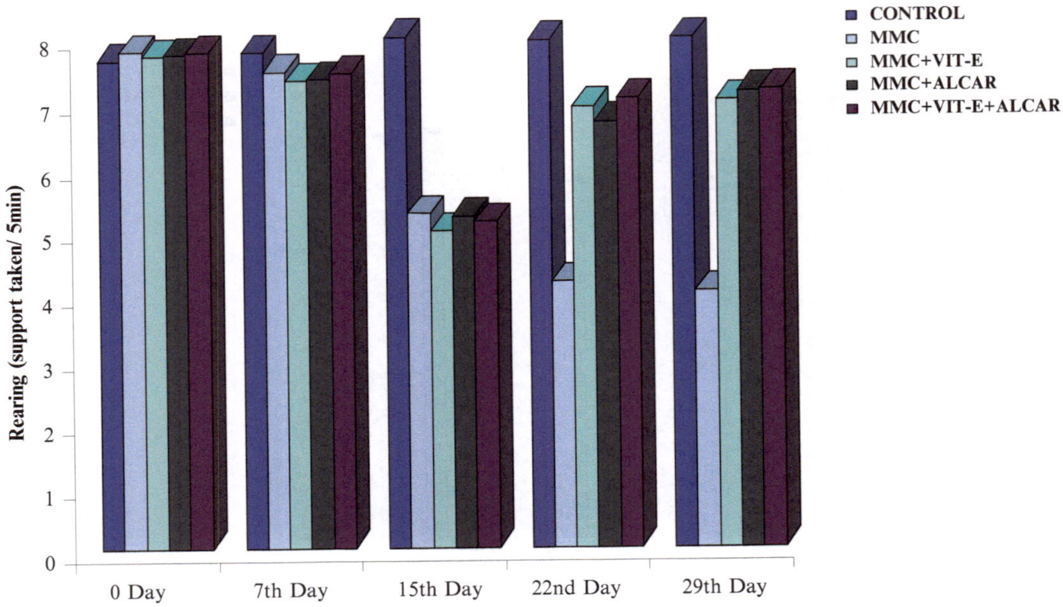

Fig. 20.4 Effect of methylmercury chloride on rearing score of rats in an open field chamber. Values represent Mean ± SEM

showing +99.52 % and +92.34 % increases in center crossing score, respectively. Combined vitamin therapy displayed +99.52 % increments in this parameter. Figure 20.5 diagrammatically depicts the results.

Discussion

The results of the present study show that methylmercury chloride induced marked decreases in locomotor activities in the open field. A salient observation was that even the lowest dose of the metal strongly decreased locomotor activity in the open-field test. In the present study, we monitored assessment of motoric behavior in young male rats orally exposed to MeHgCl and compared to age-matched control rats. We observed significant differences in all the four parameters of open-field test in this study when MeHgCl-treated rats were compared to control ones.

Mice, when exposed to a novel environment, will typically explore the new environment during the first few minutes to get acquainted with the new space and/or try to find ways to escape (Crawley 1999). In our experiment, the first 5 min of open-field testing revealed decreased rate of locomotor and exploratory activities for MeHgCl-treated rats. These data are consistent with other rodent studies reported in the literature that examined horizontal exploration after exposure to MeHg (Dore et al. 2001; Goulet et al. 2003; Kim et al. 2000; Lown et al. 1977; Pereira et al. 1999; Su and Okita 1976).

Reduced locomotor activity observed in rats in the open field suggests a lowered general arousal or increased fearfulness. In case of this test, the direction of the change found, i.e., a reduction of locomotor activity, is the same as that observed by other authors. According to some reports, prenatal MeHg exposure in rats results in reduced locomotor activity in males (Castoldi et al. 2008). In mice, however, females are the affected gender (Goulet et al. 2003).

An important issue in our study is the link between MeHg exposure and altered locomotion parameters. It has been shown that the decreased locomotor and exploratory activities reflect, at least in part, neurological damage induced by MeHg exposure (Dore et al. 2001; Gimenez-Llort et al. 2001), and our data suggest neurological damage in MeHg-exposed rats through our protocol of

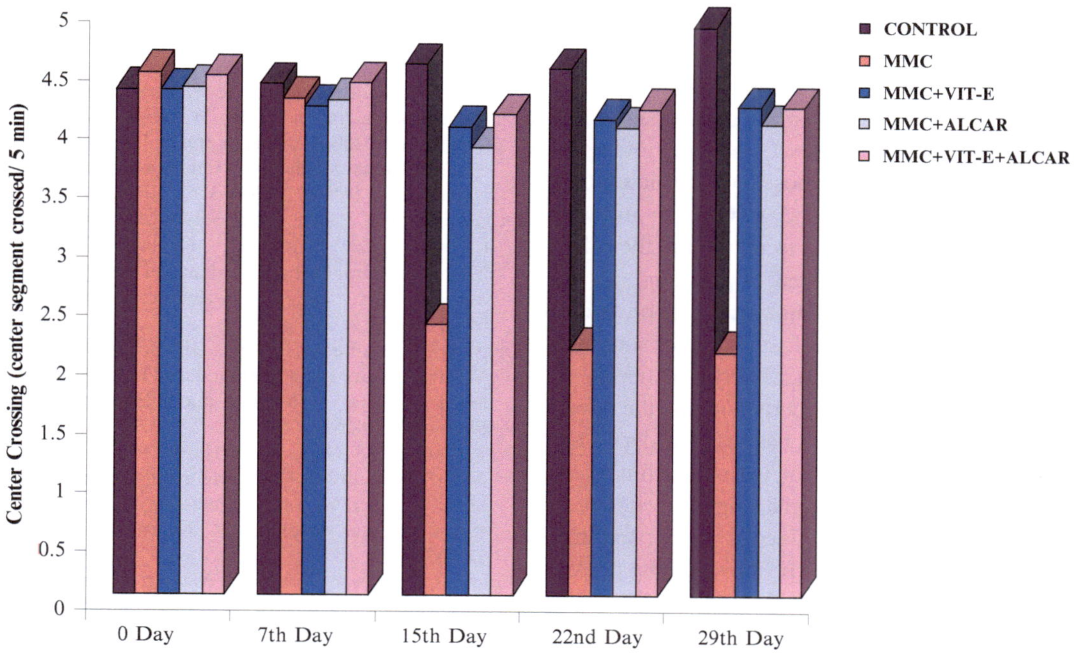

Fig. 20.5 Effects of methylmercury chloride on center crossing score of rats in an open-field chamber. Values represent mean ± SEM

intoxication. There is evidence that cerebellar cells are targeted selectively by mercury compounds in vivo (Sanfeliu et al. 2003) and that MeHg neurotoxicity affects the motor system (Grandjean et al. 1997). In fact, the relationship between MeHg-induced motor deficits and MeHg-induced cerebellar damage is a well-described phenomenon (Sakamoto et al. 1993). In this regard, reported decreased locomotor activity in animals exposed to MeHg during adulthood (Dietrich et al. 2005; Farina et al. 2005) and the early postnatal period (Manfroi et al. 2004) is evident.

In our study, the frequencies of excrement defecation and urine traces were also observed to increase by exposure to MeHgCl. It is considered as a significant factor in this test. Our results are in agreement with a study made by Norio et al. (2008). The distance walked was also reduced. There is a possibility that the distance walked in the open-field test reflects lower spontaneous locomotion activity. However, no significant difference was found in spontaneous locomotion activity evaluated in the home cage among the treatment groups. We considered that the result in the open-field test is independent of spontaneous locomotion activity and is caused presumably by emotional stress.

Anxiety is an emotional state emerging under conditions of indefinite hazards and manifesting by expectation of an unfavorable course of events, which under natural conditions help to adapt to changing environmental conditions. However, high basal anxiety can become a cause of excessive stress reactivity and injuries. The decreased activity in the open field of rats suggests an increase in emotional reactivity. The present findings indicate that vitamin E had a favorable impact on behavioral characteristics and reduced the anxiety of rats. Oral administration of ALCAR also decreased the emotional reactivity of rats in the present study, thus proving to be a good antioxidant against methylmercury-induced behavioral hazards.

Conclusion

In conclusion, it was clearly demonstrated that continuous oral administration of MeHgCl for 14 days caused behavioral changes in rats. Thus, it is evident that this behavioral test can be used as an effective tool to measure subtle motor and coordination deficits that result from exposure to moderate to low doses of neurotoxicants. We found that MeHgCl produces adverse effects in individuals that were exposed to the neurotoxicant during early adulthood. It would be interesting in future experiments to determine the effect of such exposures in aged populations, as we know that fish forms a major source of protein in the aging population, and fish is a major source of MeHg. Moreover, it is rather likely that this chemical exerts its effect differently and that the developmental or behavioral changes it induces are based on the integration of individual functions, not the interactive effect on a single function.

It can also be summarized from our findings that vitamin E and ALCAR improved the performance of rats in the open field. Thus, it can be speculated that influenced rat behavior by these antioxidants might be due to increasing attentional processes.

References

Amin-Zaki L, Majeed MA, Elhassani SB, Clarkson TW, Greenwood MR, Doherty RA (1979) Prenatal methylmercury poisoning: clinical observations over five years. Am J Dis Child 133:172–177

Bartus RT, Dean RL, Beer B, Lippa AS (1982) The cholinergic hypothesis of geriatric memory dysfunction. Science 217:408–414

Bear MF, Connors BW, Paradiso MA (1996) Chemical control of brain and behavior. In: Satterfield TS (ed) Neuroscience-exploring the brain. Williams and Wilkins, Baltimore, pp 422–423

Burbacher TM, Rodier PM, Weiss B (1990) Methylmercury developmental neurotoxicity: a comparison of effects in humans and animals. Neurotoxicol Teratol 12:191–202

Cagiano R, De Salvia MA, Persichella M, Renna G, Tattoli M, Cuomo V (1990) Behavioral changes in the offspring of rats exposed to diazepam during gestation. Eur J Pharmacol 177:67–74

Candland DK, Nagy ZM (1969) The open field: some comparative data. Ann N Y Acad Sci 159:831–851

Castoldi AF, Onishchenko N, Johansson C, Coccini T, Roda E, Vahter M, Ceccatelli S, Manzo L (2008) Neurodevelopmental toxicity of methylmercury: laboratory animal data and their contribution to human risk assessment. Regul Toxicol Pharmacol 51:215–229

Crawley JN (1999) Behavioral phenotyping of transgenic and knockout mice: experimental design and evaluation of general health, sensory functions, motor abilities, and specific behavioral tests. Brain Res 835:18

Debes F, Budtz-Jorgensen E, Weihe P, White RF, Grandjean P (2006) Impact of prenatal methylmercury exposure on neurobehavioral function at age 14 years. Neurotoxicol Teratol 28:363–375

Di Giovanni V, Cagiano R, De Salvia MA, Giustino A, Lacombo C, Renna G, Cuomo V (1993) Neurobehavioral changes produced in rats by prenatal exposure to carbon monoxide. Brain Res 616:126–131

Dietrich MO, Mantese CE, Anjos GD, Souza DO, Farina M (2005) Motor impairment induced by oral exposure to methylmercury in adult mice. Environ Toxicol Pharmacol 19:169–175

Dore FY, Goulet S, Gallagher A, Harvey PO, Cantin JF, D Aigle T, Mirault ME (2001) Neurobehavioral changes in mice treated with methylmercury at two different stages of fetal development. Neurotoxicol Teratol 23:463–472

Farina M, Cereser V, Portela LV, Mendez A, Porciuncula LO, Fornaguera J, Goncalves CA, Wofchuk ST, Rocha JBT, Souza DO (2005) Methylmercury increases S100B content in rat cerebrospinal fluid. Environ Toxicol Pharmacol 19:249–253

Gilbert SG, Grant-Webster KS (1995) Neurobehavioral effects of developmental methylmercury exposure. Environ Health Perspect 103:135–142

Gimenez-Llort L, Ahlbom E, Dare E, Vahter M, Ogren S, Ceccatelli S (2001) Prenatal exposure to methylmercury changes dopamine-modulated motor activity during early ontogeny: age and gender-dependent effects. Environ Toxicol Pharmacol 9:61–70

Goulet S, Dore FY, Mirault ME (2003) Neurobehavioral changes in mice chronically exposed to methylmercury during fetal and early postnatal development. Neurotoxicol Teratol 25:335–347

Grandjean P, Weihe P, White RF, Debes F, Araki S, Yokoyama K, Murata K, Sorensen N, Dahl R, Jorgensen PL (1997) Cognitive deficit in 7-year-old children with prenatal exposure to methylmercury. Neurotoxicol Teratol 19:417–428

Harada M (1995) Minamata disease: methylmercury poisoning in Japan caused by environmental pollution. Crit Rev Toxicol 25:1–24

Kakita K, Wakabayashi M, Su Y, Yoneoka M, Sakamoto F, Ikuta H, Takahashi H (2000) Intrauterine methylmercury intoxication consequence of the inherent brain lesions and cognitive dysfunction in maturity. Brain Res 877:322–330

Kim CY, Nakai K, Kasanuma Y, Satob H (2000) Comparison of neurobehavioral changes in three inbred strains of mice prenatally exposed to methylmercury. Neurotoxicol Teratol 22:297–403

Landrigan PJ, Sonawane B, Butler RN, Trasande L, Callan R, Droller D (2005) Early environmental origins of neurodegenerative disease in later life. Environ Health Perspect 113:1230–1233

Lown BA, Morganti JB, Stineman CH, Massaro EJ (1977) Differential effects of acute methylmercury poisoning in the mouse arising from route of administration: LD50, tissue distribution and behavior. Gen Pharmacol 8:97–101

Manfroi CB, Schwalm FD, Cereser V, Abreu F, Oliveira A, Bizarro L, Rocha JB, Frizzo ME, Souza DO, Farina M (2004) Maternal milk as methylmercury source for suckling mice: neurotoxic effects involved with the cerebellar glutamatergic system. Toxicol Sci 81:172–178

Martin P, Bateson P (1986) Measuring behavior. An introductory guide. Cambridge University Press, Cambridge

Narahashi T, Fenster CP, Quick MW, Lester RA, Marszalec W, Aistrup GL, Sattelle DB, Martin BR, Levin ED (2000) Symposium overview mechanism of action of nicotine on neuronal acetylcholine receptors, from molecule to behavior. Toxicol Sci 57:193–202

National Research Council, Committee on the Toxicological Effects of Methylmercury (2000) Toxicological effects of methylmercury. National Academy Press, Washington, DC

Newland MC, Rasmussen EB (2000) Aging unmasks adverse effects of gestation exposure to methylmercury in rats. Neurotoxicol Teratol 22:819–828

Newland MC, Reile PA, Langston JL (2004) Gestational exposure to methylmercury retards choice in transition in aging rats. Neurotoxicol Teratol 2(6):179

Norio S, Takashi O, Kunihiko N, Akiyoshi K, Tomoyuki N, Keita S, Satomi K, Miyuki S, Naoyuki K, Chieko S, Hiroshi S (2008) Effects of perinatal coexposure to methylmercury and polychlorinated biphenyls on neurobehavioral development in mice. Arch Toxicol 82:387–397

Nortan S (1989) Methods for behavioral toxicology. In: Hayes W (ed) Principles and methods of toxicology. Raven, New York, pp 553–571

Pereira ME, Morsch VM, Christofari RS, Rocha JBT (1999) Methylmercury exposure during postnatal brain growth alters behavioral response to SCH 23390 in young rats. Bull Environ Contam Toxicol 63:256–262

Rasmussen EB, Newland MC (2001) Developmental exposure to methylmercury alters behavioral sensitivity to d-amphetamine and pentobarbital in adult rats. Neurotoxicol Teratol 23:45–55

Rodier PM (1978) Behavioral teratology. In: Wilson JG, Fraser FC (eds) Handbook of teratology, vol 4. Plenum press, New York, pp 397–428

Rossi AD, Ahlbom E, Ogren SO, Nicotera P (1997) Prenatal exposure to methylmercury alters locomotor activity of male but not female rats. Exp Brain Res 117:428–436

Russel WL, Kelly EM (1982) Specific-locus mutation frequencies in mouse stem-cell spermatogonia at very low radiation dose rates. Proc Natl Acad Sci U S A 79:539–541

Sakamoto M, Nakano A, Kajiwara Y, Naruse I, Fujisaki T (1993) Effects of methyl mercury in postnatal developing rats. Environ Res 61:43–50

Sakamoto M, Kakita A, Wakabayashi K, Takahashi H, Nakano A, Akagi H (2002) Evaluation of changes in methylmercury accumulation in the developing rat brain and its effects: a study with consecutive and moderate dose exposure throughout gestation and lactation periods. Brain Res 949:51–59

Salvaterra P, Lown B, Morganti J, Massaro EJ (1973) Alterations in neurochemical and behavioral parameters in the mouse induced by low doses of methylmercury. Acta Pharmacol Toxicol 33:177–190

Sanfeliu C, Sebastia J, Cristofol R, Rodriguez-Farre E (2003) Neurotoxicity of organomercurial compounds. Neurotox Res 5:283–306

Sood PP, Bapu C, Sinha N, Rao AP (1997) Cholesterol and triglyceride fluctuations in mice tissues during methylmercury intoxication and monothiols and vitamin therapy. J Nutr Environ Med 7:155–162

Spyker JM, Smithberg M (1972) Effects of methylmercury on prenatal development in mice. Teratology 5:181–190

Spyker JM, Sparber SB, Goldberg AM (1972) Subtle consequences of methylmercury exposure: behavioral deviations in offspring of treated mothers. Science 177:621–623

Stem JM, Erskine MS, Levine S (1973) Dissociation of open-field behavior and pituitary-adrenal function. Horm Behav 4:149–162

Su MQ, Okita GT (1976) Behavioral effects on the progeny of mice treated with methylmercury. Toxicol Appl Pharmacol 38:195–205

Tobach E (1969) Experimental approaches to the study of emotional behavior. Ann N Y Acad Sci 159:621–1121

Weiss B, Stern S, Cox C, Balys M (2005) Perinatal and lifetime exposure to methylmercury in the mouse: behavioral effects. Neurotoxicology 26:675–690

21 Connection to Anxiolytic Behavior and Reflex Movements

Some testing methods used for the assessment of the emotional state in rodents are based on the conflict between exploration and aversion, that is, on the capacity of situational aversiveness to reduce or block exploratory responses. These methods include the elevated plus maze test, the black–white transition test, and the emergence-from-cage test. The elevated plus maze apparatus consists of an elevated maze with intersecting arms, of which two are open and two are closed. This test is based on the assumption that rats and mice prefer to be in an enclosed environment, compared with an open space. This test provides information concerning anxiety-like behavior in these animals. The animal is placed in the center of the maze and has free access to all arms. Entries into open and closed arms and time spent in open and closed arms are scored by incidence. This test has been validated behaviorally and pharmacologically (Pellow et al. 1985; Pellow and File 1986). Anxiogenic compounds, such as pentylenetetrazole and FG 7142, further decrease the percentage of entries into and time spent in the open arms, whereas anxiolytic drugs, such as benzodiazepines, elicit opposite effects. The elevated plus maze test has been frequently used for the assessment of emotional changes produced in rodents by developmental exposure to psychoactive compounds. Recent results obtained in rats exposed prenatally to a benzodiazepine derivative may be cited as an example (Kellogg et al. 1991). Adult male rats exposed in utero to diazepam spent significantly more time in the open arms than did rats exposed in utero to vehicle. The total amount of time spent in either the open or the closed arms, however, was not affected by prenatal drug treatment. Such data could be interpreted as indicating a decrease in the emotional reactivity of animals exposed to diazepam during gestation.

When terrestrial quadrupeds are dropped from the supine position, animals reflexively rotate their head and body and take a prone position in the air to prepare for landing upon their four limbs. This is called the air-righting reflex and is used as a model of dynamic postural arrangements. Because there is no support surface during the fall, the air-righting reflex may provide facility to investigate neuronal mechanisms of postural control based on head-based sensors free from postural control-based somatosensory input.

The present work was designed to assess the effects of low-dose methylmercury on brain control of dynamic posture and anxiety.

Behavioral Testing Procedure

After treatment, each animal of both the treated and control groups was exposed on the 29th day for 5 min in the apparatus and analyzed for anxiety and righting reflex task.

Elevated Plus Maze Test

This test, in which the animal's behavior is sensitive to the actions of anxiogenic and anxiolytic agents, is one of the most widely used tests for

S. Nabi, *Toxic Effects of Mercury*, DOI 10.1007/978-81-322-1922-4_21,
© Springer India 2014

assessing anxiety states in individuals (Lister 1987; Rogers and Cole 1994). The cross maze was elevated above the floor to a height of 50 cm and consisted of two open and two closed arms (enclosed on three sides). The rat was placed in a darkened room 5 min before the test, and the individual's behavior was tested in the maze for 5 min. The animal was placed in the center of the maze with the nose in a closed arm and measures adequately reflecting the anxiety state were recorded: the time spent in the open arms, the center, and the closed arms of the maze (data presented as the times spent in each part of the maze as percentages of the total test duration); the numbers of excursions into the open arms, center, and closed arms, again expressed as percentages; the total numbers of excursions into the open arms and center and entries into the closed arms of the maze; the numbers of transfers from one closed arm to the other; and the number of glances beneath the maze. The maze was thoroughly washed and dried after testing each individual (Figs. 21.1, 21.2, and 21.3).

Righting Reflex Test

If the rat is placed on its back, the animal quickly rights itself and assumes a normal posture. Neurological deficit is indicated by the inability to regain normal position within 5 min. Control and treated group rats were held upside down by the neck and lower back and dropped from a height of approximately 30 cm onto a cotton pad. The righting reflex was considered positive if the rat landed on all four legs, receiving a score of "1." Otherwise it was scored as "0."

Results

Effect of Methylmercury Chloride, Vitamin E, and Acetyl-L-Carnitine in Elevated Plus Maze (EPM)

The effect of methylmercury chloride treatment on the number of entries in closed and open arms in the elevated plus maze has been summarized below. There was a significant increase in occupancy in closed arm by the treated rats as compared to control rats. But the administration of vitamins markedly declined the closed arm entries, thereby revealing the ameliorative effect of these antioxidants. The data was analyzed by one-way ANOVA, followed by Student's *t*-test, and the difference between the control and treated groups was significant at $p<0.001$, 0.01, and 0.05 levels. There was +42.18 % increase in closed arm entries in MeHg-treated rats, but −10.90 % and −40.84 % decline, respectively, in vitamin E and acetyl-L-carnitine inoculated rats. Results are depicted in Fig. 21.4.

Effect of Methylmercury Chloride, Vitamin E, and Acetyl-L-Carnitine in Righting Reflex Test (RRT)

In case of righting reflex, control rats were able to assume normal posture in less than 1 s, but methylmercury chloride-treated animals took some minutes to gain the normal posture. In trial 1, there was about +53.84 % increase; in trial 2, +73.61 % increase; and in trial 3, it was +71.62 %. In vitamin E-treated rats, posture was also assumed quickly. Vitamin E showed −36.80 % recovery in trial 2 and acetyl-L-carnitine displayed −33.60 % decline in the same trial. Difference between the control and treated animals was significant at $p<0.001$, 0.01, and 0.05 levels. Diagrammatic representation of results is shown in Fig. 21.5.

Discussion

In order to study the relationship between anxiety and cognition, the elevated plus maze (EPM) is among the most popular behavioral models of anxiety. It has been used very effectively to assess neurobehavioral profile of animals under the influence of anxiogenic/anxiolytic agents (File et al. 1991). In the elevated plus maze, increased aversion of open arms are indicative of an enhanced anxiety state and our results indicate that low-dose administration of methylmercury

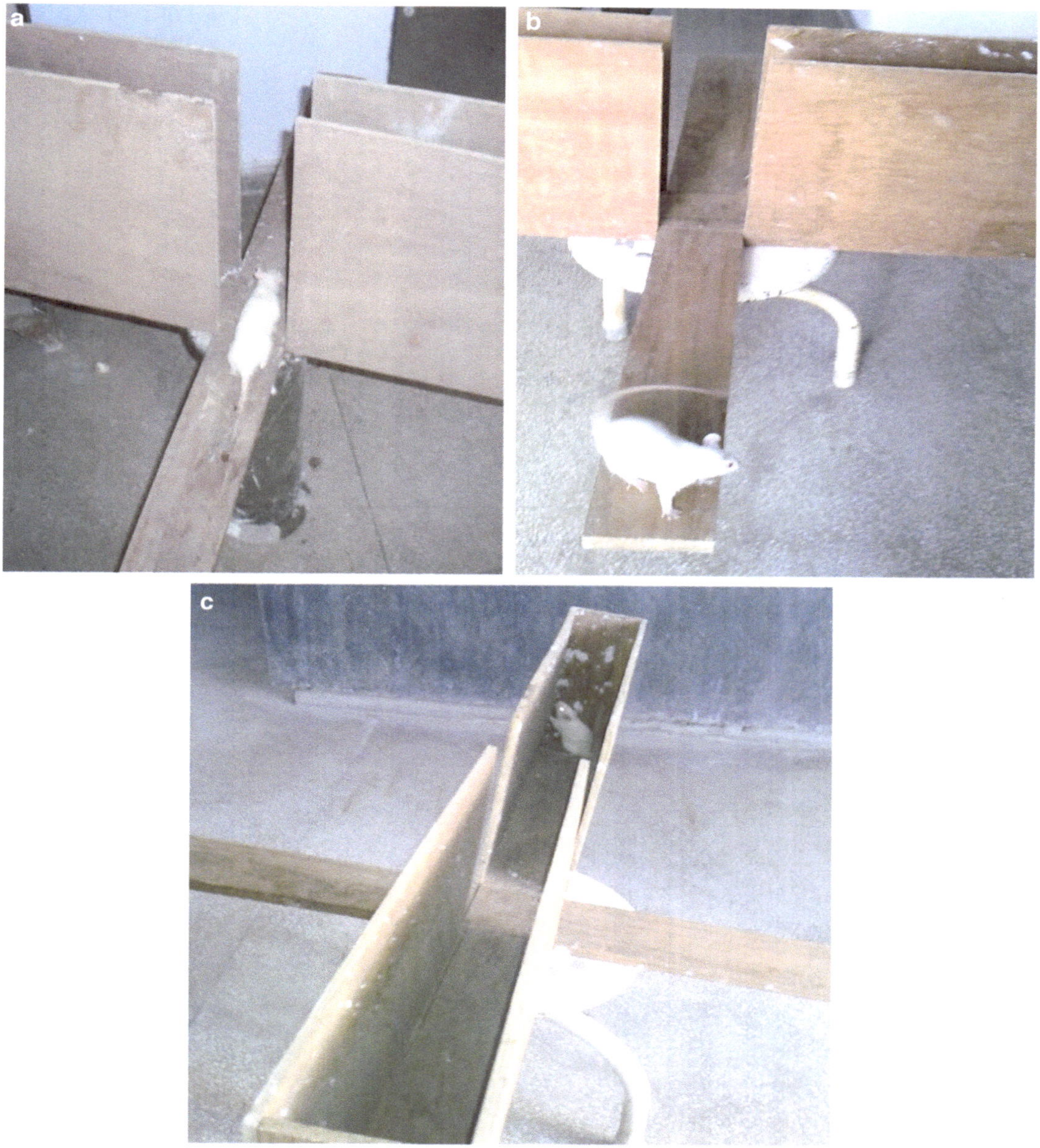

Fig. 21.1 Elevated plus maze apparatus accommodating a rat treated with MMC to assess for emotional changes. (**a**) Rat transferring from one open arm to another. (**b**) Rat glancing beneath the maze. (**c**) Rat hiding itself in closed arm showing a state of anxiety

chloride caused reduction in average number of entries and average time spent in open arms. There is profound evidence that cognitive processes and anxiety are interrelated (Goswami et al. 1996). Recently, it was shown that ghrelin, a peptide hormone, shows dose-dependent increased memory retention in the hippocampus and amygdala as well as induced anxiogenesis in these brain structures, thereby, further indicating a link between anxiogenesis and cognition (Carlini et al. 2004). Thus, our results reveal that methylmercury chloride administration increases the state of anxiety in rats, thereby causing impairment in brain cognition.

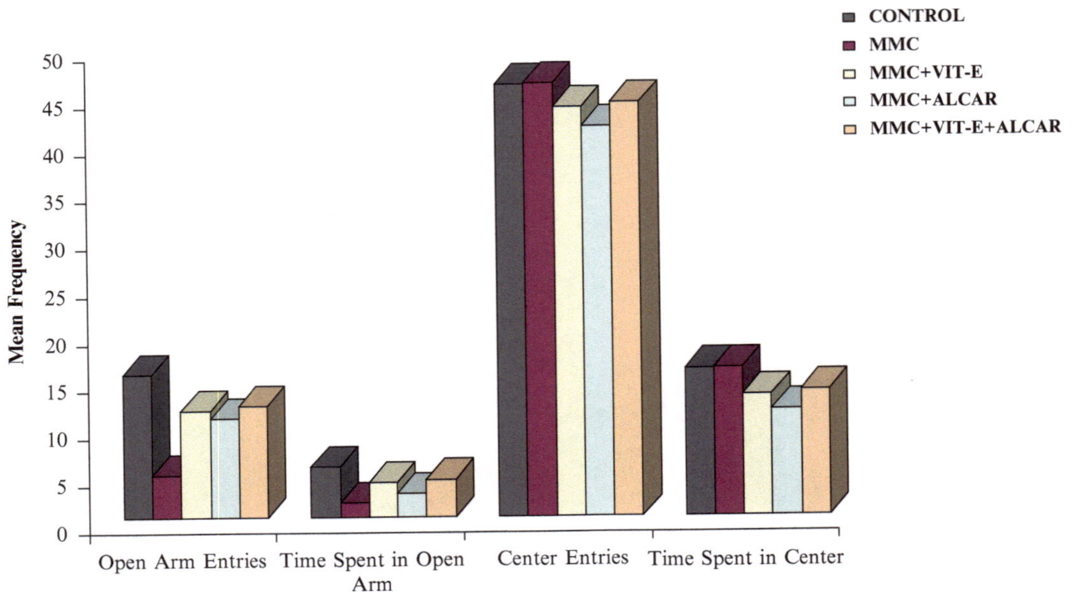

Fig. 21.2 Effects of methylmercury chloride on elevated plus maze parameters in rats. Values represent mean ± SEM.

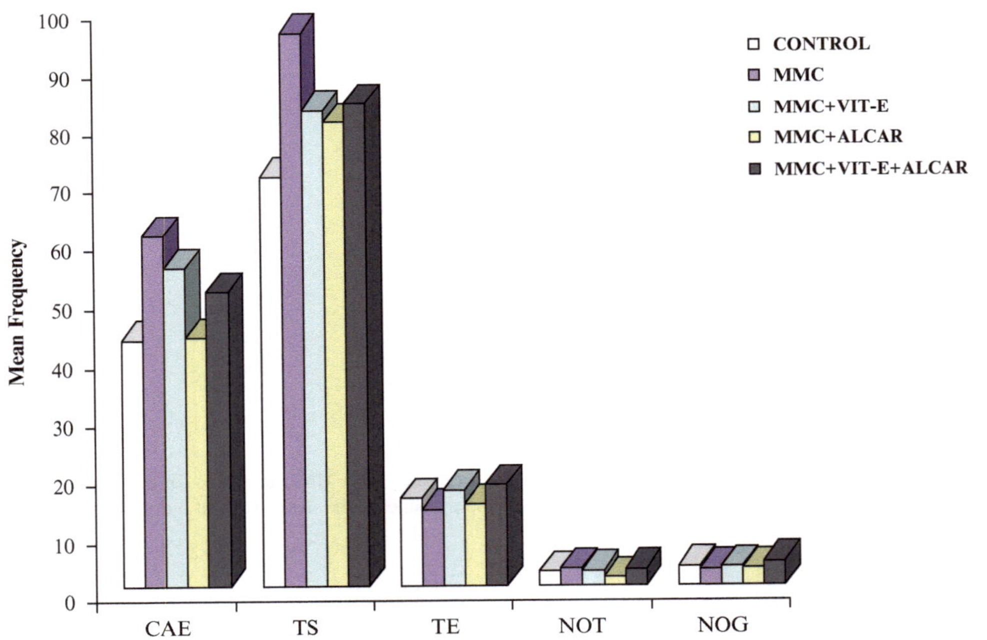

Fig. 21.3 Effects of methylmercury chloride on elevated plus maze parameters in rats. Values represent mean ± SEM. *CAE* closed arm entries, *TS* time spent in closed arm, *TE* total entries, *NOT* number of transfer, *NOG* number of glances

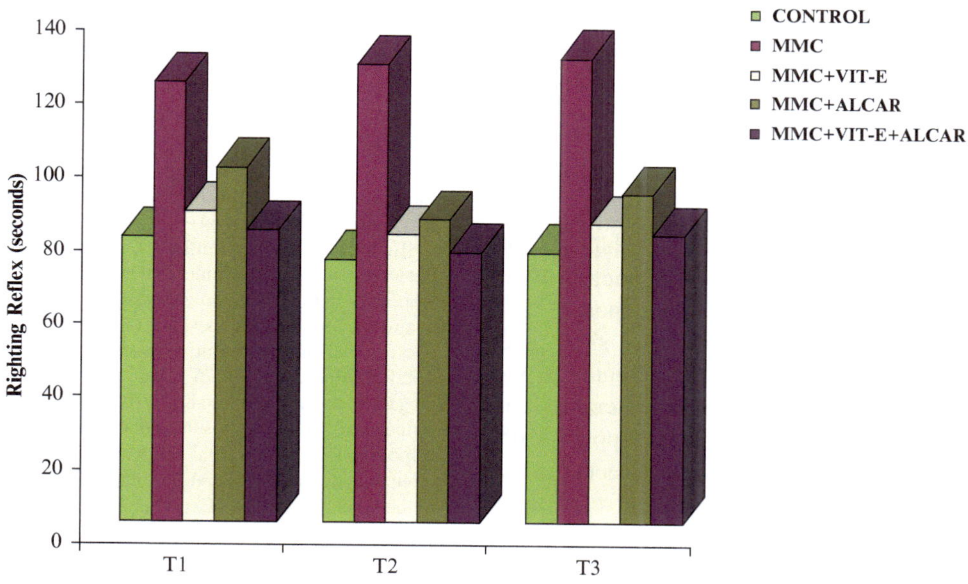

Fig. 21.4 Effects of methylmercury chloride on righting reflex of rats. Values represent mean ± SEM. *T* trial

Fig. 21.5 Righting reflex illustrated in methylmercury-treated rat

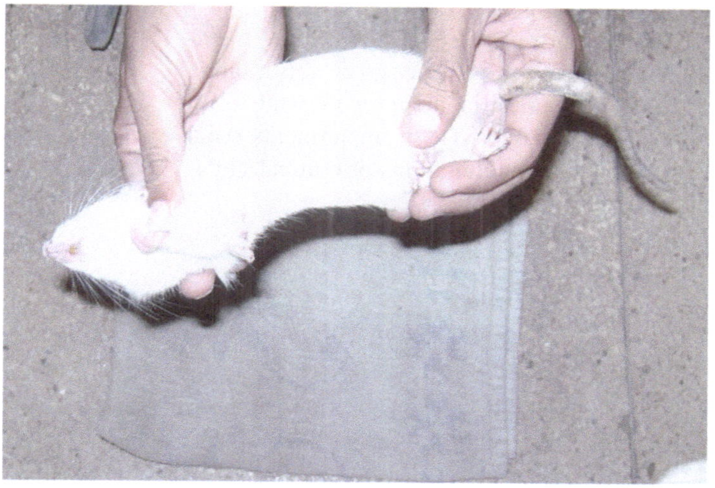

In case of righting reflex test, abnormalities were observed in righting movements in all the rats of treated group. The neural control for righting reflex is thought to be located in and around the pons and medulla (Magnus 1924). Thus, the present study confirmed that the rats with normal midbrain can perform the air-righting reflex properly. This statement was also confirmed by Ken and Takashi (2000) by a study on striated rats. They observed that only the rats with midbrain can perform air-righting reflex. Therefore, it is clear that methylmercury chloride caused damage to the mid brain of rats, thereby displaying an alteration in dynamic postures.

Conclusion

Methylmercury is a neurotoxicant that is well known to disrupt sensory-motor function with adult-onset exposure and, at high exposure levels, to produce cerebral palsy-like signs and mental retardation after developmental exposure. It is becoming evident that alterations in cognitive or intellectual function appear with low-level exposures. While these effects can be subtle, they can carry a significant economic cost. The overall pattern of neural and behavioral effects points to developmental MeHg exposure as a model of how damage to the development of frontal cortical areas can have long-lasting behavioral consequences, especially on behavioral tasks that emphasize the acquisition of a response–reinforcer relationship, choice, or perseveration.

The present results suggest that acute methylmercury exposure induces, in rats, subtle changes in short-term memory as well as in exploratory behavior. These impairments seem to be associated to alterations of cortical glutamatergic signaling.

References

Carlini VP, Varas MM, Cragnolini AB, Schioth HB, Scimonelli TN, de Barioglio SR (2004) Differential role of the hippocampus, amygdala, and dorsal raphe nucleus in regulating feeding, memory and anxiety-like behavioral responses to ghrelin. Biochem Biophys Res Commun 313:635–641

File SE, Zharkovsky A, Gulati K (1991) Effects of baclofen and nitrendipine on ethanol withdrawal responses in the rat. Neuropharmacology 30:183–190

Goswami M, Mund S, Ray A (1996) Effects of some psychotropic agents on cognitive functions in rats. Indian J Physiol Pharmacol 40:75–78

Kellogg CK, Primus RJ, Bitran D (1991) Sexually dimorphic influence of prenatal exposure to diazepam on behavioral responses to environmental challenge and on Gamma-aminobutyric acid (GABA)-Stimulated chloride uptake in the brain. J Pharmacol Exp Ther 256:259–265

Ken M, Takashi Y (2000) Abnormal air-righting reflex in striated rats. Jpn J Physio 50:163–166

Lister R (1987) The use of a plus-maze to measure anxiety in the mouse. Psychopharmacology 92:180–195

Magnus R (1924) Die Korperstellung. Springer, Berlin, p 740

Pellow S, File SE (1986) Anxiolytic and anxiogenic drug effects in exploratory activity in an elevated plus-maze: a novel test of anxiety in the rat. Pharmacol Biochem Behav 24:525–529

Pellow S, Chopin P, File SE, Briley M (1985) Validation of open-closed arm entries in an elevated plus-maze as a measure of anxiety in rat. J Neurosci Methods 14:149–167

Rogers RJ, Cole JC (1994) The elevated plus-maze: pharmacology, methodology and ethology. In: Cooper SJ, Hendrie CA (eds) Ethology and psychopharmacology. Wiley, Chichester, pp 9–44

22 Effects on Depression-Like Behavior

Depression is a heterogeneous, multifaceted disorder with symptoms manifested at the psychological, behavioral, and physiological level. This is perhaps why it is so difficult to mimic the disorder in the laboratory (American Psychiatric Association 1994). Many of the human symptoms of depression such as recurring thoughts of death or suicide or having excessive thoughts of guilt as described in the Diagnostic and Statistical Manual of the American Psychiatric Association (DSM IV) are impossible to be modeled in mice or rats. The question, therefore, remains impenetrable as to whether we can ever assume a mouse or rat is "depressed." Evolutionary theories have been proposed for psychiatric disorders (Jones and Blackshaw 2000; Nesse 2000), which would plausibly predict that also lower animal species can exhibit behaviors useful in modeling human depression. However, such hypotheses are heavily debated and are difficult to address empirically (Dubrovsky 2002; McLoughlin 2002). Another difficulty in assessing depressive states in rodents is that the underlying pathophysiology in depression is still unresolved. Further, the mode of action of clinically effective antidepressants is not yet understood beyond the fact that they primarily alter monoamine neurotransmission (Nestler et al. 2002; Frazer 1997; Richelson 2001; Blier 2001).

Despite the difficulties in translating the complexities of human affective disorders in its entire spectrum into relevant tests in mice, numerous attempts have been made to create so-called animal models of depression or at least models of some of the core aspects of depression. Such models include those paradigms where various stress and pharmacological, lesion, environmental, or genetic manipulations are applied (Cryan et al. 2002; Willner and Mitchell 2002; Cryan and Mombereau 2004; Geyer and Markou 2001; Weiss and Kilts 1998; McKinney 2001; Nestler et al. 2002).

The goal of the current experiment was to examine whether exposure to low dose of methylmercury chloride would exacerbate the effects on depression-like behavior in male rats. The two most commonly used tests for depressants and antidepressants are the Porsolt forced swim test (FST) and the tail suspension test (TST) Cryan et al. 2002. Therefore, the FST and TST behavioral assays were used in this work to elucidate the potential of MeHg to produce effects on depression-like behavior in rats.

Protocol

Behavioral Tests

Rats were subjected to two behavioral tests: the tail suspension test (TST) and the forced swim test (FST). After treatment, both control and treated rats were exposed to these tests and the immobility time was noted accordingly.

Depression: Forced Swim Test (FST)

The forced swim test was performed according to the method of Porsolt et al. (1977). A vertical

S. Nabi, *Toxic Effects of Mercury*, DOI 10.1007/978-81-322-1922-4_22,
© Springer India 2014

glass cylinder (25 cm high, 14 cm in diameter) was filled with water having a temperature of 30 °C to a depth of 20 cm. The water depth was adjusted so that the rats must swim or float without their hind limbs or tail touching the bottom. For testing, each rat was placed in the cylinder for 6 min, and the latency to float and the duration of floating (i.e., the time during which rat made only the small movements necessary to keep their heads above water) was scored. As suggested by Porsolt et al. (1977), only the data scored during the last 4 min were analyzed and presented.

In the forced swim procedure, rats are forced to swim in unescapable situation. After a period of vigorous struggling, the animal becomes immobile or makes only those movements necessary to keep its head above the water. The immobility observed in this test is considered to reflect a state of despair.

The forced swim test has a high degree of pharmacological validity as reflected by its sensitivity to major classes of antidepressants, including tricyclic antidepressants and selective serotonin reuptake inhibitors (Dalvi and Lucki 1999). It is also useful for assessing the impact of environmental and genetic manipulations on behaviors related to depression (Karolewicz and Paul 2001; Ouagazzal et al. 2003).

Depression: Tail Suspension Test (TST) [More than Just a Dryland Version of the FST]

We used the test essentially as described by Steru et al. (1985). A short piece of paper adhesive tape (about 6 cm) was attached along half the length of the tail (3 cm). The free end of the tape was attached to a 30 cm long rigid tape which hung from a horizontal bar clamped to a heavy laboratory support stand. Suspended animals were surrounded by a white wooden enclosure (45 cm high, 40 cm wide, and 40 cm deep) such that the rat's head was about 20 cm above the floor. For testing, each rat was suspended by its tail and observed for 6 min. An observer scored the total duration of a passive, "deadweight" hanging (immobility) between the periods of wriggling of the animal to avoid aversive situation.

The tail suspension test is among the most widely utilized rodent model of depression (Dalvi and Lucki 1999). In this paradigm, the rat is suspended by the tail to a tail hanger. Following an initial period of vigorous struggling, the animal gradually abates into immobility. The duration of immobility has been inferred as an index of behavioral despair (the animal has given up hope of escaping). The tail suspension test has been shown to be sensitive to various classes of antidepressants, including tricyclic antidepressants and selective serotonin reuptake inhibitors (Dalvi and Lucki 1999). Furthermore, marked genetic differences in baseline immobility were found among inbred and outbred strains of mice in this test (Dalvi and Lucki 1999; Liu and Gershenfeld 2001).

Results

Effect of Methylmercury Chloride on Immobility Period of Rats in Forced Swim Test (FST)

This paradigm evaluates animal's response to inescapable aversive situation (placement in a beaker of water), inducing an active (swimming, climbing on the walls) or inactive (floating) behavior. The latter is interpreted as a measure of depression-like behavior (Porsolt et al. 1977). Methylmercury exposed rats displayed significantly longer immobility time (passive floating without limb movements) of about +64.86 % than control animals. Vitamins significantly reduced the immobility period in rats by −56.14 % in vitamin E and −52.45 % in ALCAR, thus offering protection against methylmercury-induced increment in depression-like behavior. Combined vitamin therapy offered highest recovery of about −60.65 % (Fig. 22.1).

Effect of Methylmercury Chloride on Immobility Period of Rats in Tail Suspension Test (TST)

The effect of methylmercury in tail suspension test is shown in Fig. 22.1. The total duration of a

Fig. 22.1 Vertical glass cylinder demonstrating a rat treated with MMC and appraised for depression-like behavior in forced swim test. (**a**) Rat showing vigorous movements to escape from aversive situation. (**b**) Rat floating to keep its head above water. (**c**) Rat displaying immobility to reflect a state of despair

passive, "deadweight" hanging (immobility) was scored as a measure of depression-like behavior (Steru et al. 1985). Methylmercury chloride-intoxicated rats showed significant increase of +80.00 % in deadweight than control animals. ANOVA revealed the significance at $p<0.05$, 0.001 and 0.01. Vitamin E and acetyl-L-carnitine treatment offered significant protection of −61.48 % and −55.92 %, respectively, against methylmercury-induced cognitive deficits. Vitamins given in combined manner displayed maximal amelioration of −62.96 % (Figs. 22.2 and 22.3).

Discussion

In the present study, we also analyzed the effects of methylmercury on emotional behavior. We used the forced swim test (Porsolt et al. 1977) and tail suspension test (Steru et al. 1985), generally considered as animal models of depression. In these tests, methylmercury-induced male rats showed significantly longer immobility time that represented a behavioral despair response to an aversive situation and therefore a depression-like behavior. Depressive syndromes have been

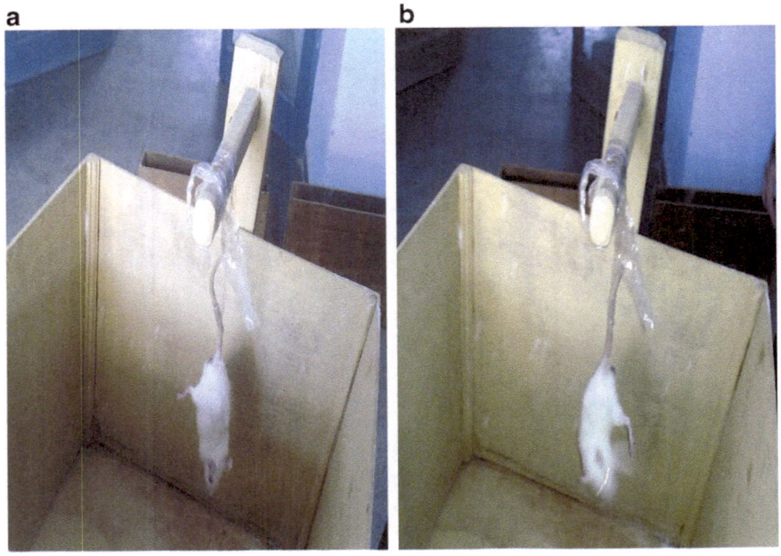

Fig. 22.2 Tail suspension apparatus hanging a MMC-treated rat evaluated for depression-like behavior. (**a**) Rat showing "passive hanging" inferred as an index of behavioral despair. (**b**) Rat showing "vigorous wriggling" to avoid aversive situation

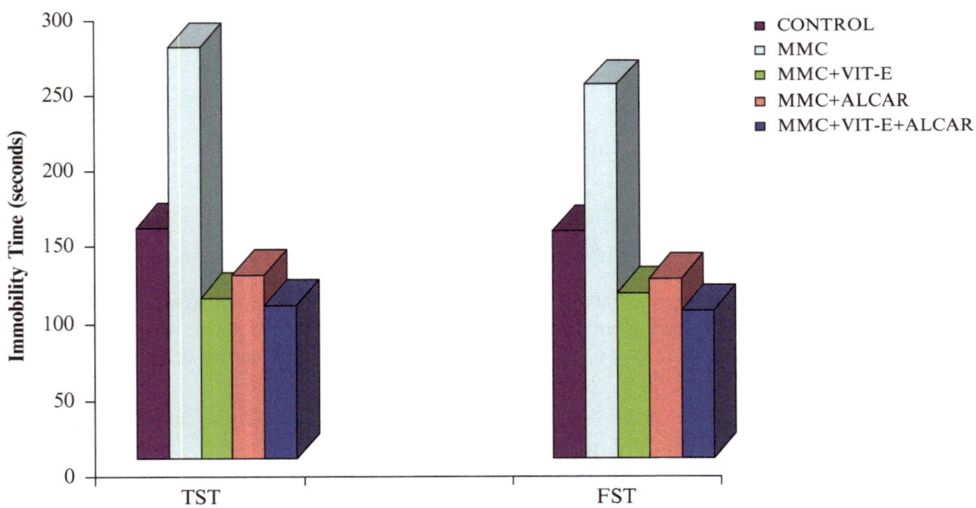

Fig. 22.3 Effects of methylmercury chloride on tail suspension test (*TST*) and forced swim test (*FST*) activity of rats. Values represent mean ± SEM

reported in humans later in life after occupational exposure to inorganic mercury (Grum et al. 2006). The neurochemical basis of methylmercury-induced behavioral alterations may be due to disturbances in a number of neurotransmitter systems, initially occurring during exposure and followed by long-lasting changes in brain functioning (Castoldi et al. 2001). Abnormalities in neuronal functioning, in turn, can be due to intracellular methylmercury toxicity, which includes alteration in Ca^{2+} homeostasis, cytoskeletal damage, and induction of oxidative stress (Sarafian and Verity 1991; Yee and Choi 1994).

Many hypotheses have been advanced to explain the physical adaptation that is the immobility response observed in the FST and TST (Cryan and Mombereau 2004). The posture of immobility in the context of the FST was originally coined "behavioral despair" by Porsolt et al. (1977), largely based on the assumption that the animals have "given up hope of escaping." In other words, the immobility represents a failure of persistence in escape-directed behavior. Other investigators have contended that the behavioral responses comprise an evolutionary preserved coping strategy Thierry et al. 1984 in which immobility behaviors represent the psychological concept of "entrapment" described in clinical depression (Dixon 1998; Gilbert and Allan 1998; Lucki 2001). Thus, the development of immobility disengages the animal from active forms of coping with stressful stimuli (Lucki 2001). Further, immobility in the TST is due to inability or reluctance to maintain effort rather than a generalized hypoactivity, as evidenced by the fact that animals can adopt this posture quickly, and drugs which may suppress activity Cryan et al. 2005 counter the immobility response. As such, this immobility may be analogous to the clinical observations that depressed patients often lack sustained expenditure of effort reflected in a pronounced psychomotor impairments (Weingartner and Silberman 1982).

Conclusion

We have shown that low level of methylmercury induces depression-like behavior in male rats. These findings point to early exposure to environmental contaminants as a possible risk factor for neurodevelopmental disorders.

The TST has great utility as a model to assess antidepressant-related behavior; however, its use is best exploited when used in battery-style fashion with other depression models such as FST. Therefore, one can conceive that these relatively robust paradigms will be useful in the future unraveling of the genetic, molecular, and neurochemical pathways relevant to depression and antidepressant action.

References

American Psychiatric Association (1994) Diagnostic and statistical manual of mental disorders, 4th edn. Author, Washington, DC

Blier P (2001) Possible neurobiological mechanisms underlying faster onset of antidepressant action. J Clin Psychiatry 62:7–11

Castoldi AF, Coccini T, Ceccatelli S, Manzo L (2001) Neurotoxicity and molecular effects of methylmercury. Brain Res Bull 55:197–203

Cryan JF, Mombereau C (2004) In search of a depressed mouse: utility of models for studying depression-related behavior in genetically modified mice. Mol Psychiatry 9:326–357

Cryan JF, Markou A, Lucki I (2002) Assessing antidepressant activity in rodents: recent developments and future needs. Trends Pharmacol Sci 23:238–245

Cryan JF, Mombereau C, Vassout A (2005) The tail suspension test as a model for assessing antidepressant activity: review of pharmacological and genetic studies in mice. Neurosci Biobehav Rev 29:571–625

Dalvi A, Lucki I (1999) Murine models of depression. Psychopharmacology 147:14–16

Dixon AK (1998) Ethological strategies for defence in animals and humans: their role in some psychiatric disorders. Br J Med Psychol 71:417–445

Dubrovsky B (2002) Evolutionary psychiatry: adaptationist and nonadaptationist conceptualizations. Prog Neuropsychopharmacol Biol Psychiatry 26:1–19

Frazer A (1997) Pharmacology of antidepressants. J Clin Psychopharmacol 17:2S–18S

Geyer M, Markou A (2001) Animal models in psychiatric disorders. In: Bloom FE, Kupfer DJ (eds) Psychopharmacology, the fourth generation of the progress. Raven, New York, pp 155–173

Gilbert P, Allan S (1998) The role of defeat and entrapment (arrested flight) in depression: an exploration of an evolutionary view. Psychol Med 28:585–598

Grum DK, Kobal AB, Arneric N, Horvat M, Zenko B, Dzeroski S, Osredkar J (2006) Personality traits in miners with past occupational elemental mercury exposure. Environ Health Perspect 114:290–296

Jones I, Blackshaw JK (2000) An evolutionary approach to psychiatry. Aust N Z J Psychiatry 34:8–13

Karolewicz B, Paul IA (2001) Group housing of mice increases immobility and antidepressant sensitivity in the forced swim and tail suspension tests. Eur J Pharmacol 415:197–201

Liu X, Gershenfeld HK (2001) Genetic differences in the tail-suspension test and its relationship to imipramine response among 11 inbred strains of mice. Biol Psychiatry 49:575–581

Lucki I (2001) A prescription to resist proscriptions for murine models of depression. Psychopharmacology 153:395–398

McKinney WT (2001) Overview of the past contributions in animal models and their changing place in psychiatry. Semin Clin Neuropsychiatry 6:68–78

McLoughlin G (2002) Is depression normal in human beings? A critique of the evolutionary perspective. Int J Ment Health Nurs 11:170–173

Nesse RM (2000) Is depression an adaptation? Arch Gen Psychiatry 57:14–20

Nestler EJ, Barrot M, DiLeone RJ, Eisch AJ, Gold SJ, Monteggia LM (2002) Neurobiology of depression review. Neuron 34:13–25

Ouagazzal AM, Moreau JL, Pauly-Evers M, Jenck F (2003) Impact of environmental housing conditions on the emotional responses of mice deficient for nociceptin/orphanin FQ peptide precursor gene. Behav Brain Res 144:111–117

Porsolt RD, Le Pichon M, Jalfre M (1977) Depression: a new animal model sensitive to antidepressant treatments. Nature 266:730–732

Richelson MDE (2001) Pharmacology of antidepressants. Mayo Clin Proc 76:511–527

Sarafian T, Verity MA (1991) Oxidative mechanisms underlying methyl mercury neurotoxicity. Int J Dev Neurosci 9:147–153

Steru L, Chermat R, Thierry B, Simon P (1985) The tail suspension test: a new method for screening antidepressants in mice. Psychopharmacology 85:367–370

Thierry B, Steru L, Chermat R, Simon P (1984) Searching-waiting strategy: a candidate for an evolutionary model of depression. Behav Neural Biol 41:180–189

Weingartner H, Silberman E (1982) Models of cognitive impairment: cognitive changes in depression. Psychopharmacol Bull 18:27–42

Weiss JM, Kilts CD (1998) Animal models of depression and schizophrenia. In: Nemeroff CB, Schatzberg AF (eds) American Psychiatric Press textbook of psychopharmacology, 2nd edn. American Psychiatric Press, Washington, DC, pp 89–131

Willner P, Mitchell PJ (2002) The validity of animal models of predisposition to depression. Behav Pharmacol 13:169–188

Yee S, Choi BH (1994) Methylmercury poisoning induces oxidative stress in the mouse brain. Exp Mol Pathol 60:188–196

Part VIII

Genetic Toxicology

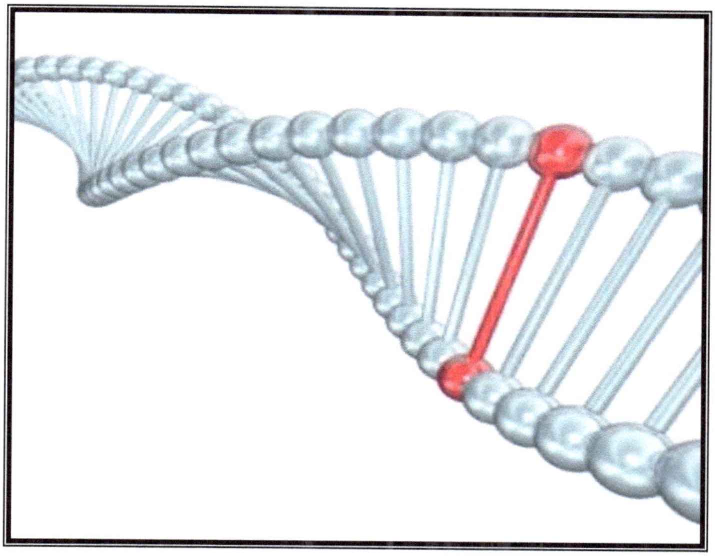

Micronucleus Test (MNT)

23

Mercury, one of the most widely diffused and hazardous organ-specific environmental contaminants, exists in a wide variety of physical and chemical states, each of which with unique characteristics of target organ specificity (Aleo et al. 2002). In nature the different forms of mercury include the metallic form, inorganic compounds, as well as alkyl, alkoxy, and aryl mercury compounds. Once introduced into the environment, mercury compounds can undergo a wide variety of transformations. In sediments, inorganic mercury ($HgCl_2$) may be converted into methyl (CH_3HgCl) and dimethyl (CH_3CH_2HgCl) forms by methanogenic bacteria. This biotransformation constitutes a serious environmental risk, given that CH_3HgCl is the most toxic of the mercury compounds and accumulates in the aquatic food chain, eventually reaching human diets (Tchounwou et al. 2003). CH_3HgCl has been an environmental concern to public health and regulatory agencies for over 50 years because of its neurotoxicity. Its association with nervous system toxicity in adults and infants near Minamata Bay, Japan, in the 1950s initiated environmental health research inquiries that continue to this day (Faustman et al. 2002). The three modern "faces" of mercury are our perceptions of risk from the exposure of billions of people to CH_3HgCl in fish, mercury vapor from amalgam tooth fillings, and CH_3CH_2HgCl in the form of thimerosal added as an antiseptic to widely used vaccines (Clarkson 2002). Mercury genotoxicity has been usually attributed to its ability to react with the sulfhydryl groups of tubulin, impairing spindle function and leading to chromosomal aberrations and polyploidy (De Flora et al. 1994). Another important mechanism of mercury genotoxicity is its ability to produce free radicals that can cause DNA damage (Schurz et al. 2000; Ehrenstein et al. 2002). In vivo studies have demonstrated a clastogenic effect of mercury on people exposed to this element in their working environment or through the consumption of contaminated food or sometimes accidentally. Increased numbers of chromosome alterations and micronuclei have been reported in people who consume contaminated fish (Amorim et al. 2000; Franchi et al. 1994) and in miners and workers of explosive factories (Al-Sabti et al. 1992; Anwar and Gabal 1991). Negative results were also obtained in some cases (Hansteen et al. 1993; Mabille et al. 1984), demonstrating that cytogenetic monitoring of peripheral blood lymphocytes in individuals exposed to mercury from different sources may not be completely specific (De Flora et al. 1994). The effects of CH_3HgCl contamination have been studied in an increasing way since the outbreaks in Japan and Iraq. Many of these studies had their focus on the neurological effects of CH_3HgCl exposure in adult animals and used high doses of this compound (1,900–30,000 ppb = μg/L) to obtain its most severe effects (National Research Council 2000). Most of the in vitro studies with lymphocytes also used high doses (250–6,250 μg/L) of mercury compounds in order to evaluate its clastogenic effects (Ogura et al. 1996; Betti et al. 1993, 1992).

S. Nabi, *Toxic Effects of Mercury*, DOI 10.1007/978-81-322-1922-4_23,
© Springer India 2014

The micronucleus test is one of the most widely applied short-term test used in genetic toxicology and has become one of the most important tests implemented by the regulatory entities of different countries to evaluate mutagenicity of, and sensitivity to, xenobiotics (OEOD 1997; EPA 1998). The experimental models proposed for these evaluations include different strains of inbred, outbred, or hybrid mice (Salamone and Mavourin 1994), transgenic mice (Recio et al. 2005), and, more recently, wildlife animal models (Da Silva et al. 2000a, b). The species employed to monitor the potential genotoxic effect must be considered as a source of variability as certain genotoxic agents have been described as species specific. For example, the effects of ionizing radiation have been highly variable when assayed with different animal species and laboratory strains (Catena et al. 1994) and differing responses have occurred between different rat strains when exposed to chemical agents such as cyclophosphamide (Hamada et al. 2001). According to Simula and Priestly (1992), these effects could be due to the influence of various factors such as the differential distribution of the compound tested within the tissues of the different species and strains of animals used in testing. The importance of strain-dependence sensitivity to different agents has been growing in the last three decades (Styles et al. 1983; Aeschbacher 1986; Sato et al. 1987, 1993). Sato et al. (1993) analyzed the micronucleus frequency in the bone marrow of different mice strains treated with base and nucleotide analogues and found that BALB/c mice were more susceptible to clastogenic effects than C_{57}BL/6 or DBA/2 strains of mice. Similar results were obtained after exposure to radiation (Bhilwade et al. 2004).

The present study was carried out to determine the changes induced in erythrocytes of rats by methylmercury chloride. Low dose was used in order to assess the genotoxic and cytotoxic effects of this compound at concentration normally found in nature, i.e., 40–500 μg/ L in the Earth's crust (Tchounwou et al. 2003).

Protocol

For genetic toxicology evaluation, three groups comprising four animals each were used in the present experimental protocol. Animals from group I served as control, while animals of groups II and III were used as experimental sets. The animals of both the experimental groups were given 2 mg/kg body weight of methylmercury chloride orally via gavage for 14 days. The animals of group II were used for the analysis of micronucleated polychromatic erythrocytes, while group III rats were examined for the assessment of chromosomal aberrations.

In case of micronucleus test, bone marrow preparations were made and stained according to the method described by Schmid (1975). The presence of micronucleated polychromatic erythrocytes was visually scored by optical microscopy using a Nikon 80i microscope fitted with a digital camera (Jenoptik, Germany). Cells were considered to be micronucleated when they found to contain neatly defined chromatin corpuscles with a diameter of less than one-third the diameter of the cell nucleus and stained equal or lighter than the nucleus of the cell from which the micronucleated cell had developed (Schmid 1975).

We prepared five slides per animal for assessing micronucleated cells in the bone marrow. Five smears prepared from blood, drawn by heart puncture from each specimen, were used for scoring peripheral blood micronucleated erythrocytes. The smears were air-dried, fixed in acetic acid–methanol = 1:3, v/v for 10 min, and stained with the May–Grünwald–Giemsa.

Labeled slides of control as well as treated animals were coded and scored in a blind manner and screened for micronuclei using light microscope under the magnification of 40× and 100×. The right area on the slide was selected for scoring where the erythrocytes were well separated, not folded, and could clearly be discriminated. Well-stained slides were chosen for the discrimination between polychromatic erythrocytes (PCEs) and normochromatic erythrocytes (NCEs). PCEs were purple to blue, while NCEs were orange to red.

The following parameters were analyzed in each specimen: (a) the frequency of PCE among the first 100 NCE observed in each slide (PCE/NCE ratio); (b) the frequency of micronucleated polychromatic erythrocytes (MNPCE) within 1,000 polychromatic erythrocytes counted; (c) the frequency of micronucleated normochromatic erythrocytes (MNNCE) within 1,000 normochromatic erythrocytes counted; (d) the frequency of dead cells (DC; erythrocytes with >3 micronuclei in the bone marrow) with in 1,000 cells; (e) the frequency of micronucleated peripheral blood erythrocytes (MNPBE) within 2,000 erythrocytes; (f) the frequency of metaphases among 1,000 cells (MI); and (g) the frequencies of cells with chromosome aberrations among 100 metaphases.

Data were expressed as mean ± SEM for each group and subjected to the statistical analysis by ANOVA (one-way) and Bonferroni post hoc test. $p<0.05$ was considered as the level of significance.

Results

Micronucleus Test: Effect of Methylmercury Chloride

Release of micronuclei (MN) as mutagenic effects of methylmercury chloride using various parameters has been summarized in Fig. 23.1. The results demonstrate microscopic analysis of changes in frequency of MN induced by 2 mg/kg body weight concentration of MeHgCl in rat bone marrow cells and peripheral blood cells. Oral administration of MeHgCl displayed an induction of MN that was significant at $p<0.05$. Chemically induced formation of MN was detectable after 14-day treatment. Erythrocytes take different stains depending on their age. Young erythrocytes appear bluish and are termed as polychromatic (Schmid 1975), whereas normochromatic erythrocytes or aged ones take red color.

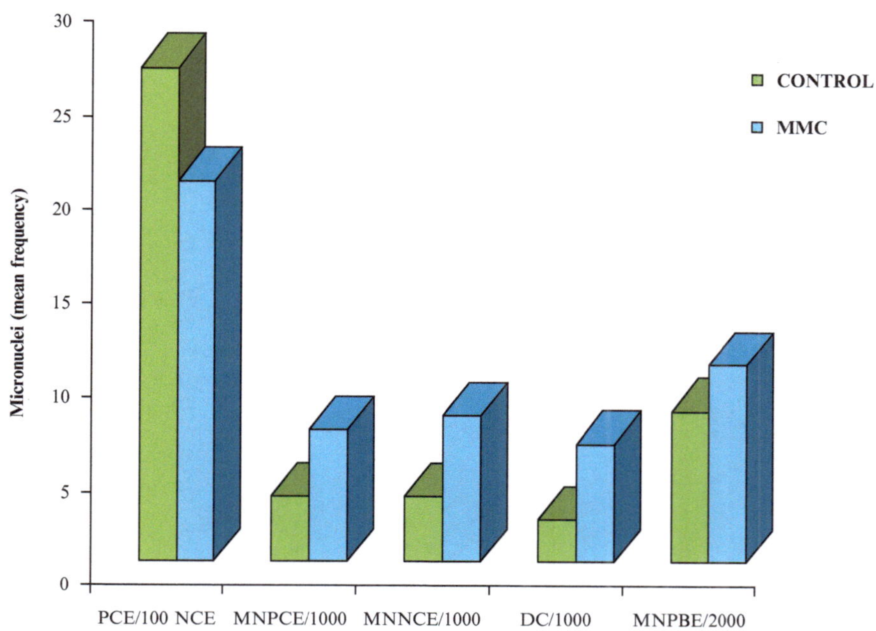

Fig. 23.1 Effect of methylmercury chloride on frequency of micronucleated polychromatic erythrocytes in rat bone marrow and peripheral blood. Values represent mean ± SEM

Discussion

The micronucleus test is widely accepted in the evaluation of clastogenic or aneugenic agents in vivo. Organomercurial compounds showed to increase micronucleus frequency in pregnant mice and their fetuses after intraperitoneal administration of o-hydroxymercuribenzoic acid anhydride, confirming placental transfer of mercury (Chorvatovicova and Kovacikova 1993; Kovachikova and Chorvatovicova 1993). Investigations on aquatic animals, as particularly exposed organisms, resulted with controversial findings. Nepomuceno et al. (1997) found a significant increase in micronuclei in fish erythrocytes after in vivo exposure to metallic mercury at concentrations of 20 and 200 mg L^{-1}Hg, but not at lower concentrations. Al-sabti (1995) described a dose-dependent increase in MN frequency after in vitro mercury and methylmercury treatment of trout hepatic cells. Some other authors did not find any significant increase in MN in fish exposed to mercury (Ayllon and Garcia-Vasquez 2000; Sanchez-Galan et al. 2001). The results cannot be compared, however, as different organisms or tissues, as well as different mercury forms and concentrations, were tested. Human exposure to mercury is proven to result in genotoxic damage. Queiroz et al. (1999) reported an increase in MN incidence in workers who were chronically exposed to mercury in concentrations considered biologically safe for the exposed population (Fig. 23.2).

The micronucleus test is often used to predict the carcinogenic potential of compounds (Sato and Tomita 2001), although the scope of this test is continuously broadening by incorporating new technologies for the detection of different genetic alterations (Zuniga-Gonzalez et al. 2001a, b; Ateeq et al. 2002; Cristaldi et al. 2004). Several authors have reported modifications to the test, including alterations in culture media, β-cytochalasin concentration, osmolarity, pH, and staining techniques, all of which are capable of introducing artifacts and producing intra- and inter-laboratory variability (Fenech et al. 2003; Mentières and Marzin 2004). Additionally, there have been reported cases of false-negative results when subjecting 5-azacitidine, diazepam, or hydroquinone to the micronucleus test using cultured LUC2 cells (Lynch and Parry 1993). Some authors have described sex as an important variable in the micronucleus test (Fenech et al. 1994; Zuniga-Gonzalez et al. 2000, 2001a, b), with males gen-

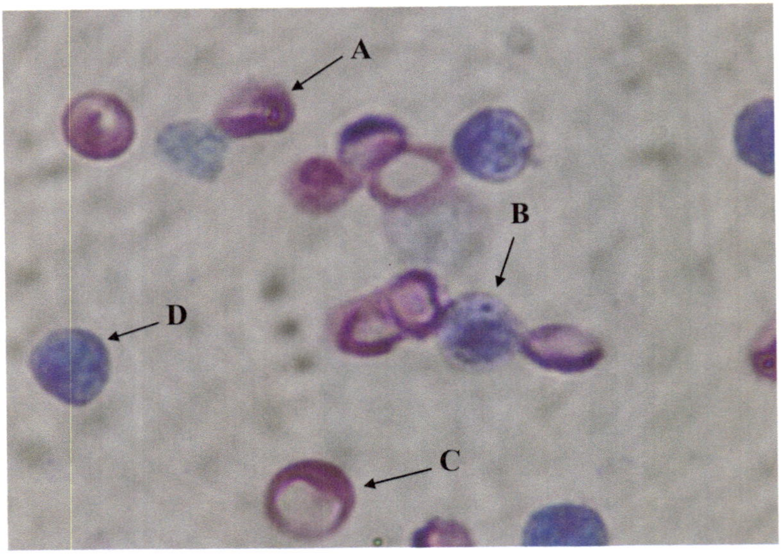

Fig. 23.2 Photomicrograph showing a plate with (**a**) micronucleated normochromatic erythrocyte, (**b**) micronucleated polychromatic erythrocyte, (**c**) normochromatic erythrocyte, (**d**) polychromatic erythrocyte

erally being more sensitive than females to the induction of micronuclei (Hayashi et al. 1982; CSGMT 1988). However, other studies have shown no sex-related differences in micronucleus test results (Vanparys et al. 1990; Mudry et al. 1994; Gimmler-Luz et al. 1997).

Our results found differences in micronucleus frequencies between the control group and the experimental group which received 2 mg/kg of methylmercury chloride. This low concentration of MMC generated micronucleated cells and these micronucleated cells contained a higher number of micronuclei showing a genotoxic damage caused by this toxic metal. The micronucleation index confirmed that MeHg may interfere in the chromosome distribution in anaphase, leading to micronuclei production. Recent works (Their et al. 2003; Bonacker et al. 2004; Stoiber et al. 2004) support the idea that mercury genotoxicity, especially micronuclei formation, may result from a disturbance of the microtubule function (tubulin assembly and kinesin-driven motility). In this case, mercury compounds may modify chromosome distribution by interfering in microtubule formation or chromosome segregation, leading to chromosome loss.

Conclusion

To conclude, our findings showed that oral exposure to low concentrations of methylmercury chloride produced a measurable genotoxic effect. A combination of the two assays proved a good choice for assessing the genotoxicity of methylmercury.

Finally, our data suggest that the difficulty to generate a differential diagnosis of MeHg genotoxicity could be avoided by the use of the above-mentioned parameters as biomarkers.

References

Aeschbacher HU (1986) Rates of micronuclei induction in different mouse strains. Mutat Res 164:109–115

Aleo MF, Morandini F, Benttoni F, Tanganelli S, Vezzola A, Giuliani R, Steimberg N, Boniotti J, Bertasi B, Losio N, Apostoli P, Mazzoleni G (2002) In vitro study of the nephrotoxic mechanism of mercuric chloride. Med Lav 93:267–278

Al-Sabti K, Lloyd DC, Edwards AA, Stegnar P (1992) A survey of lymphocyte chromosomal damage in Slovenian workers exposed to occupational clastogens. Mutat Res 280:215–223

Al-Sabti K (1995) An in vitro binucleated blocked hepatic cell technique for genotoxicity testing in fish. Mutat Res 335:109–120

Amorim AIM, Mergler D, Bahia MO, Dubeau H, Miranda D, Lebel J, Burbano RR, Lucotte M (2000) Cytogenetic damage related to low levels of methylmercury contamination in the Brazilian Amazon. An Acad Bras Cienc 72:497–507

Anwar WA, Gabal MS (1991) Cytogenetic study in workers occupationally exposed to mercury fulminate. Mutagenesis 6:189–192

Ateeq B, Abul Farah M, Niamat A, Ahmad W (2002) Induction of micronuclei and erythrocyte alterations in the Cat fish Clarias batrachus by 2, 4 dichlorophenoxyacetic acid and butachlor. Mutat Res 518:135–144

Ayllon F, Garcia-Vasquez EG (2000) Introduction of micronuclei and other nuclear abnormalities in European minnow Phoxinus phoxinus and mollie poecilia latipinna: an assessment of the fish micronucleus test. Mutat Res 467:177–186

Betti C, Davini T, Barale R (1992) Genotoxic activity of methyl mercury chloride and dimethyl mercury in human lymphocytes. Mutat Res 281:255–260

Betti C, Barale R, Pool-Zobel BL (1993) Comparative studies on cytotoxic and genotoxic effects of two organic mercury compounds in lymphocytes and gastric mucosa cells of Sprague-Dawley rats. Environ Mol Mutagen 22:172–180

Bhilwade HN, Chaubey RC, Chauhan PS (2004) Gamma Ray induced bone marrow micronucleated erythrocytes in seven strains of mouse. Mutat Res 560:19–26

Bonacker D, Stoiber T, Wang M, Bohm KJ, Prots I, Unger E, Their R, Bolt HM, Degen GH (2004) Genotoxicity of inorganic mercury salts based on disturbed microtubule function. Arch Toxicol 78:575–583

Catena C, Conti D, Villani P, Nastasi R, Archilei R, Righi E (1994) Micronuclei and 3AB in human and canine lymphocytes after *in vitro* X-irradiation. Mutat Res 312:1–8

Chorvatovicova D, Kovacikova Z (1993) Transplacental effect of organomercurial. 1. Micronucleus test. Biol Bratislava 48:271–274

Clarkson RW (2002) The three modern faces of mercury. Environ Health Perspect 110:11–23

Collaborative Study Group for the Micronucleus Test (CSGMT) (1988) Strain difference in the micronucleus test. Mutat Res 204:307–316

Cristaldi M, Ieradi LA, Udroiu I, Zilli R (2004) Comparative evaluation of background micronucleus frequencies in domestic mammals. Mutat Res 559:1–9

Da Silva J, de Freitas TRO, Heuser V, Marinho JR, Bittencourt F, Cerski CTS, Kliemann LM, Erdtmann B (2000a) Effects of chronic exposure to coal in wild rodent (*Ctenomys torquatus*) evaluated by multiple methods and tissues. Mutat Res 470:35–39

Da Silva J, de Freitas TRO, Heuser V, Marinho JR, Erdtman B (2000b) Genotoxicity biomonitoring in coal regions using wild rodent *Ctenomys torquatus* by comet assay and micronucleus test. Environ Mol Mutagen 35:270–278

De Flora S, Benniceli C, Bagnasco M (1994) Genotoxicity of mercury compounds. A review. Mutat Res 31:57–79

Ehrenstein C, Shu P, Wickenheiser EB, Hirner AV, Dolfen M, Emons H, Obe G (2002) Methyl mercury uptake and associations with the induction of chromosomal aberrations in Chinese hamster ovary (CHO) cells. Chem Biol Interact 141:259–274

Environmental Protection Agency (EPA) (1998) Health effects test guidelines OPPTS 870.5395. Mammalian erythrocyte micronucleus test. National Service Center for Environmental Publications (NSCEP), Washington, DC, pp 1–12

Faustman EM, Ponce RA, Ou YC, Mendonza MA, Lewandowski R, Kavanagh T (2002) Investigations of methylmercury-induced alterations in neurogenesis. Environ Health Perspect 110:859–864

Fenech M, Neville S, Rinaldi J (1994) Sex is an important variable affecting spontaneous micronucleus frequency in cytokinesis-blocked lymphocytes. Mutat Res 313:203–207

Fenech M, Chang M, Kirsch-Volders M, Holland N, Bonassi S, Zeiger E (2003) HUMN project: detailed description of the scoring criteria for the cytokinesis-block micronucleus assay using isolated human lymphocyte cultures. Mutat Res 534:65–75

Franchi E, Loprieno G, Ballardin M, Petrozzi L, Migliore L (1994) Cytogenetic monitoring of fishermen with environmental mercury exposure. Mutat Res 320:23–29

Gimmler-Luz MC, Rodrigues de Andrade HH, Tozzo Marafon-Bayer A (1997) Benzidine and diaminobenzidine induced micronuclei in mice after intraperitoneal and oral single or multiple treatments. Braz J Genet 20:247–252

Hamada S, Yamasaki K, Nakanishi S, Omori T, Serikawa T, Hayashi M (2001) Evaluation of general suitability of the rat for the micronucleus assay: the effect of cyclophosphamide in 14 rats. Mutat Res 495:127–134

Hansteen H, Ellingsen DG, Clausen KO, Kjuus H (1993) Chromosome aberrations in chloralkali workers previously exposed to mercury vapour. Scand J Work Environ Health 19:375–381

Hayashi M, Sofuni T, Ishidate M (1982) High sensitivity in micronucleus induction of a mouse strain (MS). Mutat Res 105:253–256

Kovachikova Z, Chorvatovicova D (1993) Transplacental effect of organomercurial. 2. Alanine and aspartate aminotransferase activities in liver. Biol Bratislava 48:275–277

Lynch A, Parry J (1993) The cytochalasin-B micronucleus/kinetochore assay *in vitro*: Studies with 10 suspected aneugens. Mutat Res 287:71–86

Mabille V, Roels H, Jacquet P, Léonard A, Lauweris R (1984) Cytogenetic examination of leucocytes of workers exposed to mercury vapour. Int Arch Occup Environ Health 53:257–260

Mentieres S, Marzin D (2004) Apoptosis may contribute to false positive results in the in vitro micronucleus test performed in extreme osmolality, ionic strength and pH conditions. Mutat Res 560:101–118

Mudry MD, Carballo M, Labal de Vinuesa ML, Gonzalez Cid M, Larripa I (1994) Mutagenic bioassay of certain pharmacological drugs: III. Metronidazole (MTZ). Mutat Res 305:127–132

National Research Council, Committee on the Toxicological Effects of Methylmercury (2000) Toxicological effects of methylmercury. National Academy Press, Washington, DC

Nepomuceno JC, Ferrari I, Spano MA, Centeno AJ (1997) Detection of micronuclei in peripheral erythrocytes of cyprinus carpio exposed to metallic mercury. Environ Mol Mutagen 30:293–297

Ogura H, Takeuchi T, Morimoto K (1996) A comparison of the 8-hydroxydeoxyguanosine, chromosome aberrations and micronucleus techniques for the assessment of the genotoxicity of mercury compounds in human blood lymphocytes. Mutat Res 340:175–182

Organization for Economic Operation and Development (OEOD) (1997) Guidelines for testing of chemicals n. 474. Mammalian erythrocyte micronucleus test. OEOD, Paris, pp 1–10

Queiroz MLS, Bincoletto C, Quadros MR, De Capitani EM (1999) Presence of micronuclei in lymphocytes of mercury exposed workers. Immunophamacol Immunotoxicol 21:141–150

Recio L, Bauer A, Faiola B (2005) Use of genetically modified mouse models to assess pathways of benzene-induced bone marrow cytotoxicity and genotoxicity. Chem Biol Interact 153–154:159–164

Salamone MF, Mavourin KH (1994) Bone marrow micronucleus assay: a review of the mouse stocks used and their published mean spontaneous micronucleus frequencies. Environ Mol Mutagen 23:239–273

Sanchez-Galan S, Linde AR, Ayllon F, Garcia-Vazquez E (2001) Induction of micronuclei in eel (Anguilla anguilla) by heavy metals. Ecotoxicol Environ Safe 49:139–143

Sato S, Tomita I (2001) Short-term screening method for the prediction of carcinogenicity or chemical substances: current status and problems of an *in vivo* rodent micronucleus assay. J Health Sci 47:1–8

Sato S, Kitajima H, Konishi S, Takizawa H, Inui N (1987) Mouse strain differences in the induction of micronuclei by polycyclic aromatic hydrocarbons. Mutat Res 192:185–189

Sato S, Tazikawa H, Inui N (1993) Mouse strain differences in induction of micronuclei by base analogues and nucleosides. Mutat Res 301:45–49

Schmid W (1975) The micronucleus test. Mutat Res 31:9–15

Schurz F, Sabater-Vilar M, Fink-Gremmels J (2000) Mutagenicity of mercury chloride and mechanisms of cellular defence: the role of metal-binding proteins. Mutagenesis 15:525–530

Simula AP, Priestly BG (1992) Species differences in the genotoxicity of cyclophosphamide and styrene in three *in vivo* assays. Mutat Res 271:49–58

Stoiber T, Bonacker D, Bohm KJ, Bolt HM, Their R, Degen GH, Unger E (2004) Disturbed microtubule function and induction of micronuclei by chelate complexes of mercury (II). Mutat Res 563:97–106

Styles JA, Richardson CR, Burlinson B (1983) A comparison of incidence of micronuclei in blood and bone marrow in 3 strains of mouse dosed with cyclophosphamide or hexamethylphosphoramide (HMPA). Mutat Res 122:143–147

Tchounwou PB, Ayensu WK, Ninashvili N, Sutton D (2003) Environmental exposure to mercury and its toxicopathologic implications for public health. Environ Toxicol 18:149–175

Their R, Bonacker D, Stoiber T, Bohm KJ, Wang M, Unger E, Bolt HM, Degen G (2003) Interaction of metal salts with cytoskeletal motor protein systems. Toxicol Lett 140–141:75–81

Vanparys P, Vermeiren F, Sysmans M, Temmerman R (1990) The micronucleus assay as a test for the detection of aneugenic activity. Mutat Res 244:95–103

Zuniga-Gonzalez G, Torres-Bugarín O, Luna-Aguirre J, Gonzalez-Rodríguez A, Zamora-Perez A, Gomez-Meda BC, Ventura-Aguilar AJ, Ramos-Ibarra ML, Ramos-Mora A, Ortiz GG, Gallegos-Arreola MP (2000) Spontaneous micronuclei in peripheral blood erythrocytes from 54 animal species (mammals, reptiles and birds): part two. Mutat Res 467:99–103

Zuniga-Gonzalez G, Torres-Bugarin O, Ramos-Ibarra ML, Zamora-Perez A, Gomez-Meda BC, Ventura-Aguilar AJ, Ramos-Mora A, Ortiz GG, Alvarez-Moya C, Gonzalez-Rodriguez A, Luna-Aguirre J, Gallegos-Arreola MP (2001a) Variation of micronucleated erythrocytes in peripheral blood of *Sciurus aureogaster* in relation to age: an increment of micronucleated polychromatic erythrocytes after the administration of colchicine. Environ Mol Mutagen 37:173–177

Zuniga-Gonzalez G, Torres-Bugarin O, Zamora-Perez A, Gomez-Meda BC, Ramos-Ibarra ML, Martinez-Gonzalez S, Gonzalez-Rodriguez A, Luna-Aguirre J, Ramos-Mora A, Ontiveros-Lira D, Gallegos-Arreola MP (2001b) Differences in the number of micronucleated erythrocytes among young and adult animals including humans. Spontaneous micronuclei in 43 species. Mutat Res 494:161–167

24 Chromosomal Aberrations

Mercury, one of the most widely diffused and hazardous organ-specific environmental contaminants, exists in a wide variety of physical and chemical states, each of which with unique characteristics of target organ specificity (Aleo et al. 2002). In nature the different forms of mercury include the metallic form, inorganic compounds, as well as alkyl, alkoxy, and aryl mercury compounds. Once introduced into the environment, mercury compounds can undergo a wide variety of transformations. In sediments, inorganic mercury ($HgCl_2$) may be converted into methyl (CH_3HgCl) and dimethyl (CH_3CH_2HgCl) forms by methanogenic bacteria. This biotransformation constitutes a serious environmental risk, given that CH_3HgCl is the most toxic of the mercury compounds and accumulates in the aquatic food chain, eventually reaching human diets (Tchounwou et al. 2003). CH_3HgCl has been an environmental concern to public health and regulatory agencies for over 50 years because of its neurotoxicity. Its association with nervous system toxicity in adults and infants near Minamata Bay, Japan, in the 1950s initiated environmental health research inquiries that continue to this day (Faustman et al. 2002). The three modern "faces" of mercury are our perceptions of risk from the exposure of billions of people to CH_3HgCl in fish, mercury vapor from amalgam tooth fillings, and CH_3CH_2HgCl in the form of thimerosal added as an antiseptic to widely used vaccines (Clarkson 2002). Mercury genotoxicity has been usually attributed to its ability to react with the sulfhydryl groups of tubulin, impairing spindle function and leading to chromosomal aberrations and polyploidy (De Flora et al. 1994). Another important mechanism of mercury genotoxicity is its ability to produce free radicals that can cause DNA damage (Schurz et al. 2000; Ehrenstein et al. 2002). In vivo studies have demonstrated a clastogenic effect of mercury on people exposed to this element in their working environment or through the consumption of contaminated food or sometimes accidentally. Increased numbers of chromosome alterations and micronuclei have been reported in people who consume contaminated fish (Amorim et al. 2000; Franchi et al. 1994) and in miners and workers of explosive factories (Al-Sabti et al. 1992; Anwar and Gabal 1991). Negative results were also obtained in some cases (Hansteen et al. 1993; Mabille et al. 1984), demonstrating that cytogenetic monitoring of peripheral blood lymphocytes in individuals exposed to mercury from different sources may not be completely specific (De Flora et al. 1994). The effects of CH_3HgCl contamination have been studied in an increasing way since the outbreaks in Japan and Iraq. Many of these studies had their focus on the neurological effects of CH_3HgCl exposure in adult animals and used high doses of this compound (1,900–30,000 ppb = μg/L) to obtain its most severe effects (National Research Council 2000). Most of the in vitro studies with lymphocytes also used high doses (250–6,250 μg/L) of mercury compounds in order to evaluate its clastogenic effects (Ogura et al. 1996; Betti et al. 1992, 1993).

S. Nabi, *Toxic Effects of Mercury*, DOI 10.1007/978-81-322-1922-4_24,
© Springer India 2014

Protocol

For genetic toxicology evaluation, three groups comprising four animals each were used in the present experimental protocol. Animals from group I served as control, while animals of groups II and III were used as experimental sets. The animals of both the experimental groups were given 2 mg/kg body weight of methylmercury chloride orally via gavage for 14 days. The animals of group II were used for the analysis of micronucleated polychromatic erythrocytes, while group III rats were examined for the assessment of chromosomal aberrations.

Following the oral administration of toxic metal, rats were sacrificed on the scheduled day by cervical dislocation and immediately dissected to obtain bone marrow for cytogenetic preparations. For chromosomal aberration, 120 min prior to killing, colchicine (4 mg/kg) was administered to the animals. The time of sacrifice was decided on preliminary observations for scoring as optimal aberrations. The slides of bone marrow cells were prepared and stained according to the routine hypotonic solution–acetic acid–methanol–air-drying–Giemsa schedule for metaphase plate analysis (Preston et al. 1987).

We prepared five slides per animal for assessing micronucleated cells in the bone marrow. Five smears prepared from blood, drawn by heart puncture from each specimen, were used for scoring peripheral blood micronucleated erythrocytes. The smears were air-dried, fixed in acetic acid–methanol=1:3, v/v for 10 min, and stained with the May–Grünwald–Giemsa.

Well-spread metaphases were selected from five slides per animal for scoring the frequencies of cells with gaps and breaks, cells with at least one chromosome showing early separation of the centromere (early segregation), polyploidy, and extremely damaged with nuclear material visibly different from normal patterns of cell division stages.

Results

The data obtained from approximately 400 metaphases analyzed per treatment showed a significant increase in the frequency of chromatid gaps and chromosomal alterations from methylmercury chloride at low concentration.

Chromosomal Aberrations: Methylmercury Connection

The presented results indicate that MeHgCl is both cytotoxic and clastogenic. The dose- and time-dependent total chromosomal aberrations are summarized below. There was a significant increase in aberrations after 14-day treatment with this toxic metal.

Figures 24.1 and 24.2a, b illustrates the changes in morphology following treatment with clastogen MeHgCl. As shown in the figure, this agent caused chromosome damage, which includes chromosomal, and chromatid breakage and gaps.

Discussion

Numerous studies deal with the evaluation of genotoxic effects of mercury and mercury compounds on mammalian cells in culture and experimental animals, as well as in exposed human populations. Results point to an inhibition of DNA synthesis (Choi and Kim 1984), DNA damage (Grover et al. 2001), inhibition of spindle microtubule assembly (De Flora et al. 1994), reduction in the frequency of mitosis (Bahia et al. 1999), endoreduplication (Lee et al. 1997), and chromosomal damages (Amorim et al. 2000; Akiyama et al. 2001). Because of mercury accumulation in aquatic predators and marine fish and mammals, these organisms are often investigated for mercury-induced genotoxic effects (Fig. 24.2).

Mercury compounds induce the disturbance of the mitotic spindle (Their et al. 2003) resulting in

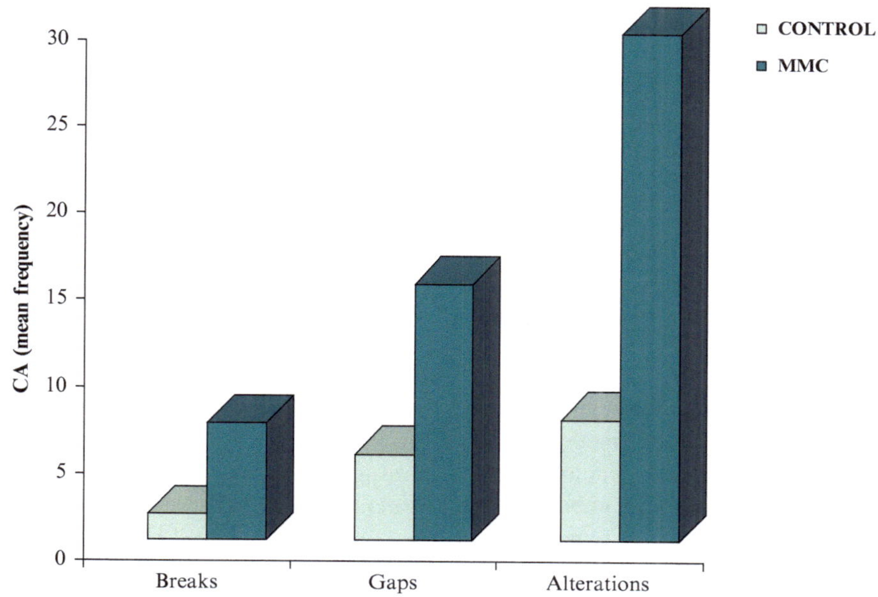

Fig. 24.1 Effect of methylmercury chloride on chromosomal aberration in rat bone marrow. Values represent mean ± SEM

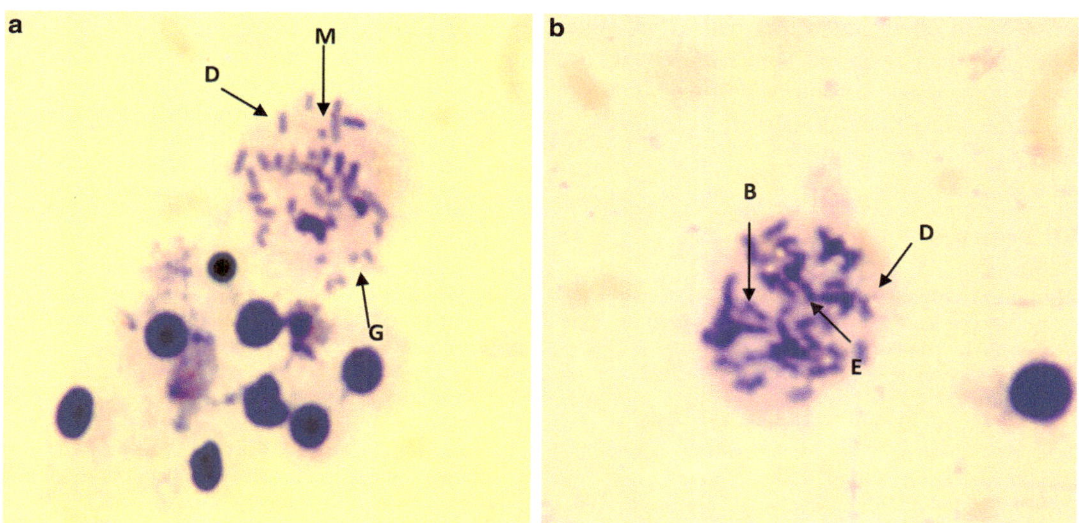

Fig. 24.2 Photomicrograph of rat bone marrow showing aberrated plates with (**a**): chromosome gap (*G*), minute (*M*), and deletion (*D*) (**b**): exchange (*E*) chromosome break (*B*) deletion (*D*)

aneuploidy and polyploidy, which is due to its affinity to sulfhydryl groups in the spindle apparatus (Andersen et al. 1983). In the present study, a significant dose-related increase in the number of cells showing chromosome aberrations was observed after treatment with CH_3HgCl. However, such increase was not linearly related to dose. This effect can be explained by the fact that CH_3HgCl acts similarly to x-rays (Betti et al. 1993), increasing DNA damage in a dose-dependent manner until a plateau is reached, with a decrease in damage being observed even though the doses continue to increase. This effect reflects two different phenomena: a dose-dependent increase in the proportion of normal cells whose DNA is damaged and a dose-dependent decrease of the probability that such cells can survive higher exposures (Hall 2000). Similar results were observed on TK6 cells after exposition to CH_3HgCl (Bahia et al. 1999).

Conclusion

To conclude, our findings showed that oral exposure to low concentrations of methylmercury chloride produced a measurable genotoxic effect. Based on mercury's known ability to bind sulfhydryl groups, several hypotheses were raised about potential molecular mechanisms for the metal genotoxicity. Mercury may be involved in four main processes that lead to genotoxicity: generation of free radicals and oxidative stress, action on microtubules, influence on DNA repair mechanisms, and direct interaction with DNA molecules. All data reviewed here contributed to a better knowledge of the widespread concern about the safety limits of mercury exposure.

References

Akiyama M, Oshima H, Nakamura M (2001) Genotoxicity of mercury used in chromosome aberration tests. Toxicol In Vitro 15:463–467

Aleo MF, Morandini F, Benttoni F, Tanganelli S, Vezzola A, Giuliani R, Steimberg N, Boniotti J, Bertasi B, Losio N, Apostoli P, Mazzoleni G (2002) In vitro study of the nephrotoxic mechanism of mercuric chloride. Med Lav 93:267–278

Al-Sabti K, Lloyd DC, Edwards AA, Stegnar P (1992) A survey of lymphocyte chromosomal damage in Slovenian workers exposed to occupational clastogens. Mutat Res 280:215–223

Amorim AIM, Mergler D, Bahia MO, Dubeau H, Miranda D, Lebel J, Burbano RR, Lucotte M (2000) Cytogenetic damage related to low levels of methylmercury contamination in the Brazilian Amazon. An Acad Bras Cienc 72:497–507

Andersen O, Ronne M, Nordberg G (1983) Effects of inorganic metal salts on chromosome length in human lymphocytes. Hereditas 98:65–70

Anwar WA, Gabal MS (1991) Cytogenetic study in workers occupationally exposed to mercury fulminate. Mutagenesis 6:189–192

Bahia MO, Amorim MIM, Burbano RR, Vincent S, Dubeau H (1999) Genotoxic effects of mercury on *in vitro* cultures of human cells. An Acad Bras Cienc 71:437–443

Betti C, Davini T, Barale R (1992) Genotoxic activity of methyl mercury chloride and dimethyl mercury in human lymphocytes. Mutat Res 281:255–260

Betti C, Barale R, Pool-Zobel BL (1993) Comparative studies on cytotoxic and genotoxic effects of two organic mercury compounds in lymphocytes and gastric mucosa cells of Sprague-Dawley rats. Environ Mol Mutagen 22:172–180

Choi H, Kim C (1984) The comparative effects of methylmercury chloride and mercuric chloride upon DNA synthesis in mouse fetal astrocytes in vitro. Exp Mol Pathol 41:371–376

Clarkson RW (2002) The three modern faces of mercury. Environ Health Perspect 110:11–23

De Flora S, Benniceli C, Bagnasco M (1994) Genotoxicity of mercury compounds. A review. Mutat Res 31:57–79

Ehrenstein C, Shu P, Wickenheiser EB, Hirner AV, Dolfen M, Emons H, Obe G (2002) Methyl mercury uptake and associations with the induction of chromosomal aberrations in Chinese hamster ovary (CHO) cells. Chem Biol Interact 141:259–274

Faustman EM, Ponce RA, Ou YC, Mendonza MA, Lewandowski R, Kavanagh T (2002) Investigations of methylmercury-induced alterations in neurogenesis. Environ Health Perspect 110:859–864

Franchi E, Loprieno G, Ballardin M, Petrozzi L, Migliore L (1994) Cytogenetic monitoring of fishermen with environmental mercury exposure. Mutat Res 320:23–29

Grover P, Saleha Banu B, Dana Devi K, Begum S (2001) In vivo genotoxic effects of mercuric chloride in rat peripheral blood leucocytes using comet assay. Toxicology 167:191–197

Hall EJ (2000) Cell survival curves. In: Hall EJ (ed) Radiobiology for the radiologist, 5th edn. Lippincott, Williams and Wilkins, Philadelphia, pp 32–50

Hansteen H, Ellingsen DG, Clausen KO, Kjuus H (1993) Chromosome aberrations in chloralkali workers previously exposed to mercury vapour. Scand J Work Environ Health 19:375–381

Lee CH, Lin RH, Liu SH, Lin-Shiau SY (1997) Distinct genotoxicity of phenylmercury acetate in human lymphocytes as compared with other mercury compounds. Mutat Res 392:269–276

Mabille V, Roels H, Jacquet P, Léonard A, Lauweris R (1984) Cytogenetic examination of leucocytes of workers exposed to mercury vapour. Int Arch Occup Environ Health 53:257–260

National Research Council, Committee on the Toxicological Effects of Methylmercury (2000)

Toxicological effects of methylmercury. National Academy Press, Washington, DC

Ogura H, Takeuchi T, Morimoto K (1996) A comparison of the 8-hydroxydeoxyguanosine, chromosome aberrations and micronucleus techniques for the assessment of the genotoxicity of mercury compounds in human blood lymphocytes. Mutat Res 340:175–182

Preston RJ, Dean BJ, Galloway S, Holden H, McFee AF, Shelby M (1987) Mammalian in vivo cytogenetic assay: analysis of chromosomal aberrations in bone marrow cells. Mutat Res 189:157–165

Schurz F, Sabater-Vilar M, Fink-Gremmels J (2000) Mutagenicity of mercury chloride and mechanisms of cellular defence: the role of metal-binding proteins. Mutagenesis 15:525–530

Tchounwou PB, Ayensu WK, Ninashvili N, Sutton D (2003) Environmental exposure to mercury and its toxicopathologic implications for public health. Environ Toxicol 18:149–175

Their R, Bonacker D, Stoiber T, Bohm KJ, Wang M, Unger E, Bolt HM, Degen G (2003) Interaction of metal salts with cytoskeletal motor protein systems. Toxicol Lett 140–141:75–81

Part IX

Methylmercury Toxicity and Linked Diseases

25 Methylmercury and Minamata Disease

Health is a state of complete physical, mental and social well-being, and not merely the absence of disease or infirmity."

(Unknown Author)

Minamata disease, sometimes referred to as Chisso-Minamata disease, is a neurological syndrome caused by severe mercury poisoning. Symptoms include ataxia, numbness in the hands and feet, general muscle weakness, narrowing of the field of vision, and damage to hearing and speech. In extreme cases, insanity, paralysis, coma, and death follow within weeks of the onset of symptoms. A congenital form of the disease can also affect fetuses in the womb.

Minamata disease was first discovered in Minamata city in Kumamoto Prefecture, Japan, in 1956. It was caused by the release of methylmercury in the industrial wastewater from the Chisso Corporation's chemical factory, which continued from 1932 to 1968. This highly toxic chemical bioaccumulated in shellfish and fish in Minamata Bay and the Shiranui Sea, which when eaten by the local populace resulted in mercury poisoning. While cat, dog, pig, and human deaths continued over more than 30 years, the government and company did little to prevent the pollution. The animal effects were severe enough in cats that they came to be called "dancing cat fever" (Withrow and Vail 2007).

As of March 2001, 2,265 victims had been officially recognized (1,784 of whom had died) (Official Government Figure, 2001), and over 10,000 had received financial compensation from Chisso (Minamata Disease Archives). By 2004, Chisso Corporation had paid $86 million in compensation and in the same year was ordered to clean up its contamination (Jane Hightower 2008). On March 29, 2010, a settlement was reached to compensate as-yet uncertified victims (Asahi Shimbun News 2010).

A second outbreak of Minamata disease occurred in Niigata Prefecture in 1965. The original Minamata disease and Niigata Minamata disease are considered two of the four big pollution diseases of Japan (Fig. 25.1).

1908–1955

The Chisso Corporation first opened a chemical factory in Minamata in 1908. Initially producing fertilizers, the factory followed the nationwide expansion of Japan's chemical industry, branching out into production of acetylene, acetaldehyde, acetic acid, vinyl chloride, and octanol, among others. The Minamata factory became the most advanced in all of Japan, both before and after World War II. The waste products resulting from the manufacture of these chemicals were released into the Minamata Bay through the factory wastewater. These pollutants had an environmental impact. Fisheries were damaged in terms of reduced catches, and in response, Chisso reached two separate compensation

S. Nabi, *Toxic Effects of Mercury*, DOI 10.1007/978-81-322-1922-4_25,
© Springer India 2014

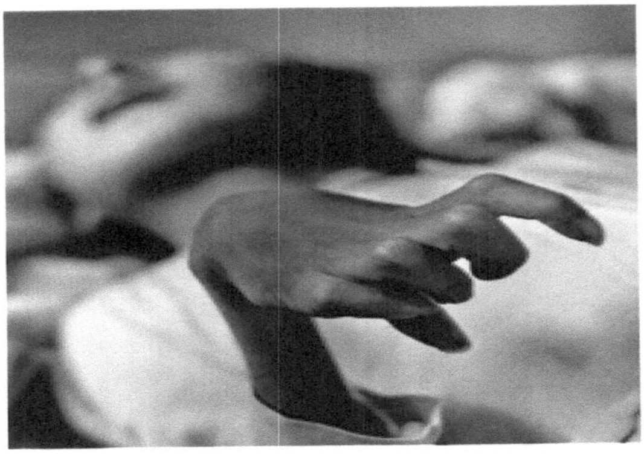

Fig. 25.1 The crippled hand of a Minamata disease victim (Source: English Wikipedia)

agreements with the fishery cooperative in 1926 and 1943 (Harada, p 15). The rapid expansion of the Minamata factory spurred on the local economy, and as Chisso prospered, so did Minamata. This fact, combined with the lack of other industries, meant that Chisso had great influence in Minamata. At one point, over half of the tax revenue of Minamata City authority came from Chisso and its employees, and the company and its subsidiaries were responsible for creating a quarter of all jobs in Minamata (George, pp 35–36). Minamata was even dubbed Chisso's "castle town," in reference to the capital cities of feudal lords who ruled Japan during the Edo period (George, p 26).

The Chisso-Minamata factory first started acetaldehyde production in 1932, producing 210 tons that year. By 1951, production had jumped to 6,000 tons per year and reached a peak of 45,245 tons in 1960 (Report of the Social Scientific Study Group on Minamata Disease). Throughout, the Chisso factory's output amounted to between a quarter and a third of Japan's total acetaldehyde production. The chemical reaction used to produce the acetaldehyde used mercury sulfate as a catalyst. A side reaction of the catalytic cycle led to the production of a small amount of an organic mercury compound, namely, methylmercury (Information on Mercury). This highly toxic compound was released into the Minamata Bay from the start of production in 1932 until 1968, when this production method was discontinued.

1956–1959

On April 21, 1956, a 5-year-old girl was examined at the Chisso Corporation's factory hospital in Minamata, Japan, a town on the west coast of the southern island of Kyushu. The physicians were puzzled by her symptoms: difficulty walking, difficulty speaking, and convulsions. Two days later her younger sister also began to exhibit the same symptoms and she too was hospitalized. The girls' mother informed doctors that her neighbor's daughter was also experiencing similar problems. After a house-to-house investigation, eight further patients were discovered and hospitalized. On May 1, the hospital director reported to the local public health office the discovery of an "epidemic of an unknown disease of the central nervous system," marking the official discovery of Minamata disease (Harada, p 10).

To investigate the epidemic, the city government and various medical practitioners formed the Strange Disease Countermeasures Committee (奇病対策委員会 *Kibyō Taisaku Iinkai*[?]) at the end of May 1956. Owing to the localized nature of the disease, it was suspected to be contagious, and as a precaution, patients were isolated and their homes disinfected. Although contagion was later disproved, this initial response contributed to the stigmatization and discrimination experienced by Minamata victims from the local community. During its investigations, the committee uncovered surprising anecdotal evidence of the strange behavior of cats and other wildlife in the areas surrounding patients' homes. From around 1950 onward, cats had been seen to have convulsions, go mad, and die. Locals called it the "cat dancing disease" (猫踊り病 *neko odori byō*[?]), owing to their erratic movement (Withrow and Vail 2007). Crows had fallen from the sky, seaweed no longer grew on the sea bed, and fish floated dead on the surface of the sea. As the extent of the outbreak was understood, the committee invited researchers from the Kumamoto University to help in the research effort (Nicol 2012).

The Kumamoto University Research Group was formed on August 24, 1956. Researchers from the School of Medicine began visiting Minamata regularly and admitted patients to the university hospital for detailed examinations. A more complete picture of the symptoms exhibited by patients was gradually uncovered. The disease developed without any prior warning, with patients complaining of a loss of sensation and numbness in their hands and feet. They became unable to grasp small objects or fasten buttons. They could not run or walk without stumbling, their voices changed in pitch, and many patients complained of difficulties seeing, hearing, and swallowing. In general, these symptoms deteriorated and were followed by severe convulsions, coma, and eventually death. By October 1956, 40 patients had been discovered, 14 of whom had died: an alarming mortality rate of 36.7 % (Harada, pp 23–24).

Finding the Cause

Researchers from the Kumamoto University also began to focus on the cause of the strange disease. They found that the victims, often members of the same family, were clustered in fishing hamlets along the shore of Minamata Bay. The staple food of victims was invariably fish and shellfish from Minamata Bay. The cats in the local area, which tended to eat scraps from the family table, had died with symptoms similar to those now discovered in humans. This led the researchers to believe that the outbreak was caused by some kind of food poisoning, with contaminated fish and shellfish being the prime suspects.

On November 4, the research group announced its initial findings: "Minamata disease is rather considered to be poisoning by a heavy metal… presumably it enters the human body mainly through fish and shellfish" (Harada, pp 26–27).

Identification of Mercury

As soon as the investigation identified a heavy metal as the causal substance, the wastewater from the Chisso plant was immediately suspected as the origin. The company's own tests revealed that its wastewater contained many heavy metals in concentrations sufficiently high to bring about serious environmental degradation including lead, mercury, manganese, arsenic, selenium, thallium, and copper. Identifying which particular poison was responsible for the disease proved to be extremely difficult and time-consuming. During the years 1957 and 1958, many different theories were proposed by different researchers. At first, manganese was thought to be the causal substance due to the high concentrations found in fish and the organs of the deceased. Thallium, selenium, and a multiple contaminant theory were also proposed, but it was not until March 1958, when visiting British neurologist Douglas McAlpine suggested that Minamata symptoms resembled those of organic mercury poisoning, that the focus of the investigation centered on mercury.

In February 1959, the mercury distribution in Minamata Bay was investigated. The results shocked the researchers involved. Large quantities of mercury were detected in fish, shellfish, and sludge from the bay. The highest concentrations centered around the Chisso factory wastewater canal in Hyakken Harbour and decreased going out to sea, clearly identifying the plant as the source of contamination. Pollution was so heavy at the mouth of the wastewater canal that a figure of 2 kg of mercury per ton of sediment was measured: a level that would be economically viable to mine. Indeed, Chisso did later set up a subsidiary to reclaim and sell the mercury recovered from the sludge (Harada, p 50).

Hair samples were taken from the victims of the disease and also from the Minamata population in general. In patients, the maximum mercury level recorded was 705 ppm (parts per million), indicating very heavy exposure, and in non-symptomatic Minamata residents, the level was 191 ppm. This compared to an average level of 4 ppm for people living outside the Minamata area (Harada, p 50).

On November 12, 1959, the Ministry of Health and Welfare's Minamata Food Poisoning Subcommittee published its results:

> Minamata disease is a poisoning disease that affects mainly the central nervous system and is caused by the consumption of large quantities of

fish and shellfish living in Minamata Bay and its surroundings, the major causative agent being some sort of organic mercury compound. (Harada, p 52)

1959

During the investigation by researchers at the Kumamoto University, the causal substance had been identified as a heavy metal, and it was widely presumed that the Chisso plant was the source of the contamination. Chisso was coming under closer scrutiny, and in order to deflect criticism, the wastewater output route was changed. Chisso knew of the environmental damage caused by its wastewater and was well aware that it was the prime suspect in the Minamata disease investigation. Despite this, from September 1958, instead of discharging its waste into the Hyakken Harbour (the focus of investigation and source of original contamination), it discharged wastewater directly into Minamata River. The immediate effect was the death of fish at the mouth of the river, and from that point on, new Minamata disease victims began to appear in other fishing villages up and down the coast of the Shiranui Sea, as the pollution spread over an even greater area (Harada, pp 38–39) (Fig. 25.2).

Chisso failed to cooperate with the investigation team from Kumamoto University. It withheld information on its industrial processes, leaving researchers to speculate what products the factory was producing and by what methods (Ui, Chapter 4-Section IV). The Chisso factory's hospital director, Hajime Hosokawa, established a laboratory in the research division of the plant to carry out his own experiments into Minamata disease in July 1959. Food to which factory wastewater had been added was fed to healthy cats. Seventy-eight days into the experiment, cat 400 exhibited symptoms of Minamata disease, and pathological examinations confirmed a diagnosis of organic mercury poisoning. The company did not reveal these significant results to the investigators and ordered Hosokawa to stop his research (George, pp 60–61).

In an attempt to undermine Kumamoto University researchers' organic mercury theory, Chisso and other parties with a vested interest that the factory remain open (including the Ministry of International Trade and Industry and the Japan Chemical Industry Association) funded research into alternative causes of the disease, other than its own waste (The Stockholm Appeal).

Compensation of Fishermen and Patients, 1959

Polluting wastewater had damaged the fisheries around Minamata ever since the opening of the Chisso factory in 1908. The *Minamata Fishing Cooperative* had managed to win small payments of "sympathy money" (見舞い金 *mimaikin*?) from the company in 1926 and again in 1943, but

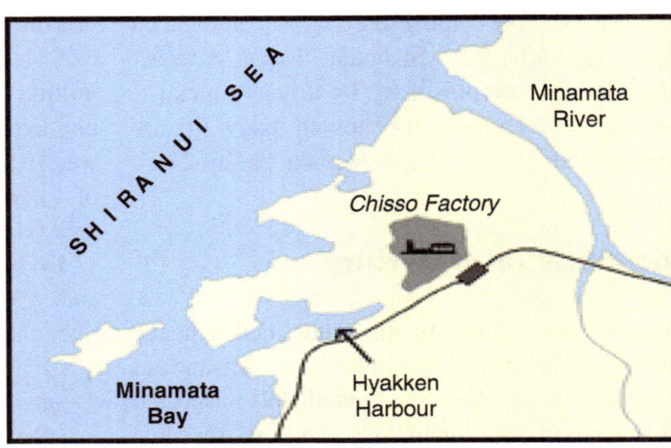

Fig. 25.2 The Chisso factory and its wastewater routes (Source: English Wikipedia)

after the outbreak of Minamata disease, the fishing situation was becoming critical. Fishing catches had declined by 91 % between the years 1953 and 1957. The Kumamoto prefectural government issued a partial ban on the *sale* of fish caught in the heavily polluted Minamata Bay, but not an all-out ban, which would have legally obliged it to compensate the fishermen. The fishing cooperative protested against Chisso and angrily forced their way into the factory on August 6 and 12, demanding compensation. A committee was set up by Minamata Mayor Todomu Nakamura to mediate between the two sides, but this committee was stacked heavily in the company's favor. On August 29, the fishing cooperative agreed to the mediation committee's proposal, stating: "In order to end the anxiety of the citizens, we swallow our tears and accept." The company paid the cooperative JPY20 million (USD 55,600) and set up a JPY15 million (USD 41,700) fund to promote the recovery of fishing.

Since the change of route of wastewater output in 1958, pollution had spread up and down the Shiranui Sea, damaging fisheries there too. Emboldened by the success of the small Minamata cooperative, the *Kumamoto Prefectural Alliance of Fishing Cooperatives* also decided to seek compensation from Chisso. On October 17, 1,500 fishermen from the alliance descended on the factory to demand negotiations. When this produced no results, the alliance members took their campaign to Tokyo, securing an official visit to Minamata by members of the Japanese Diet. During the visit on November 2, alliance members forced their way into the factory and rioted, causing many injuries and JPY10 million (USD 27,800) worth of damage. The violence was covered widely in the media, bringing the nation's attention to the Minamata issue for the first time since the outbreak began. Another mediation committee was set up and an agreement hammered out and signed on December 17. JPY25 million "sympathy money" was paid to the alliance, and a JPY65 million fishing recovery fund established.

In 1959, the victims of Minamata disease were in a much weaker position than the fishermen. The recently formed *Minamata Disease Patients Families Mutual Aid Society* was much more divided than the fishing cooperatives. Patients' families were the victim of discrimination and ostracism from the local community. Local people felt that the company (and their city that depended upon it) was facing economic ruin. To some patients, this ostracism by the community represented a greater fear than the disease itself. After beginning a sit-in at the factory gates in November 1959, the patients asked Kumamoto Prefecture Governor Hirosaku Teramoto to include the patients' request for compensation with the mediation that was ongoing with the prefectural fishing alliance. Chisso agreed and after a few weeks' further negotiation, another "sympathy money" agreement was signed. Patients who were certified by a Ministry of Health and Welfare committee would be compensated: adult patients received JPY100,000 (USD 278) per year, children JPY30,000 (USD 83) per year, and families of dead patients would receive a one-off JPY320,000 (USD 889) payment (Fig. 25.3).

Wastewater Treatment

On October 21, 1959, Chisso was ordered by the Ministry of International Trade and Industry to switch back its wastewater drainage from the Minamata River to Hyakken Harbour and to speed up the installation of wastewater treatment systems at the factory. Chisso installed a Cyclator purification system on December 19, 1959, and opened it with a special ceremony. Chisso's president Kiichi Yoshioka drank a glass of water supposedly treated through the Cyclator to demonstrate that it was safe. In fact, the wastewater from the acetaldehyde plant, which the company knew still contained mercury and led to Minamata disease when fed to cats, was not treated through the Cyclator at the time. Testimony at a later Niigata Minamata disease trial proved that Chisso knew the Cyclator to be completely ineffective: "…the purification tank was installed as a social solution and did nothing to remove organic mercury" (Harada, p 56).

The deception was successful and almost all parties involved in Minamata disease were duped into believing that the factory's wastewater had

Fig. 25.3 Protestors at the gates of the Chisso factory (W. E. Smith) (Source: English Wikipedia)

been made safe from December 1959 onward. This widespread assumption meant that doctors were not expecting new patients to appear, resulting in numerous problems in the years to follow, as the pollution continued. In most people's minds, the issue of Minamata disease had been resolved.

1959–1969

The years between the first set of "sympathy money" agreements in 1959 and the start of the first legal action to be taken against Chisso in 1969 are often called the "ten years of silence." In fact, much activity on the part of the patients and fishermen took place during this period, but nothing had a significant impact on the actions of the company or the coverage of Minamata in the national media.

Continued Pollution

Despite the almost universal assumption to the contrary, the wastewater treatment facilities installed in December 1959 had no effect on the level of organic mercury being released into the Shiranui Sea. The pollution and the disease it caused continued to spread. The Kumamoto and Kagoshima prefectural governments conducted a joint survey in late 1960 and early 1961 into the level of mercury in the hair of people living around the Shiranui Sea. The results confirmed that organic mercury had spread all around the inland sea and that people were still being poisoned by contaminated fish. Hundreds of people were discovered to have levels greater than 50 ppm of mercury in their hair, the level at which people are likely to experience nerve damage. The highest result recorded was that of a woman from Goshonoura Island who had 920 ppm in her sample.

The prefectural governments did not publish the results and did nothing in response to these surveys. The participants who had donated hair samples were not informed of their result, even when they requested it. A follow-up study 10 years later discovered that many had died from "unknown causes" (George, pp 144–145).

Congenital Minamata Disease

Local doctors and medical officials had noticed for a long time an abnormally high frequency of cerebral palsy and other infantile disorders in the

Minamata area. In 1961, a number of medical professionals including Masazumi Harada (later to receive an honor from the United Nations for his body of work on Minamata disease) set about reexamining children diagnosed with cerebral palsy. The symptoms of the children closely mirrored those of adult Minamata disease patients, but many of their mothers did not exhibit symptoms. The fact that these children had been born after the initial outbreak and had never been fed contaminated fish also led their mothers to believe they were not victims. At that time, the medical establishment believed the placenta would protect the fetus from toxins in the bloodstream, which is indeed the case with most chemicals. What was not known at the time was that exactly the opposite is the case with methylmercury: the placenta removes it from the mother's bloodstream and concentrates the chemical in the fetus.

After several years of study and the autopsies of two children, the doctors announced that these children were suffering from an as yet unrecognized congenital form of Minamata disease. The certification committee convened on November 29, 1962, and agreed that the 2 dead children and the 16 children still alive should be certified as patients and therefore liable for "sympathy" payments from Chisso, in line with the 1959 agreement (Harada, pp 68–77).

Outbreak of Niigata Minamata Disease

Minamata disease broke out again in 1965, this time along the banks of the Agano River in Niigata Prefecture. The polluting factory (owned by Showa Denko) employed a chemical process using a mercury catalyst very similar to that used by Chisso in Minamata. As in Minamata, from the autumn of 1964 to the spring of 1965, cats living along the banks of the Agano River had been seen to go mad and die. Before long, patients appeared with identical symptoms to patients living on the Shiranui Sea, and the outbreak was made public on June 12, 1965. Researchers from the Kumamoto University Research Group and Hajime Hosokawa (who had retired from Chisso in 1962) used their experience from Minamata and applied it to the Niigata outbreak. In September 1966, a report was issued proving Showa Denko's pollution to be the cause of this second Minamata disease.

Unlike the patients in Minamata, the victims of Showa Denko's pollution lived a considerable distance from the factory and had no particular link to the company. As a result the local community was much more supportive of patients' groups, and a lawsuit was filed against the company in March 1968, only 3 years after discovery.

The events in Niigata catalyzed a change in response to the original Minamata incident. The scientific research carried out in Niigata forced a reexamination of that done in Minamata, and the decision of Niigata patients to sue the polluting company allowed the same response to be considered in Minamata. Masazumi Harada has said that, "It may sound strange, but if this second Minamata disease had not broken out, the medical and social progress achieved by now in Kumamoto… would have been impossible" (Harada, p 90).

Around this time two other pollution-related diseases were also grabbing headlines in Japan. Victims of Yokkaichi asthma and itai-itai disease were forming citizens' groups and filed lawsuits against the polluting companies in September 1967 and March 1968, respectively. As a group, these diseases came to be known as the four big pollution diseases of Japan (George, pp 174–175).

Slowly but surely, the mood in Minamata and Japan as a whole was shifting. Minamata patients found the public gradually becoming more receptive and sympathetic as the decade wore on. This culminated in 1968 with the establishment in Minamata of the *Citizens' Council for Minamata Disease Countermeasures*, which was to become the chief citizens' support group to the Minamata patients. A founding member of the citizens' council was Michiko Ishimure, a local housewife and poet who later that year published *Pure Land, Poisoned Sea: Our Minamata Disease* (苦海浄土—わが水俣病 *Kugai Jōdo: Waga Minamatabyō*), a book of poetic essays that received national acclaim.

1969–1973

Finally on September 26, 1968 – 12 years after the discovery of the disease (and 4 months after Chisso had stopped production of acetaldehyde using its mercury catalyst) – the government issued an official conclusion as to the cause of Minamata disease:

> Minamata disease is a disease of the central nervous system, a poisoning caused by long-term consumption, in large amounts, of fish and shellfish from Minamata Bay. The causative agent is methylmercury. Methylmercury produced in the acetaldehyde acetic acid facility of Shin Nihon Chisso's Minamata factory was discharged in factory wastewater… Minamata disease patients last appeared in 1960, and the outbreak has ended. This is presumed to be because consumption of fish and shellfish from Minamata Bay was banned in the fall of 1957, and the fact that the factory had waste-treatment facilities in place from January 1960.

The conclusion contained many factual errors: eating fish and shellfish from other areas of the Shiranui Sea, *not just* Minamata Bay, could cause the disease; eating small amounts, *as well as* large amounts, of contaminated fish over a long time also produced symptoms; and the outbreak had not in fact "ended" in 1960, nor had mercury-removing wastewater facilities been installed in January 1960. Nevertheless, the government announcement brought a feeling of relief to a great many victims and their families. Many felt vindicated in their long struggle to force Chisso to accept responsibility for causing the disease and expressed thanks that their plight had been recognized by their social superiors. The struggle now focused on to what extent the victims should be compensated (George, pp 187–190).

Struggle for a New Agreement

In light of the government announcement, the patients of the Mutual Aid Society decided to ask for a new compensation agreement with Chisso and submitted the demand on October 6. The company replied that it was unable to judge what would be a fair compensation and asked the national government to set up a binding arbitration committee to decide. This proposal split the members of the patients' society, many of whom were extremely wary of entrusting their fate to a third party, as they had done in 1959 with unfortunate results. At a meeting on April 5, 1969, the opposing views within the society could not be reconciled, and the organization split into the *Arbitration Group* (who were willing to accept binding arbitration) and the *Litigation Group* (who decided to sue the company). That summer, Chisso sent gifts to the families who opted for arbitration rather than litigation.

An arbitration committee was duly set up by the Ministry of Health and Welfare on April 25, but it took almost a year to draw up a draft compensation plan. A newspaper leak in March 1970 revealed that the committee would ask Chisso to pay only JPY2 million (USD 5,600) for dead patients and JPY140,000 to JPY200,000 (USD 390–560) per year to surviving patients. The Arbitration Group were dismayed by the sums on offer. They petitioned the committee, together with patients and supporters of the Litigation Group, for a fairer deal. The arbitration committee announced their compensation plan on May 25 in a disorderly session at the Ministry of Health and Welfare in Tokyo. Thirteen protesters were arrested. Instead of accepting the agreement as they had promised, the Arbitration Group asked for increases. The committee was forced to revise its plan, and the patients waited inside the Ministry building for 2 days while they did so. The final agreement was signed on May 27. Payments for deaths ranged from JPY1.7 million to JPY4 million (USD 4,700–11,100), one-time payments from JPY1 million to JPY4.2 million (USD 2,760–11,660), and annual payments of between JPY170,000 and 380,000 (USD470–1,100) for surviving patients. On the day of the signing, the Minamata Citizens' Council held a protest outside the Minamata factory gates. One of the Chisso trade unions held an 8 h strike in protest at the poor treatment of the Arbitration Group by their own company (George, pp. 191–202) (Fig. 25.4).

The Litigation Group, representing 41 certified patients (17 already deceased) in 28 families,

Fig. 25.4 Minamata patients and family members hold photographs of their dead family members during a demonstration (W. E. Smith) (Source: English Wikipedia)

submitted their suit against Chisso in the Kumamoto District Court on June 14, 1969. The leader of the group, Eizō Watanabe (a former leader of the Mutual Aid Society), declared that, "Today, and from this day forth, we are fighting against the power of the state." Those who decided to sue the company came under fierce pressure to drop their lawsuits against the company. One woman was visited personally by a Chisso executive and harassed by her neighbors. She was ignored; her family's fishing boat used without permission, their fishing nets cut, and human feces thrown at her in the street (George, p. 205).

The Litigation Group and their lawyers were helped substantially by an informal national network of citizens' groups that sprung up around the country in 1969. The Associations to Indict [Those Responsible for] Minamata disease (水俣病を告発する会 *Minamata-byō o Kokuhatsu Suru Kai*) were instrumental in raising awareness and funds for the lawsuit. The Kumamoto branch in particular was especially helpful to the case. In September 1969, they set up a Trial Research Group, which included law professors, medical researchers (including Masazumi Harada), sociologists, and even the housewife and poet Michiko Ishimure to provide useful material to the lawyers to improve their legal arguments. In fact, their report *Corporate Responsibility for Minamata Disease: Chisso's Illegal Acts* (Corporate Responsibility for Minamata Disease), published in August 1970, formed the basis of the ultimately successful lawsuit (George, pp 241–249).

The trial lasted almost 4 years. The Litigation Group lawyers sought to prove Chisso's corporate negligence. Three main legal points had to be overcome to win the case. First, the lawyers had to show that methylmercury caused Minamata disease and that the company's factory was the source of pollution. The extensive research by the Kumamoto University and the government conclusion meant that this point was proved quite easily. Second, could and should the company have anticipated the effect of its wastewater and should it have taken steps to prevent the tragedy (i.e., was the company negligent in its duty of care)? Third, was the "sympathy money" agreement of 1959, which forbade the patients from claiming any further compensation, a legally binding contract?

The trial heard from patients and their families, but the most important testimony came from Chisso executives and employees. The most dramatic testimony came from Hajime Hosokawa who spoke on July 4, 1970, from his hospital bed where he was dying of cancer. He explained his experiments with cats, including the infamous "cat 400," which developed Minamata disease after being fed factory wastewater. He also spoke of his opposition to the 1958 change in wastewater output route from Hyakken Harbour to Minamata River. His testimony was backed up by a colleague who also told how company officials had ordered them to halt their cat experiments in the autumn of 1959. Hajime Hosokawa died 3 months after giving his testimony. Former factory manager Eiichi Nishida admitted that the company put profits ahead of safety, resulting in dangerous working conditions and a lack of care with mercury. Former Chisso President Kiichi Yoshioka admitted that the company promoted a theory of dumped World War II explosives even though it knew it to be unfounded.

The verdict handed down on March 20, 1973, represented a complete victory for the patients of the Litigation Group:

> The defendant's factory was a leading chemical plant with the most advanced technology and ... should have assured the safety of its wastewater. The defendant could have prevented the occurrence of Minamata disease or at least have kept it at a minimum. We cannot find that the defendant took any of the precautionary measures called for in this situation whatsoever. The presumption that the defendant had been negligent from beginning to end in discharging wastewater from its acetaldehyde plant is amply supported. The defendant cannot escape liability for negligence.

The "sympathy money" agreement was found to be invalid, and Chisso was ordered to make one-time payments of JPY18 million (USD 66,000) for each deceased patient and from JPY16 million to JPY 18 million (USD 59,000–66,000) for each surviving patient. The total compensation of JPY937 million (USD 3.4 million) was the largest sum ever awarded by a Japanese court (Harada, pp 156–157 and George, p 208).

Uncertified Patients' Fight to Be Recognized

While the struggles of the arbitration and litigation groups against Chisso were continuing, a new group of Minamata disease sufferers emerged. In order to qualify for compensation under the 1959 agreement, patients had to be officially recognized by various ad hoc certification committees according to their symptoms. However, in an effort to limit the liability and financial burden on the company, these committees were sticking to a rigid interpretation of Minamata disease. They required that patients must exhibit all symptoms of Hunter–Russell syndrome – the standard diagnosis of organic mercury poisoning at the time – which originated from an industrial accident in the United Kingdom in 1940. The committee certified only patients exhibiting explicit symptoms of the British syndrome, rather than basing their diagnosis on the disease in Japan. This resulted in many applicants being rejected by the committee, leaving them confused and frustrated (*Mercury Poisoning of Thousands Confirmed by* Jonathan Watts).

Victims

As of March 2001, 2,265 victims have been officially certified (1,784 of whom have died) (official government figure as of March 2001), and over 10,000 people have received financial compensation from Chisso (*Minamata Disease Archives*), although they are not recognized as official victims. The issue of quantifying the impact of Minamata disease is complicated, as a full epidemiological study has never been conducted and patients were ever recognized only if they voluntarily applied to a Certification Council in order to seek financial compensation (*Mercury Poisoning of Thousands Confirmed by* Jonathan Watts). Many victims of Minamata disease faced discrimination and ostracism from the local community if they came out into the open about their symptoms. Some people feared the disease to be contagious, and many local people were fiercely loyal to Chisso, depending on the company for their livelihoods. In this atmosphere, sufferers were reluctant to come forward and seek certification. Despite these factors, over 17,000 people have applied to the Council for certification. Also, in recognizing an applicant as a Minamata disease sufferer, the Certification Council qualified that patient to receive financial compensation from Chisso. As such, the Council has always been under immense pressure to reject claimants and minimize the financial burden placed on Chisso. Rather than being a Council of medical recognition, the decisions of the Council were always affected by the economic and political factors surrounding Minamata and the Chisso Corporation. Furthermore, compensation of the victims led to continued strife in the community, including unfounded accusations that some of the people who sought compensation did not actually suffer from the disease (*Ten Things to Know about Minamata Disease* by Soshisha).

Democratizing Effects

According to Timothy S. George, the environmental protests that surrounded the disease appeared to aid in the democratization of Japan (George 2001). When the first cases were reported and subsequently suppressed, the rights of the victims were not recognized, and they were given no compensation. Instead, the afflicted were ostracized from their community due to ignorance about the disease, as people were afraid that it was contagious.

The people directly impacted by the pollution of Minamata Bay were not originally allowed to participate in actions that would affect their future. Disease victims, fishing families, and company employees were excluded from the debate. Progress occurred when Minamata victims were finally allowed to come to a meeting to discuss the issue. As a result, postwar Japan took a small step toward democracy.

Through the evolution of public sentiments, the victims and environmental protesters were able to acquire standing and proceed more effectively in their cause. The involvement of the press also aided the process of democratization because it caused more people to become aware of the facts of Minamata disease and the pollution that caused it.

Although the environmental protests did result in Japan becoming more democratized, it did not completely rid Japan of the system that first suppressed the fishermen and victims of Minamata disease.

Media

Photographic documentation of Minamata started in the early 1960s. One photographer who arrived in 1960 was Shisei Kuwabara, straight from university and photo school, who had his photographs published in *Weekly Asahi* as early as May 1960. The first exhibition of his photographs of Minamata was held in the Fuji Photo Salon in Tokyo in 1962, and the first of his book-length anthologies, *Minamata Disease*, was published in Japan in 1965. He has returned to Minamata many times since (*Shisei Kuwabara* – "Minamata").

However, it was a dramatic photographic essay by W. Eugene Smith that brought world attention to Minamata disease. He and his Japanese wife lived in Minamata from 1971 to 1973. The most famous and striking photo of the essay, *Tomoko Uemura in Her Bath* (1972), shows Ryoko Uemura, holding her severely deformed daughter, Tomoko, in a Japanese bath chamber. Tomoko was poisoned by methylmercury while still in the womb. The photo was very widely published. It was posed by Smith with the cooperation of Ryoko and Tomoko in order to dramatically illustrate the consequences of the disease. It has subsequently been withdrawn from circulation at the request of Tomoko's family and therefore does not appear in recent anthologies of Smith's works (*Tomoko Uemura, R.I.P.* by Jim Hughes). Smith and his wife were extremely dedicated to the cause of the victims of Minamata disease, closely documenting their struggle for recognition and right to compensation. Smith was himself attacked and seriously injured by Chisso employees in an incident in Goi, Ichihara City, near Tokyo on January 7, 1972, in an attempt to stop the photographer from further revealing the issue to the world (Smith, pp 94–95). The 54-year-old Smith survived the attack, but his sight in one eye deteriorated, and his health never fully recovered before his death in 1978.

Japanese photographer Takeshi Ishikawa, who assisted Smith in Minamata, has since exhibited his own photographs documenting the disease. His photographs cover the years 1971 to the present day with Minamata victims as his subjects (Hirano 2012a).

Fig. 25.5 Memorial at the Minamata disease municipal museum (Source: English Wikipedia)

The prominent Japanese documentary filmmaker Noriaki Tsuchimoto made a series of films, starting with *Minamata: The Victims and Their World* (1971), documenting the incident and siding with the victims in their struggle against Chisso and the government.

In Popular Culture

Toshiko Akiyoshi, touched by the plight of the fishing village, wrote a jazz suite, "Minamata," that was to be the central piece of the Toshiko Akiyoshi–Lew Tabackin Big Band's 1976 album on RCA, *Insights*. The piece was constructed in three parts, to musically reflect the tragedy – "Peaceful Village," "Prosperity & Consequence," and "Epilogue." Akiyoshi used Japanese vocalists to sing the Japanese lyrics of a tone poem that were part of the composition. The album won many awards in jazz circles, including *Downbeat*'s best album award, largely on the strength of this piece, which brought some further attention on the tragedy (Scott Yanow 1978; Insights (Toshiko Akiyoshi–Lew Tabackin Big Band)).

The song "Kepone Factory" on Dead Kennedys' *In God We Trust, Inc.*, makes reference to the disaster in its chorus.

The song "The Disease of the Dancing Cats" by the band Bush on "The Science of Things" album is in reference to the disaster (Fig. 25.5).

Minamata Disease Today

Minamata disease remains an important issue in contemporary Japanese society. Lawsuits against Chisso and the prefectural and national governments are still continuing, and many regard the government responses to date as inadequate (Minamata's latest chapter, 2006). The company's "historical overview" in its current website makes no mention of their role in the mass contamination of Minamata and the dreadful aftermath. Their 2004 Annual Report however reports an equivalent of about USD50 million (5,820 million yen) in "Minamata Disease Compensation Liabilities." From 2000 to 2003, the company also reported total compensation liabilities of over US$170 million. Their 2000 accounts also show that the Japanese and Kumamoto prefectural governments waived an enormous US$560 million in related liabilities. Their FY2004 and FY2005 reports refer to Minamata disease as "Mad Hatter's Disease," a term coined from the mercury poisoning experienced by hatmakers of the last few centuries (cf Mad Hatter) (FY2004 and FY2005 financial results).

A memorial service was held at the Minamata Disease Municipal Museum on May 1, 2006, to mark 50 years since the official discovery of the disease. Despite bad weather, the service was attended by over 600 people, including Chisso Chairman Shunkichi Goto and Environment Minister Yuriko Koike (Memorial service marks Minamata tragedy's 50th year, 2006).

On Monday, March 29, 2010, a group of 2,123 uncertified victims reached a settlement with the government of Japan, the Kumamoto Prefectural government, and Chisso Corporation to receive individual lump-sum payments of 2.1 million yen and monthly medical allowances (Asahi Shimbun News 2010; Hirano 2012b).

Most congenital patients are now in their 40s and 50s and their health is deteriorating. Their parents, who are often their only source of care, are into their 70s or 80s or already deceased. Often these patients find themselves tied to their own homes and the care of their family, effectively isolated from the local community. Some welfare facilities for patients do exist. One notable example is *Hot House* (ほっとはうす *Hotto Hausu*?), a vocational training center for congenital patients as well as other disabled people in the Minamata area. Hot House members are also involved in raising awareness of Minamata disease, often attending conferences and seminars as well as making regular visits to elementary schools throughout Kumamoto Prefecture (Takeko Kato 2006).

References

Asahi Shimbun News (2010) Agreement reached to settle Minamata suit, 31 March 2010. Retrieved 1 Apr 2010

George TS (2001) Minamata: pollution and the struggle for democracy in postwar Japan. Harvard University Asia Center, Cambridge. ISBN 0-674-00785-9

Hightower J (2008) Diagnosis mercury: money, politics and poison. Island Press, Washington, DC, p 77

Hirano K (2012a) Life with Minamata disease in photos, Japan Times, 15 Nov 2012, p 3

Hirano K (2012b) Kyodo news, Mercury pact falls short on Minamata, Japan Times, 1 Mar 2012, p 3

Nicol CW (2012) Minamata: a saga of suffering and hope. Japan Times, p 10

Takeko Kato (2006) Advanced welfare should arise from Minamata. Asahi Shimbun, 10 May 2006. Retrieved 29 Oct 2006

Withrow SJ, Vail DM (2007) Withrow and MacEwen's small animal clinical oncology. Elsevier, St. Louis, pp 73–74. ISBN 0721605583

Yanow S (1978) Record Review Magazine. 2(3). Downbeat. http://www.downbeat.com/default.asp?sect=stories&subsect=story_detail&sid=704; liner notes for Toshiko Akiyoshi/Lew Tabackin Big Band LP, Insights 1976 RCA VICTOR AFL1-2678 STEREO

Methylmercury and Alzheimer's Disease

26

Alzheimer's is the cleverest thief, because she not only steals from you, but she steals the very thing you need to remember what's been stolen.

(Jarod Kintz)

Methylmercury, a pollutant produced by various industrial activities, is a potent neurotoxin that has now caused serious contamination issues within our oceans. As a fat-soluble molecule, methylmercury enters the food chain and accumulates in the flesh of the fish that then may end up in our supermarkets. Consuming larger, longer living fish on a regular basis is now known to pose a serious health hazard, especially for children and pregnant women who are consequently advised to limit (or even avoid) the intake of some species such as fresh tuna or marlin.

The accumulation of mercury within the body can have profound long-term effects on the nervous system and has been linked to a variety of conditions including Alzheimer's disease where it is believed to play a part in nerve cell death. Lipoproteins, such as high-density lipoprotein (HDL) and low-density lipoprotein (LDL), are combinations of lipids (fat) and proteins that function to transport fat around via the blood, a function that is generally associated with cholesterol and, therefore, cardiovascular health. However, approximately one in seven people carries a gene that causes their body to produce a particular lipoprotein called apoE4, known to play a significant role in the development of Alzheimer's disease. Those who inherit the apoE4 gene from one parent are three times more likely than average to develop Alzheimer's disease, with those who inherit the gene from both parents having a tenfold risk of developing the disease (Donix et al. 2010). There are multiple hypotheses as to why those carrying the apoE4 gene are more likely to develop Alzheimer's than those who carry the apoE3 or apoE2 genes; one such hypothesis regards the role that these lipoproteins play in mercury transport within the body, as mercury accumulation in the brain has been linked to the progression of Alzheimer's. Like all proteins, apolipoprotein is made of chains of amino acids. Cysteine is of particular relevance, as this amino acid contains sulfur, a member of a class of substances called "mercaptans," the Latin name for "mercury capture." Because apoE2, the protective form of apoE, contains two cysteine amino acids, it is particularly efficient at removing mercury from the system. In contrast, apoE3 has only one cysteine, and apoE4 none, making it the most ineffective at removing excess mercury from the body.

Given that fish oils are thought to offer protection against neuronal death and therefore the onset of dementia, it seems that ingesting them in high doses may negate any beneficial therapeutic effects unless they are highly purified to ensure all heavy metals are removed. The growing omega 3 market means there are more products of differing qualities and strengths, and the processes used to isolate and purify oils can also

S. Nabi, *Toxic Effects of Mercury*, DOI 10.1007/978-81-322-1922-4_26,
© Springer India 2014

differ quite significantly. It would certainly be advisable to choose fish oil supplements that have been purified under pharmaceutical grade conditions to ensure the product not only offers the best possible health benefits but can also guarantee to be contaminant free.

Alzheimer's disease (AD) rarely occurs in early forms between the age of 30 and 65 (5–10 %) and frequently in late forms above the age of 65. On average the duration of the disease is 6–10 years, although duration of survival decreases with increasing age. In the USA, Alzheimer's disease causes costs abounding to an estimated 90 billion dollars (Ernst and Hay 1997). It ranks fourth among all death causes, meanwhile infesting 4.5 million citizens (Helmuth 2003). According to estimations, a total of 16 million individuals will be affected by the year 2050 (Helmuth 2003; Brookmeyer and Gray 2000). In recent years, the incidence of Alzheimer's disease has been on the rise. At least 30–50 % of all individuals above the age of 85 are affected in industrialized countries (Breteler et al. 1992). With ever increasing life spans, Alzheimer's disease will be one of the major public health problems of coming decades.

The central pathogenetic mechanism is neurodegeneration and inflammatory processes, which in turn produce oxidative stress that accelerates neuron damage. The neurodegeneration starts with a hyperphosphorylation of the tau protein due to as yet unknown reasons. This in turn leads to a breakdown of microtubules which form the cytoskeleton of the neuron and are essential for the neuron's metabolism and functioning. The vital processes of the neuron are disturbed and finally neuron death ensues. The deposits cannot be adequately removed and form neurofibrillary tangles which in turn accelerate the inflammatory cascade and the positive feedback circle that leads to the progression of the disease. Nerve cell degeneration produces damages in the cholinergic projective systems of the basal prefrontal brain, in the entorhinal cortex, and the hippocampus at early stages (Dickson 2001; Wenk 2003; Arendt 2002). The neuronal losses are highest in the nucleus basalis of Meynert (NBM) and reach more than 90 % at advanced affection stages (Braak et al. 2000; Sassin et al. 2000). Due to the concomitant reduction of the cholinergic activity of the cerebrum, which normally determines the activity status of the cortex, memory performance is significantly impaired despite the fact that the cerebral cortex does not show much damage (Dickson 2001; Wenk 2003; Arendt 2002). In the course of Alzheimer's disease, considerable and unusual amounts of extracellular protein accretions are traceable. Fiber mass consists of insoluble amyloid-b-protein (Ab). Therefore, it cannot be removed by antibodies. Accretion of Ab causes induction of inflammatory processes and an increased creation of free oxygen radicals which are further enhanced through elevated homocysteine and metals (White et al. 2001; Butterfield 2002; Jellinger 2003; Bush 2003a, b). This might explain the neurotoxicity of amyloid accretions.

The cause of Alzheimer's disease is yet unknown. About 3–5 % of all cases are genetically determined, suggesting a multicausal model for the disease. Studies on migration suggest that exogenous factors might be responsible for triggering this pathological positive feedback circle (Hendrie et al. 1995; Grant 1999; Osuntokun et al. 1992). The amount of neurofibrillary tangles found in eminently affected brain regions in Alzheimer's disease correlates with the severity of Alzheimer's disease (Braak et al. 1995, 1999, 1997; Dickson et al. 1992). Minor neurofibrillary nerve cell changes may occur as early as 50 years before onset of clinical symptoms (Braak et al. 1999). Thus, age is not the cause but only one factor for its clinical manifestation (Braak et al. 1997). Interestingly, neurofibrillary tangles in low amounts are already found in about 20 % of individuals aged 20–30 years without clinical symptoms of Alzheimer's disease (Braak et al. 1997). In the age group 70–80 years, 90 % of the individuals display neurofibrillary tangles in their brains. In this cohort, 35 % have highest numbers of histological detectable neurofibrillary tangles and subsequently suffer clinically detectable from Alzheimer's disease (Braak et al. 1997). Thus, if an exogenous factor contributes to the development of neurofibrillary tangles and consequently Alzheimer's disease, this factor must be present in a great portion of the public only in industrial developed countries. In the past 20 years, a number of studies were published suggesting a potential pathogenetic role of inorganic mercury in Alzheimer's disease.

Major human sources of mercury include fish consumption (Clarkson et al. 2003; Schober et al. 2003), dental amalgams (Mutter et al. 2004), and vaccines (Clarkson et al. 2003). Regular fish consumption and intake of omega 3 fatty acids reduce the risk of developing Alzheimer's disease (Grant 1997; Grant et al. 2002; Clarke et al. 1998; Kalmijn et al. 1997; Morris et al. 2003). Selenium, which is found in fish, is essential for the function of glutathione peroxidase regenerating glutathione, which is an important antioxidant and detoxification enzyme. Furthermore, selenium disposes mercury directly by tightly binding it to mercury selenite, which is nontoxic. This is shown by autopsy studies that assess the ratio of selenium and mercury in several organs (Drasch et al. 2000). Methylmercury in fish, being bound to cysteine, appears to be by far less toxic than hitherto assumed and is about 20 times less toxic than methylmercury chloride usually used in experiments (Harris et al. 2003). For that reason, methylmercury found in fish seems not to be involved in the pathogenesis of Alzheimer's disease. Inorganic mercury (found in dental amalgam) or ethylmercury (found in vaccines) may play a major role.

Experimental Mercury Effects and Alzheimer's Disease

Inhibition and Deterioration of Neurotubulin

It was shown that both organic (Falconer et al. 1994) and inorganic mercury (Duhr et al. 1993) cause those biochemical changes in tubulin structures which can be found in brains of patients with Alzheimer's disease (Duhr et al. 1993). In healthy human brain tissue cultures, only mercury, even in lower concentrations, but not aluminum, lead, zinc, or iron, was able to inhibit binding to guanosine triphosphate (GTP), which is necessary for tubulin synthesis and thus for neuron function (Duhr et al. 1993). Mercury inhibits ADP-ribosylation of tubulin and actin (Palkiewicz et al. 1994). This process leads to an inhibition of polymerization of tubulin to microtubulin. As a result, neurofibrillary tangles and senile plaques are formed. Living rats exposed to mercury vapor (250 + 300 μg/m^3) four times a day exhibit the same molecular changes in their brain tissue as those caused in human brain cell cultures after 14 days (Pendergrass et al. 1997). These changes are similar to those found postmortem in brains of patients with Alzheimer's disease (Duhr et al. 1993; Pendergrass and Haley 1995, 1996). Tubulin is assumed to be the most vulnerable protein for mercury, because administration of very low doses of inorganic mercury does not inhibit other GTP- or ATP-binding proteins (Pendergrass and Haley 1995, 1996). Tubulin has at least 14 sulfhydryl groups which bind mercury with high affinity resulting in functional losses of tubulin and creation of neurofibrillary tangles. Since human nerve cells do not regenerate, any blocking of neurotubulin is particularly grave.

Creation of Neurofibrillary Tangles and Amyloid

Administration of very low doses of inorganic mercury (0.18 μM) has been shown to promote hyperphosphorylation of tau protein in neuronal cell cultures within 24 h (Olivieri et al. 2000). Hyperphosphorylation of tau is the first biochemical change to be observed in the development of Alzheimer's disease and results in formation of neurofibrillary tangles and failure of nerve cell functions. Administration of mercury to nerve cells provokes also production of amyloid-b 40–42 (Olivieri et al. 2000).

Glutathione Consumption and Increased Oxidative Stress

Within 30 min., low doses of inorganic mercury reduce glutathione concentration by increasing oxidative stress in a cell culture model. Addition of melatonin is able to protect the nerve cells from the damaging impact of mercury (Olivieri et al. 2000). Melatonin is an antioxidant and in addition has the ability to bind and eliminate metals (Limson et al. 1998). Although cobalt also has been reported to decrease glutathione concentration in neuronal cell cultures and to release secretion of amyloid-b (Olivieri et al. 2002), it is not able to hyperphosphorylate tau protein and build up neurofibrillary tangles (Olivieri et al. 2001). Changes brought about by cobalt were only observable above concentrations 1,700 higher than those of mercury (300 µM cobalt vs. 180 nM mercury) (Olivieri et al. 2001, 2002).

Neurodegeneration Through Mercury

Leong et al. 2001 demonstrated axon degeneration and formation of neurofibrillary tangles in animal neuronal cell cultures within minutes and with lowest amounts of inorganic mercury (2 µl 100 nM in 2 ml neuronal cell culture nourishing solution). This neurodegenerative effect was not demonstrable with other metals like aluminum, lead, cadmium, or manganese (Leong et al. 2001). In neuronal stem cells, inorganic mercury of 2 and 5 µg/ml impaired tubulin functions for 48 h (Cedrola et al. 2003). It caused apoptosis, programmed cell death of nerve cells, and induced expression of heat shock proteins (Cedrola et al. 2003).

Comparison with Mercury Concentrations in Human Brain Tissues

Mercury load in the brain of patients with Alzheimer's disease was specified at 20 and 178 ng/g (Ehmann et al. 1986; Thompson et al. 1988; Saxe et al. 1999). This amounts to a molar mercury concentration of 0.1–0.89 µmol. In the above-mentioned experimental studies on nerve cells, exclusive administration of mercury of a final concentration of 0.0001 µMol (2 µl 0.1 µMolar mercury in 2 ml nourishing solution) resulted in axon degeneration and creation of neurofibrillary tangles (Leong et al. 2001). Addition of 0.18 µmol mercury leads to secretion of amyloid-b 40 and 42, increased oxidative stress, and hyperphosphorylation of the tau protein (Olivieri et al. 2000, 2002).

Increase of Glutamate Toxicity

It is assumed that toxicity of the excitatory neurotransmitter glutamate plays a role in neuronal death in neurodegenerative diseases (Bleich et al. 2003). Glutamate is toxic when it accumulates and when protective mechanisms fail. One such protective mechanism is the enzyme glutamine synthetase primarily found in astrocytes (Robinson 2001). Mercury inhibits its reuptake of glutamate in the astrocytes and other cells of the nervous system (Aschner et al. 2000; Brookes 1992) resulting in extracellular accumulation of glutamate. In addition, mercury and lead inhibit the enzyme glutamine synthetase which converts glutamate to nontoxic glutamine (Sierra and Tiffany-Castiglioni 1991). Inorganic mercury (Hg++) appears to inhibit GS to a larger degree than methylmercury (Allen et al. 2001). It has been shown that glutamine synthetase is reduced in the brain of patients with Alzheimer's disease (Robinson 2001; Butterfield et al. 1997), whereas the concentration of glutamine synthetase in the liquor is increased (Gunnersen and Haley 1992). Glutamine synthetase concentration in the liquor, stemming from enhanced degradation of astrocytes having a high concentration of glutamine synthetase, could thus have diagnostic relevance for Alzheimer's disease (Gunnersen and Haley 1992; Tumani et al. 1999).

Enzyme Inhibition

Creatine kinase (CK) is an enzyme crucial for energy production in all body cells. Its function is reduced in patients with Alzheimer's disease (David et al. 1998). Since it possesses many sulfhydryl groups, similar to glutamine synthetase and tubulin, mercury inhibits its functions (Haley 2002).

With respect to Alzheimer's disease, protein kinase plays an important role in the production of normal amyloid. Whenever these are inhibited, the enzyme b-secretase modulates the metabolisms of APP (amyloid precursor protein) in a way that amyloid-b is increasingly produced (Olivieri et al. 2000). Protein kinase C is inhibited by mercury both in vitro and in the brain tissue (Matsuoka et al. 2000; Rajanna et al. 1995).

Synergistic Effects of Other Metals

The Alzheimer's disease-typical neuronal changes (hyperphosphorylation of tau protein, occurrence of neurofibrillary tangles, amyloid-b, tubulin inhibition, axon degeneration, increase of glutamine synthetase in the liquor) found in nerve cells and animals may not be caused by other metals (lead, cadmium, aluminum, copper, zinc, iron, chrome, manganese), but other metals may potentize mercury effects by contributing to oxidative stress (Haley 2002; Duhr et al. 1993; Olivieri et al. 2001; Leong et al. 2001).

Experimental Effects of Estrogens

Estrogen is able to compensate the damaging effects of mercury in a cell model, when it is concurrently administered (Olivieri et al. 2002). This could provide an explanation for the findings of some studies showing reduced risk of developing Alzheimer's disease with high-dose estrogen replacement (Olivieri et al. 2002; Henderson 2000; Henderson et al. 1994; Paganini-Hill and Henderson 1994), which, however, is not a consistent finding (Rapp et al. 2003).

Mercury in Brains of Patients with Alzheimer's Disease

Ehmann et al. 1986 examined 81 brain specimens from 14 patients with Alzheimer's disease and 147 specimens from 28 controls with the same age. Out of 17 target elements, the biggest differences were found for mercury and bromine levels in the cerebral brain tissue of patients with Alzheimer's disease (3.4 ± 3.7 ng/g vs. 17.5 ± 1.3 ng/g, $p < 0.05$). In the gray matter, they found more mercury (patients with Alzheimer's disease: 42.7 ng/g vs. 14.7 ng/g; controls: 29.0 ng/g vs. 20.5) (Ehmann et al. 1986). In the nucleus basalis of Meynert, mercury concentration was four times higher in 14 patients with Alzheimer's disease compared with 15 controls (Thompson et al. 1988). Other elements were significantly increased, too (iron, sodium, and zinc) (Thompson et al. 1988). In tissue specimens from temporal lobes of 10 patients with Alzheimer's disease and 12 controls, there were significant increases of mercury concentrations in the microsomes of the brain cells and nonsignificantly increased mercury values in other brain fractions (temporal lobe, mitochondria, and cell nuclei) (Wenstrup et al. 1990). The total mercury content in the temporal lobes of the patients with Alzheimer's disease was 176 ng/g, as compared to 69.6 ng/g of the controls (Wenstrup et al. 1990).

In the pituitary, significant differences were found for mercury content between 43 patients with Alzheimer's disease and 15 controls (Cornett et al. 1998). Other studies found nonsignificantly increased mercury values in the olfactory region of the amygdala and amygdala and hippocampus, respectively (Cornett et al. 1998), but nonsignificantly decreased mercury values in the cerebellum and rhinencephalon (Cornett et al. 1998). Another study found no elevated mercury levels in brains of patients with Alzheimer's disease compared to controls (Fung et al. 1997).

Saxe et al. 1999 were unable to find significant mercury increases in an autopsy study of specific brain regions of 68 patients with Alzheimer's disease (mercury on average 20.3–61.1 ng/g depending on the area) when compared to 33 controls with the same age (mercury on average 30.1–88.9 ng/g). Also, there was neither a correlation between number and duration of amalgam filling and mercury concentration in the Alzheimer's disease group nor in the control group (Saxe et al. 1999). This is astonishing because other human autopsy studies show such correlations (Drasch et al. 1994, 1992; Guzzi et al. 2002;

Nylander 1986; Nylander et al. 1987; Nylander and Weiner 1991; Eggleston and Nylander 1987; Weiner and Nylander 1993). Interestingly, in patients unaffected by Alzheimer's disease, mercury concentration in the olfactory region was twice as high as in controls (88.9 ng/g vs. 41.7 g/ng) (Saxe et al. 1999). In sum, autopsy studies examining mercury load in the brains of patients with Alzheimer's disease, although suggestive, are not consistent. One potential confounding factor might be the loosely defined staging of patients with Alzheimer's disease when autopsied.

Mercury in Living Patients with Alzheimer's Disease

Mercury blood concentration in 33 patients with Alzheimer's disease was twice as high (2.64 μg/L) as that of 45 depressed patients (1.20 μg/L) and 65 patients without psychiatric disorders (1.09 μg/L) ($p<0.0005$) (Hock et al. 1998). In the early form of Alzheimer's disease (13 persons younger than 65 years), difference was even higher (3.32 μg/L, $p=0.0002$). Correlation with blood mercury concentration and the amyloid-b-protein (Ab) in the liquor cerebrospinalis was significant ($p=0.0015$). These findings were confirmed (Basun et al. 1991), yet disconfirmed (Fung et al. 1995). The average mercury load in the urine of nine patients with Alzheimer's disease (2.96 ± 1.13 μg/L) as compared to nine controls (1.86 ± 0.9 μg/L) was not significantly different (Fung et al. 1995). This obviously is a power problem, as the effect size of this difference (standardized mean difference) is 1.1 and thus large.

Several studies measuring the nails and hair of patients with Alzheimer's disease found less mercury in the patients compared to controls (Vance et al. 1988, 1990). This could be attributable to the fact that nail and hair merely reflect a recent mercury exposure. Patients with Alzheimer's disease would be less exposed to environmental mercury than controls because they are accommodated in foster homes (Vance et al. 1990). Mercury concentrations in the blood may only be elevated when nerve cell damage is greatest (presumably at intermediate stages), suggesting a curvilinear association between Alzheimer's disease and mercury excretion, a possibility that researchers have eschewed so having normally supposed and looked for a linear relationship.

Dental Condition and the Risk of Developing Alzheimer's Disease

A recent analysis of 10,263 individuals from Canada yielded a distinct association between dental condition and the risk for Alzheimer's disease. The fewer the number of teeth, the higher the risk (Lund et al. 2003). The authors took this as evidence for the fact that amalgam fillings are not causal for Alzheimer's disease (Lund et al. 2003). Alzheimer's disease takes about 30–50 years to clinically manifest itself (Braak et al. 1997). Patients with fewer teeth previously had poorer dental conditions and had therefore presumably been provided with mercury containing amalgam over a longer period. Thus, they are likely to have been exposed to mercury vapor in a vulnerable phase and to a larger extent than persons still having teeth in advanced years.

Metal Chelation as Potential Therapy of Alzheimer's Disease?

If mercury is involved in the genesis of Alzheimer's disease, preventive and possibly therapeutic strategies may be developed provided neurodegeneration has not progressed too much.

It has been suggested that aluminum and iron play an important role in Alzheimer's disease (Kruck 1993). Therefore, the chelator desferrioxamine, which has the capacity to bind iron, aluminum, and to a lesser degree mercury, was tested in clinical trials (McLachlan et al. 1993; Crapper McLachlan et al. 1991). Recently, clioquinol, formerly approved as an antibiotic in Japan, has been successfully applied in animal and clinical studies (Bush 2002; Helmuth 2002) to treat Alzheimer's disease due to its capacity as a chelating agent to bind copper and zinc. Chelating agents, which bind copper and zinc usually also, have the capacity to bind mercury.

Implications

Converging findings from experimental, epidemiological, and clinical studies identify inorganic mercury as one of the potential exogenous factors responsible for Alzheimer's disease. The main source for inorganic mercury in habitants of industrial developed countries is dental amalgam (Mutter et al. 2004; Lorscheider et al. 1995). Other metals and noxes might have synergistic effects with mercury. Due to the complex relationships and restricted technical measurement equipment in some studies, individual findings appear contradictory and a lot remains to be clarified. This is a situation analogous to other hotly debated areas, like the association of smoking and cancer or estrogen replacement and myocardial infarction.

Obviously, definite knowledge about the causal role of mercury in Alzheimer's disease may only be derived from large, long-term prospective epidemiological studies examining occurrence of Alzheimer's disease in subjects exposed to the risk of mercury in amalgam and other sources, compared with those at lower risk. Since apolipoprotein E could constitute an important protective or risk factor, it should be monitored in future studies. Additionally, clinical studies with chelating agents measuring mercury excretion could give indirect evidence and would offer a therapeutic strategy for Alzheimer's disease at early stages. The findings reviewed here should have made plausible that the potential association between mercury and Alzheimer's disease should be one of prime importance, since the public health impact is enormous.

References

Allen JW, Mutkus LA, Aschner M (2001) Mercuric chloride, but not methylmercury, inhibits glutamine synthetase activity in primary cultures of cortical astrocytes. Brain Res 891:148–157

Arendt Th (2002) Neuronale pathologie. In: Beyreuther K, Einhäupl KM, Förstl H, Kurz A (eds) Demenzen. Thieme, Stuttgart, pp 106–117

Aschner M, Yao CP, Allen JW, Tan KH (2000) Methylmercury alters glutamate transport in astrocytes. Neurochem Int 37:188–206

Basun H, Forssell LG, Wetterberg L, Winblad B (1991) Metals and trace elements in plasma and cerebrospinal fluid in normal aging and Alzheimer's disease. J Neural Transm [P-D Sect] 3:231–258

Bleich S, Romer K, Wiltfang J, Kornhuber J (2003) Glutamate and the glutamate receptor system: a target for drug action. Int J Geriatr Psychiatry 18:S33–S40

Braak H, Braak E, Yilmazer D, Bohl J (1995) Age-related changes of the human cerebral cortex. In: Cruz-Sanchez FF, Ravid R, Cuzner ML (eds) Neuropathological diagnostic criteria for brain banking. Biomedical Health Research IOS Press, Amsterdam, pp 14–19

Braak H, Griffing K, Braak E (1997) Neuroanatomy of Alzheimer's disease. Alzheimer's Res 3:235–247

Braak E, Griffing K, Arai K, Bohl J, Bratzke H, Braak H (1999) Neuropathology of Alzheimer's disease: what is new since A. Alzheimer? Eur Arch Psychiatry Clin Neurosci 249:14–22

Braak H, Del Tredici K, Schultz C, Braak E (2000) Vulnerability of select neuronal types to Alzheimer's disease. Ann N Y Acad Sci 924:53–61

Breteler MM, Claus JJ, van Duijn CM, Launer LJ, Hofman A (1992) Epidemiology of Alzheimer's. Epidemiol Rev 14:59–82

Brookes N (1992) In vitro evidence for the role of glutamate in the CNS toxicity of mercury. Toxicology 76:245–256

Brookmeyer R, Gray S (2000) Methods for projecting the incidence and prevalence of chronic diseases in aging populations: application to Alzheimer's. Stat Med 19:1481–1493

Bush AI (2002) Metal complexing agents as therapies for Alzheimer's disease. Neurobiol Aging 23:1031–1038

Bush AI (2003a) Copper, zinc, and the metallobiology of Alzheimer's disease. Alzheimer Dis Assoc Disord 17:147–150

Bush AI (2003b) The metallobiology of Alzheimer's disease. Trends Neurosci 26:207–214

Butterfield DA (2002) Amyloid beta-peptide (1-42)-induced oxidative stress and neurotoxicity: implications for neurodegeneration in Alzheimer's disease brain. A review. Free Radic Res 36:1307–1313

Butterfield DA, Hensley K, Cole P et al (1997) Oxidatively induced structural alteration of glutamine synthetase assessed by analysis of spin label incorporation kinetics: relevance to Alzheimer's disease. J Neurochem 68:2451–2457

Cedrola S, Guzzi G, Ferrari D et al (2003) Inorganic mercury changes the fate of murine CNS stem cells. FASEB J 17:869–871

Clarke R, Smith AD, Jobst KA, Refsum H, Sutton L, Ueland PM (1998) Folate, vitamin B12, and serum total homocysteine levels in confirmed Alzheimer disease. Arch Neurol 55:1449–1455

Clarkson TW, Magos L, Myers GJ (2003) The toxicology of mercury-current exposures and clinical manifestations. N Engl J Med 349:1731–1737

Cornett CR, Markesbery WR, Ehmann WD (1998) Imbalances of trace elements related to oxidative damage in Alzheimer's disease brain. Neurotoxicology 19:339–346

Crapper McLachlan DR, Dalton AJ, Kruck TP et al (1991) Intramuscular desferrioxamine in patients with Alzheimer's disease. Lancet 337:1304–1308

David S, Shoemaker M, Haley BE (1998) Abnormal properties of creatine kinase in Alzheimer's disease brain: correlation of reduced enzyme activity and active site photolabeling with aberrant cytosol-membrane partitioning. Brain Res Mol Brain Res 54:276–287

Dickson DW (2001) Neuropathology of Alzheimer's disease and other dementias. Clin Geriatr Med 17:209–228

Dickson DW, Crystal HA, Mattiace LA et al (1992) Identification of normal and pathological aging in prospectively studied nondemented elderly humans. Neurobiol Aging 13:179–189

Donix M, Poettrich K, Weiss PH, Werner A, von Kummer R, Fink GR, Holthoff VA (2010) Age-dependent differences in the neural mechanisms supporting long-term declarative memories. Arch Clin Neuropsychol 25:383–395

Drasch G, Schupp I, Riedl G, Günther G (1992) Einfluß von Amalgamfüllungen auf die Quecksilberkonzentration in menschlichen Organen. Dtsch Zahnarztl Z 47:490–496

Drasch G, Schupp I, Hofl H, Reinke R, Roider G (1994) Mercury burden of human fetal and infant tissues. Eur J Pediatr 153:607–610

Drasch G, der Mail S, Schlosser C, Roider G (2000) Content of non-mercury-associated selenium in human tissues. Biol Trace Elem Res 77:219–230

Duhr EF, Pendergrass JC, Slevin JT, Haley BE (1993) HgEDTA complex inhibits GTP interactions with the E-site of brain beta-tubulin. Toxicol Appl Pharmacol 122:273–280

Eggleston DW, Nylander M (1987) Correlation of dental amalgam with mercury in brain tissue. J Prosthet Dent 58:704–707

Ehmann WD, Markesbery WR, Alauddin M, Hossain TI, Brubaker EH (1986) Brain trace elements in Alzheimer's disease. Neurotoxicology 7:195–206

Ernst RL, Hay JW (1997) Economic research on Alzheimer disease: a review of the Alzheimer. Dis Assoc Disord 11(Suppl 6):135–145

Falconer MM, Vaillant A, Reuhl KR, Laferriere N, Brown DL (1994) The molecular basis of microtubule stability in neurons. Neurotoxicology 15:109–122

Fung YK, Meade AG, Rack EP et al (1995) Determination of blood mercury concentrations in Alzheimer's patients. J Toxicol Clin Toxicol 33:243–247

Fung YK, Meade AG, Rack EP, Blotcky AJ (1997) Brain mercury in neurodegenerative disorders. J Toxicol Clin Toxicol 35:49–54

Grant WB (1997) Dietary Links to Alzheimer's disease. Alzheimer's Disease Review 2:42–55

Grant WB (1999) Dietary links to Alzheimer's disease: 1999 update. J Alzheimer's Dis 1:197–201

Grant WB, Campbell A, Itzhaki RF, Savory J (2002) The significance of environmental factors in the etiology of Alzheimer's disease. J Alzheimers Dis 4:179–189

Gunnersen D, Haley B (1992) Detection of glutamine synthetase in the cerebrospinal fluid of Alzheimer diseased patients: a potential diagnostic biochemical marker. Proc Natl Acad Sci U S A 89:11949–11953

Guzzi G, Grandi M, Cattaneo C (2002) Should amalgam fillings be removed? Lancet 360:2081

Haley B (2002) The relationship of toxic effects of mercury to exacerbation of the medical condition classified as Alzheimer's disease. http://www.fda.gov/ohrms/dockets/dailys/02/Sep02/091602/80027dd5.pdf. Accessed 16 May 2004

Harris HH, Pickering IJ, George GN (2003) The chemical form of mercury in fish. Science 301:1203

Helmuth L (2002) New therapies. New Alzheimer's treatments that may ease the mind. Science 297:1260–1262

Helmuth L (2003) Detangling Alzheimer's disease. New insights into the biological bases of the most common cause of dementia are pointing to better diagnostics and possible therapeutics. Sci Aging Knowledge Environ 2003:oa2

Henderson VW (2000) Oestrogens and dementia. Novartis Found Symp 230:254–265

Henderson VW, Paganini-Hill A, Emanuel CK, Dunn ME, Buckwalter JG (1994) Estrogen replacement therapy in older women. Comparisons between Alzheimer's disease cases and nondemented control subjects. Arch Neurol 51:896–900

Hendrie HC, Osuntokun BO, Hall KS et al (1995) Prevalence of Alzheimer's disease and dementia in two communities: Nigerian Africans and African Americans. Am J Psychiatry 152:1485–1492

Jellinger KA (2003) General aspects of neurodegeneration. J Neural Transm Suppl 65:101–144

Kalmijn S, Launer LJ, Ott A, Witteman JC, Hofman A, Breteler MM (1997) Dietary fat intake and the risk of incident dementia in the Rotterdam Study. Ann Neurol 42:776–782

Leong CC, Syed NI, Lorscheider FL (2001) Retrograde degeneration of neurite membrane structural integrity of nerve growth cones following in vitro exposure to mercury. Neuroreport 12:733–737

Limson J, Nyokong T, Daya S (1998) The interaction of melatonin and its precursors with aluminium, cadmium, copper, iron, lead, and zinc: an adsorptive voltammetric study. J Pineal Res 24:15–21

Lorscheider FL, Vimy MJ, Summers AO (1995) Mercury exposure from "silver" tooth fillings: emerging evidence questions a traditional dental paradigm. FASEB J 9:504–508

Lund JP, Mojon P, Pho M, Feine JS (2003) Alzheimer's disease and edentulism. Age Ageing 32:228–229

Matsuoka M, Wispriyono B, Iryo Y, Igisu H (2000) Mercury chloride activates c-Jun N-terminal kinase and induces c-jun expression in LLC-PK1 cells. Toxicol Sci 53:361–368

McLachlan DR, Smith WL, Kruck TP (1993) Desferrioxamine and Alzheimer's disease: video home behavior assessment of clinical course and measures of brain aluminum. Ther Drug Monit 15:602–607

Morris MC, Evans DA, Bienias JL et al (2003) Consumption of fish and n-3 fatty acids and risk of incident Alzheimer disease. Arch Neurol 60:940–946

Mutter J, Naumann J, Sadaghiani C, Walach H, Drasch G (2004) Amalgam studies: disregarding basic principles of mercury toxicity. Int J Hyg Environ Health 205(8):435

Nylander M (1986) Mercury in pituitary glands of dentists. Lancet 1:442

Nylander M, Weiner J (1991) Mercury and selenium concentrations and their interrelations in organs from dental staff and the general population. Br J Ind Med 48:729–734

Nylander M, Friberg L, Lind B (1987) Mercury concentrations in the human brain and kidneys in relation to exposure from dental amalgam fillings. Swed Dent J 11:179–187

Olivieri G, Brack C, Muller-Spahn F, Stahelin HB, Herrmann M, Renard P, Brockhaus M, Hock C (2000) Mercury induces cell cytotoxicity and oxidative stress and increases beta-amyloid secretion and tau phosphorylation in SHSY5Y neuroblastoma cells. J Neurochem 74:231–236

Olivieri G, Hess C, Savaskan E et al (2001) Melatonin protects SHSY5Y neuroblastoma cells from cobalt-induced oxidative stress, neurotoxicity and increased beta-amyloid secretion. J Pineal Res 31:320–325

Olivieri G, Novakovic M, Savaskan E et al (2002) The effects of beta-estradiol on SHSY5Y neuroblastoma cells during heavy metal induced oxidative stress, neurotoxicity and beta-amyloid secretion. Neuroscience 113:849–855

Osuntokun BO, Hendrie HC, Ogunniyi AO et al (1992) Cross-cultural studies in Alzheimer's disease. Ethn Dis 2:352–357

Paganini-Hill A, Henderson VW (1994) Estrogen deficiency and risk of Alzheimer's disease in women. Am J Epidemiol 140:256–261

Palkiewicz P, Zwiers H, Lorscheider FL (1994) ADP-ribosylation of brain neuronal proteins is altered by in vitro and in vivo exposure to inorganic mercury. J Neurochem 62:2049–2052

Pendergrass JC, Haley BE (1995) Mercury-EDTA complex specifically blocks brain beta-tubulin- GTP interactions: similarity to observations in Alzheimer's disease. In: Friberg LT, Scrauzer GN (eds) Status quo and perspectives of amalgam and other dental materials. Georg Thieme Verlag, Stuttgart, pp 98–105

Pendergrass JC, Haley BE (1996) Inhibition of brain tubulin-guanosine 5′-triphosphate interactions by mercury: similarity to observations in Alzheimer's diseased brain. In: Sigel H, Sigel A (eds) Metal ions in biological systems, vol 34. Marcel Dekker, New York, pp 461–478

Pendergrass JC, Haley BE, Vimy MJ, Winfield SA, Lorscheider FL (1997) Mercury vapor inhalation inhibits binding of GTP to tubulin in rat brain: similarity to a molecular lesion in Alzheimer's diseased brain. Neurotoxicology 18:315–324

Rajanna B, Chetty CS, Rajanna S, Hall E, Fail S, Yallapragada PR (1995) Modulation of protein kinase C by heavy metals. Toxicol Lett 81:197–203

Rapp SR, Espeland MA, Shumaker SA et al (2003) Effect of estrogen plus progestin on global cognitive function in postmenopausal women: the women's health initiative memory study: a randomized controlled trial. JAMA 289:2663–2672

Robinson SR (2001) Changes in the cellular distribution of glutamine synthetase in Alzheimer's disease. J Neurosci Res 66:972–980

Sassin I, Schultz C, Thal DR et al (2000) Evolution of Alzheimer's disease-related cytoskeletal changes in the basal nucleus of Meynert. Acta Neuropathol 100:259–269

Saxe SR, Wekstein MW, Kryscio RJ et al (1999) Alzheimer's disease, dental amalgam and mercury. J Am Dent Assoc 130:191–199

Schober SE, Sinks TH, Jones RL et al (2003) Blood mercury levels in US children and women of childbearing age, 1999–2000. JAMA 289:1667–1674

Sierra EM, Tiffany-Castiglioni E (1991) Reduction of glutamine synthetase activity in astroglia exposed in culture to low levels of inorganic lead. Toxicology 65:295–304

Thompson CM, Markesbery WR, Ehmann WD, Mao YX, Vance DE (1988) Regional brain trace-element studies in Alzheimer's disease. Neurotoxicology 9:1–7

Tumani H, Shen G, Peter JB, Bruck W (1999) Glutamine synthetase in cerebrospinal fluid, serum, and brain: a diagnostic marker for Alzheimer disease? Arch Neurol 56:1241–1246

Vance DE, Ehmann WD, Markesbery WR (1988) Trace element imbalances in hair and nails of Alzheimer's disease patients. Neurotoxicology 9:197–208

Vance DE, Ehmann WD, Markesbery WR (1990) A search for longitudinal variations in trace element levels in nails of Alzheimer's disease patients. Biol Trace Elem Res 26–27:461–470

Weiner JA, Nylander M (1993) The relationship between mercury concentration in human organs and different predictor variables. Sci Total Environ 138:101–115

Wenk GL (2003) Neuropathologic changes in Alzheimer's disease. J Clin Psychiatry 64:7–10

Wenstrup D, Ehmann WD, Markesbery WR (1990) Trace element imbalances in isolated subcellular fractions of Alzheimer's disease. Brain Res 533:125–131

White AR, Huang X, Jobling MF et al (2001) Homocysteine potentiates copper- and amyloid beta peptide-mediated toxicity in primary neuronal cultures: possible risk factors in the Alzheimer's-type neurodegenerative pathways. J Neurochem 76:1509–1520

27 Methylmercury and Parkinson's Disease

With Parkinson's, it's like you're in the middle of the street and you're stuck there in cement shoes and you know a bus is coming at you, but you don't know when. You think you can hear it rumbling, but you have a lot of time to think. And so you just don't live that moment of the bus hitting you until it happens. There are all kinds of room in that space.

(Michael Fox)

There has been a huge increase in the incidence of degenerative neurological conditions in virtually all Western countries over the last two decades. The increase in Parkinson's and other motor neuron disease has been over 50 %. The primary cause appears to be increased exposures to toxic pollutants such as toxic metals, pesticides, etc., resulting in brain inflammation and oxidative damage of free radicals.

Dental amalgam fillings are the largest source of mercury in most people with daily exposures documented to commonly be above government health guidelines (World Health Organization 1991). This is due to continuous vaporization of mercury from amalgam in the mouth, along with galvanic currents from mixed metals in the mouth that deposit the mercury in the gums and oral cavity. Due to the high daily mercury exposure and excretion into home and business sewers of those with amalgam, dental amalgam is also the largest source of the high levels of mercury found in all sewers and sewer sludge and, thus, according to government studies a significant source of mercury in rivers, lakes, bays, fish, and crops. People also get significant exposure from vaccinations, fish, and dental office vapor.

When amalgam was placed into teeth of monkeys and rats, within 1 year mercury was found to have accumulated in the brain, trigeminal ganglia, spinal ganglia, kidneys, liver, lungs, hormone glands, and lymph glands. People also commonly get exposures to mercury and other toxic metals such as lead, arsenic, nickel, and aluminum from food, water, and other sources. All of these are highly neurotoxic and are documented to cause neurological damage which can result in chronic neurological conditions over time, as well as ADHD, mood, and behavioral disorders. A study found that those with occupational exposure to lead, arsenic, or copper have more than double the incidence of Parkinson's than normal.

Mercury is one of the most toxic substances in existence and is known to bioaccumulate in the body of people and animals that have chronic exposure. Mercury exposure is cumulative and comes primarily from four main sources: silver (mercury) dental fillings, food (mainly fish), vaccinations, and occupational exposure. Whereas mercury exposure from fish is primarily methylmercury and mercury from vaccinations is thimerosal (ethylmercury), mer-

S. Nabi, *Toxic Effects of Mercury*, DOI 10.1007/978-81-322-1922-4_27,
© Springer India 2014

cury from occupational exposure and dental fillings is primarily from elemental mercury vapor. Developmental and neurological conditions occur at lower levels of exposure from mercury vapor than from inorganic mercury or methylmercury. Mercury in amalgam fillings, because of its high vapor pressure and galvanic action with other metals in the mouth, has been found to be continuously vaporized and released into the body and has been found to be directly correlated to the number of amalgam surfaces and the largest source of mercury in the majority of people (World Health Organization 1991), typically between 60 and 90 % of the total. The level of daily exposure of those with several amalgam fillings commonly exceeds the US EPA health guideline for daily mercury exposure of 0.1 ug/kg body weight/day, and the oral mercury level commonly exceeds the mercury MRL of the US ATSDR of 0.2 ug/cubic meter of air. When amalgam fillings are replaced, levels of mercury in the blood, urine, saliva, and feces typically rise temporarily but decline between 60 and 90 % within 6–9 months.

The main factors determining whether chronic conditions are induced by metals appear to be exposure and genetic susceptibility, which determines individual's immune sensitivity and ability to detoxify metals. Very low levels of exposure have been found to seriously affect relatively large groups of individuals who are immune sensitive to toxic metals or have an inability to detoxify metals due to such as deficient sulfoxidation or metallothionein function or other inhibited enzymatic processes related to detoxification or excretion of metals II mechanisms by which mercury causes neurological conditions found in Parkinson's and neurodegenerative diseases.

Programmed cell death (apoptosis) is documented to be a major factor in degenerative neurological conditions like ALS, Alzheimer's, MS, Parkinson's, etc. Some of the factors documented to be involved in apoptosis of neurons and immune cells include inducement of the inflammatory cytokine tumor necrosis factor-alpha (TNFa), reactive oxygen species, and oxidative stress; reduced glutathione levels and liver enzyme effects; inhibition of protein kinase C and cytochrome P450; nitric oxide and peroxynitrite toxicity; excitotoxicity and lipid peroxidation; excess free cysteine levels; excess glutamate toxicity; excess dopamine toxicity; amyloid-beta generation (Olivieri et al. 2000); increased calcium influx toxicity; and DNA fragmentation (Aschner et al. 1997; Ariza et al. 1998) and mitochondrial membrane dysfunction. Mitochondrial DNA mutations or dysfunction is fairly common, found in at least 1 in every 200 people, and toxicity effects affect this population more than those with less susceptibility to mitochondrial dysfunction. This has been found to be a factor in conditions like Parkinson's. The mechanisms by which mercury causes (often synergistically along with other toxic exposures) all of these conditions and neuronal apoptosis will be documented.

TNFa (tumor necrosis factor-alpha) is a cytokine that controls a wide range of immune cell response in mammals, including cell death (apoptosis) in neuronal and immune cells. This process is involved in inflammatory and degenerative neurological conditions like ALS, MS, Parkinson's, rheumatoid arthritis, etc. Cell signaling mechanisms like sphingolipids are part of the control mechanism for the TNFa apoptosis mechanism. Glutathione is an amino acid that is a normal cellular mechanism for controlling apoptosis. When glutathione is depleted in the brain, reactive oxidative species increased, and CNS and cell signaling mechanisms are disrupted by toxic exposures such as mercury, neuronal cell apoptosis results and neurological damage.

> Mercury has been shown to induce TNFa and deplete glutathione, causing inflammatory effects and cellular apoptosis in neuronal and immune cells.

Mercury's biochemical damage at the cellular level includes DNA damage, inhibition of DNA, and RNA synthesis (Aschner et al. 1997; Ariza et al. 1998); alteration of protein structure (Aschner et al. 1997); alteration of the transport of calcium (Olivieri et al. 2000); inhibition of glucose transport and of enzyme function, protein transport, and other essential nutrient transport; induction of free radical formation; depletion of cellular glutathione (necessary for detoxification processes); inhibition of glutathione peroxidase enzyme; inhibition of glutamate uptake; inducement of peroxynitrite and lipid peroxidation damage, which causes abnormal migration of neurons in the cerebral cortex and immune system damage; and inducement of inflammatory cytokines.

Oxidative stress and reactive oxygen species (ROS) have been implicated as major factors in neurological disorders including stroke, Parkinson's disease (PD), Alzheimer's, ALS, etc. Mercury-induced lipid peroxidation has been found to be a major factor in mercury's neurotoxicity, along with leading to decreased levels of glutathione peroxidation and superoxide dismutase (SOD). Only a few micrograms of mercury severely disturb cellular function and inhibit nerve growth. Exposure to mercury results in metalloprotein compounds that have genetic effects, having both structural and catalytic effects on gene expression. Mercury inhibits sulfur ligands in MT and in the case of intestinal cell membranes inactivates MT that normally binds cuprous ions, thus allowing buildup of copper to toxic levels in many and malfunction of the Zn-Cu SOD function. Mercury also causes displacement of zinc in MT and SOD, which has been shown to be a factor in neurotoxicity and neuronal diseases. Some of the processes affected by such metalloprotein control of genes include cellular respiration, metabolism, enzymatic processes, metal-specific homeostasis, and adrenal stress response systems. Significant physiological changes occur when metal ion concentrations exceed threshold levels. Such metalloprotein formation also appears to have a relation to autoimmune reactions in significant numbers of people. Increased formation of reactive oxygen species (ROS) has also been found to increase formation of advanced glycation end products (AGEs) that have been found to cause activation of glial cells to produce superoxide and nitric oxide, and they can be considered part of a vicious cycle, which finally leads to neuronal cell death in the substantia nigra in PD.

Mercury exposure causes high levels of oxidative stress/reactive oxygen species (ROS), which has been found to be a major factor in apoptosis and neurological disease including dopamine or glutamate-related apoptosis. Mercury and quinones form conjugates with thiol compounds such as glutathione and cysteine and cause depletion of glutathione, which is necessary to mitigate reactive damage. Such conjugates are found to be highest in the brain substantia nigra with similar conjugates formed with L-Dopa and dopamine in Parkinson's disease. Mercury depletion of GSH and damage to cellular mitochondria and the increased lipid peroxidation in protein and DNA oxidation in the brain appear to be a major factor in Parkinson's disease. Exposure to mercury vapor and methylmercury is well documented to commonly cause conditions involving tremor and/or ataxia, with populations exposed to mercury experiencing tremor on average proportional to exposure level. One study found higher than average levels of mercury in the blood, urine, and hair of Parkinson's disease patients. Another study found blood and urine mercury levels to be very strongly related to Parkinson's with odds ratios of approx. 20 at high levels of Hg exposure. Other studies that reviewed occupational exposure data found that occupational exposure to manganese and copper has high odds ratios in relation to PD, as well as multiple exposures to these and lead, but one study noted that this effect was only seen for exposure of over 20 years. Occupational exposure to mercury has been found to cause Parkinson's. One study found the EDTA chelation was effective in reducing some of the effects.

Glutamate is the most abundant amino acid in the body and in the CNS acts as an excitory neurotransmitter, which also causes inflow of calcium. Astrocytes, a type of cell in the brain and CNS with the task of keeping clean the area around nerve cells, have a function of neutralizing excess glutamate by transforming it to glutamic acid. If astrocytes are not able to rapidly neutralize excess glutamate, then a buildup of glutamate and calcium occurs, causing swelling and neurotoxic effects. Mercury and other toxic metals inhibit astrocyte function in the brain and CNS, causing increased glutamate and calcium-related neurotoxicity which is responsible for much of the fibromyalgia symptoms. This is also a factor in conditions such as CFS, Parkinson's, and ALS.

Parkinson's disease involves the aggregation of alpha-synuclein to form fibrils, which are the major constituent of intracellular protein inclusions (Lewy bodies and Lewy neurites) in dopaminergic neurons of the substantia nigra. Occupational exposure to specific metals, especially manganese, copper, lead, iron, mercury, and aluminum, appears to be a risk factor for Parkinson's disease based on epidemiological studies. Elevated levels of several of these metals have also been reported in the substantia nigra of Parkinson's disease subjects.

Exposure to aluminum hydroxide in vaccines also appears to sometimes cause symptoms similar to Parkinson's or other neurological conditions.

Na (+), K (+)-ATPase is a transmembrane protein that transports sodium and potassium ions across cell membranes during an activity cycle that uses the energy released by ATP hydrolysis. Mercury is documented to inhibit Na (+), K (+)-ATPase function at very low levels of exposure (Rajanna et al. 1990). Studies have found that in Parkinson's cases, there was an elevation in plasma serum digoxin and a reduction in serum magnesium, RBC membrane Na (+)-K + ATPase activity. The activity of all serum free radical scavenging enzymes, concentration of glutathione, alpha-tocopherol, iron binding capacity, and ceruloplasmin decreased significantly in PD, while the concentration of serum lipid peroxidation products and nitric oxide increased. The inhibition of Na + -K + ATPase can contribute to increase in intracellular calcium and decrease in magnesium, which can result in (1) defective neurotransmitter transport mechanism, (2) neuronal degeneration and apoptosis, (3) mitochondrial dysfunction, and (4) defective Golgi body function and protein processing dysfunction. It is documented in this paper that mercury is a cause of most of these conditions seen in Parkinson's.

Many studies of patients with major neurological or degenerative diseases have found evidence amalgam fillings may play a major role in development of conditions such as Alzheimer's, ALS, MS, Parkinson's, etc. Mercury exposure causes high levels of oxidative stress/reactive oxygen species (ROS), which has been found to be a major factor in neurological disease. Mercury and quinones form conjugates with thiol compounds such as glutathione and cysteine and cause depletion of glutathione, which is necessary to mitigate reactive damage. Such conjugates are found to be highest in the brain substantia nigra with similar conjugates formed with L-Dopa and dopamine in Parkinson's disease. Mercury depletion of GSH and damage to cellular mitochondria and the increased lipid peroxidation in protein and DNA oxidation in the brain appear to be a major factor in Parkinson's disease.

An EKM system for evaluating nerve and muscle function ability using a set of five measures (precision, imprecision, tremor, Fitts' constant, and irregularity) and tested on a group of Cree Indians with mercury exposure from fish eating. Ninety-six participants, including 30 control subjects, 36 Cree subjects exposed to mercury, 21 subjects with Parkinson's disease, 6 with presumed cerebellar deficit, and 3 with essential tremor, participated in the study. An ANOVA on the three largest groups generated significant results for tremor, Fitts' constant, and irregularity between the Cree and the control subjects and on Fitts' constant and irregularity between the subjects with Parkinson's disease and the control subjects. Three subgroups of the same mean age composed of six subjects each were selected. One was composed of Cree subjects with the highest level of mercury exposure, another with Cree subjects having a low level of mercury exposure, and a third with control subjects. An ANOVA on these three groups revealed a significant difference between both groups of Cree subjects and the control group for Fitts' constant and irregularity. These preliminary results suggest that the EKM system is able to discriminate the performance of different groups of subjects and found significant evidence that mercury exposure is related to nerve and muscle function conditions such as tremor and Parkinson's (Beuter et al. 1999).

Though mercury vapor and organic mercury readily cross the blood–brain barrier, mercury has been found to be taken up into neurons of the brain and CNS without having to cross the blood–brain barrier, since mercury has been found to be taken up and transported along nerve axons as well through calcium and sodium channels and along the olfactory path. Exposure to inorganic mercury has significant effects on blood parameters and liver function. Studies have found that in a dose-dependent manner, mercury exposure causes reductions in oxygen consumption and availability, perfusion flow, biliary secretion, hepatic ATP concentration, and cytochrome P450 liver content, while increasing blood hemolysis products and tissue calcium content and inducing heme oxygenase, porphyria, and platelet aggregation through interfering with the sodium pump.

Studies have found mercury and lead cause autoantibodies to neuronal proteins, neurofilaments, and myelin basic protein (MBP). Mercury and cadmium also have been found to interfere with zinc binding to MBP, which affects MS symptoms since zinc stabilizes the association of MBP with brain myelin. MS has also been found to commonly be related to inflammatory activity in the CNS such as that caused by the reactive oxygen species and cytokine generation caused by mercury and other toxic metals. Antioxidants like lipoic acid which counteract such free radical activity have been found to alleviate symptoms and decrease demyelination. A group of metal-exposed MS patients with amalgam fillings were found to have lower levels of red blood cells, hemoglobin, hematocrit, thyroxine, T cells, and CD8+ suppressor immune cells than a group of MS patients with amalgam replaced and more exacerbations of MS than those without. Immune and autoimmune mechanisms are thus seen to be a major factor in neurotoxicity of metals. Mercury penetrates and damages the blood–brain barrier allowing penetration of the barrier by other substances that are neurotoxic. Such damage to the blood–brain barrier's function has been found to be a major factor in chronic neurological diseases such as MS, and studies have found mercury-related mental effects to be indistinguishable from those of MS patients. MS patients have been found to have much higher levels of mercury in cerebrospinal fluid compared to controls. Large German studies including studies at German universities have found that MS patients usually have high levels of mercury body burden, with one study finding 300 % higher than controls. Most recovered after mercury detox with some requiring additional treatment for viruses and intestinal dysbiosis. Similarly, thousands of MS patients have been documented to have recovered or significantly improved after amalgam replacement.

Mercury has been found to accumulate preferentially in the primary motor function-related areas such as the brain stem, cerebellum, rhombencephalon, dorsal root ganglia, and anterior horn motor neurons, which enervate the skeletal muscles. There is considerable indication this may be a factor in development of ALS and other neurodegenerative conditions. Treatment using IV glutathione, vitamin C, and minerals has been found to be very effective in the stabilizing and amelioration of some of these chronic neurological conditions by neurologists such as Perlmutter in Florida.

Low levels of toxic metals have been found to inhibit dihydropteridine reductase, which affects the neural system function by inhibiting brain transmitters through its effect on phenylalanine, tyrosine, and tryptophan transport into neurons. This was found to cause severe impaired amine synthesis and hypokinesis. Tetrahydrobiopterin, which is essential in production of neurotransmitters, is significantly decreased in patients with Alzheimer's, Parkinson's, and MS. Such patients have abnormal inhibition of neurotransmitter production. Supplements which inhibit breach of the blood–brain barrier such as bioflavonoids have been found to slow such neurological damage.

Clinical tests of patients with MND, ALS, Parkinson's, Alzheimer's, lupus (SLE), and rheumatoid arthritis have found that the patients generally have elevated plasma cysteine to sulfate ratios, with the average being 500 % higher than controls, and in general being poor sulfur oxidizers. Mercury has been shown to diminish and block sulfur oxidation and thus reducing glutathione levels which is the part of this process involved in detoxifying and excretion of toxics like mercury. Glutathione is produced through the sulfur oxidation side of this process. Low levels of available glutathione have been shown to increase mercury retention and increase toxic effects, while high levels of free cysteine have been demonstrated to make toxicity due to inorganic mercury more severe. Mercury has also been found to play a part in neuronal problems through blockage of the P-450 enzymatic process. Other toxic metals and toxics such as pesticides have also been found to cause the types of damage seen in Parkinson's and to exposure to have positive correlation to Parkinson's. Another exposure that affects some appears to be hexane. There are synergistic effects of various toxics that result in conditions like Parkinson's. Determination of one's factors by history assessment and tests is a first step in improving the condition.

One genetic difference found in animals and humans is cellular retention differences for metals related to the ability to excrete mercury. For example, it has been found that individuals with genetic blood factor type apoE4 do not excrete mercury readily and bioaccumulate mercury, resulting in susceptibility to chronic autoimmune conditions such as Alzheimer's, Parkinson's, etc., as early as age 40, whereas those with type apoE2 readily excrete mercury and are less susceptible. Those with type apoE3 are intermediate to the other two types.

The Huggins Clinic using total dental revision (TDR) has successfully treated over a thousand patients with chronic autoimmune conditions like MS, Parkinson's, lupus, ALS, AD, diabetes, etc., including himself with the population of over 1,000 (approx. 85 %) who experienced significant improvement in MS. Jaw bone cavitations were found to be common significant factors in some of these conditions such as Parkinson's.

Huggins Total Dental Revision Protocol

(a) History questionnaire and panel of tests.
(b) Replace amalgam fillings starting with filling with highest negative current or highest negative quadrant, with supportive vitamin/mineral supplements.
(c) Extract all root canalled teeth using proper finish protocol.
(d) Test and treat cavitations and amalgam tattoos where relevant.
(e) Supportive supplementation and periodic monitoring tests evaluate the need for further treatment (not usually needed).
(f) Avoid acute exposures/challenges to the immune system on a weekly 7/14/21 day pattern.

Tests Suggested by Huggins/Levy (35) for Evaluation and Treatment of Mercury Toxicity

(a) Hair element test (low hair mercury level does not indicate low body level) (more than three essential minerals out of normal range indicates likely metals toxicity)
(b) CBC blood test with differential and platelet count
(c) Blood serum profile
(d) Urinary mercury (for person with average exposure with amalgam fillings, average mercury level is 3–4 ppm; lower test level than this likely means person is a poor excretor and accumulating mercury, often mercury toxic)
(e) Fractionated porphyrin urine test (note test results sensitive to light, temperature, shaking)
(f) Individual tooth electric currents (replace high negative current teeth first)
(g) Patient questionnaire on exposure and symptom history
(h) Specific gravity of urine (test for pituitary function, s.g. >1.022 normal; s.g. <1.008 consistent with depression and suicidal tendencies)

Note: During initial exposure to mercury, the body marshals immune system and other measures to try to deal with the challenge, so many test indicators will be high; after prolonged exposure, the body and immune system inevitably lose the battle and measures to combat the challenge decrease – so some test indicator scores decline. Chronic conditions are common during this phase. Also high mercury exposures with low hair mercury or urine mercury level usually indicate body is retaining mercury and likely toxicity problem. In such cases where (calcium >1,100 or <300 ppm) and low test mercury, manganese, zinc, potassium; mercury toxicity likely and hard to treat since retaining mercury.

Test Results Indicating Mercury/ Metals Toxicity

(a) White blood cell count >7,500 or <4,500
(b) Hematocrit >50 % or <40 %
(c) Lymphocyte count >2,800 or <1,800 years
(d) Blood protein level >7.5 g/100 mL
(e) Triglycerides >150 mg %mL
(f) BUN >18 or <12
(g) Hair mercury >1.5 ppm or <.4 ppm
(h) Oxyhemoglobin level <55 % saturated
(i) Carboxyhemoglobin >2.5 % saturated
(j) T lymphocyte count <2,000
(k) DNA damage/cancer
(l) TSH >1 ug
(m) Hair aluminum >10 ppm
(n) Hair nickel >1.5 ppm
(o) Hair manganese >0.3 ppm
(p) Immune reactive to mercury, nickel, aluminum, etc.
(q) High hemoglobin and hematocrit and high alkaline phosphatase (alkphos) and lactic dehydrogenase (LDA) during initial phases of exposure; with low/marginal hemoglobin and hematocrit plus low oxyhemoglobin during long-term chronic fatigue phase

Note: After treatment of many cases of chronic autoimmune conditions such as MS, ALS, Parkinson's, Alzheimer's, CFS, lupus, rheumatoid arthritis, etc., it has been observed that often mercury along with root canal toxicity or cavitation toxicity is a major factor in these conditions, and most with these conditions improve after TDR if protocol is followed carefully.

Medical studies and doctors treating fibromyalgia have found that supplements which cause a decrease in glutamate or protect against its effects have a positive effect on fibromyalgia. Some that have been found to be effective in treating metals-related autoimmune conditions such as Parkinson's include vit B6, coenzyme Q10, methylcobalamin (B12), L-carnitine, choline, ginseng, Ginkgo biloba, vitamins C and E, nicotine, octacosanol, phosphatidylserine, omega 3 fatty acids (fish and flaxseed oil), turmeric, lipoic acid, proteolytic enzymes, and Hydergine. Reduced glutathione (GSH) and N-acetyl cysteine (NAC) have been found to be protective against cellular apoptosis seen in Parkinson's and other neurodegenerative conditions. High levels of vitamins C and E along with zinc have also been found protective against oxidative stress and some effects of mercury toxicity including for Parkinson's. CoQ10 at 600 mg per day was found effective at reducing Parkinson's effects. IGF-1 treatments have also been found to alleviate some of the symptoms of ALS. There is also evidence that melatonin and curcumin may have beneficial effects on reducing metal toxicity. Turmeric/curcumin has been found to reduce some of the toxic and inflammatory effects of toxic metals. Lithium supplements (lithium carbonate and lithium orotate) have been found to be effective in protecting neurons and brain function from oxidative and excitotoxic effects. A recent study demonstrated that combined treatment with lithium and valproic acid elicits synergistic neuroprotective effects against glutamate excitotoxicity in cultured brain neurons.

Doctors affiliated with the Life Enhancement Foundation have developed a diet and supplementation protocol to reduce Parkinson's effects and delay the start time of daily levodopa therapy. Dietary considerations include avoidance of alcohol, sugar, red meats, cow's milk products, gluten, fried foods, aspartame, MSG, and pesticides.

Some clinics have found root canals, cavitations, and amalgam tattoos to also be a factor in such autoimmune conditions and that treatment of them improves prognosis in recovery from these conditions.

References

Ariza ME, Bijur GN, Williams MV (1998) Lead and mercury metagenesis: role of H_2O_2, superoxide dismutase, and xanthine oxidase. Environ Mol Mutagen 31:352–361

Aschner M, Cherian MG, Klaassen CD, Palmiter RD, Erickson JC, Bush AI (1997) Metallothioneins in brain: the role in physiology and pathology. Toxicol Appl Pharmacol 142:229–242

Beuter A, De Geoffroy A, Edwards R (1999) Analysis of rapid alternating movements in Cree subjects exposed to methylmercury and in subjects with neurological deficits. Environ Res 8(0):64

Olivieri G, Brack C, Muller-Spahn F, Stahelin HB, Herrmann M, Renard P, Brockhaus M, Hock C (2000) Mercury induces cell cytotoxicity and oxidative stress and increases beta-amyloid secretion and tau phosphorylation in SHSY5Y neuroblastoma cells. J Neurochem 74:231–236

Rajanna B, Hobson M, Harris L, Ware L, Chetty CS (1990) Effects of cadmium and mercury on Na+, K + ATPases and the uptake of ^{3}H-dopamine in rat brain synaptosomes. Arch Physiol Biochem 98:291–296

World Health Organization (1991) Inorganic mercury, vol 118. Distribution and Sales Service, International Programme on Chemical Safety, Geneva

28 Methylmercury and Thimerosal

> "I think that the biological case against Thimerosal is so dramatically overwhelming anymore that only a very foolish or a very dishonest person with the credentials to understand this research would say that Thimerosal wasn't most likely the cause of autism."
>
> Boyd Haley Ph.D. (2006)

There are two, very different, types of mercury which people should know about: *methylmercury* and *ethylmercury*.

Mercury is a naturally occurring element found in the Earth's crust, air, soil, and water. Since the Earth's formation, volcanic eruptions, weathering of rocks, and burning coal have caused mercury to be released into the environment. Once released, certain types of bacteria in the environment can change mercury into *methylmercury*. Methylmercury makes its way through the food chain in fish, animals, and humans. At high levels, it can be toxic to people.

Thimerosal contains a different form of mercury called *ethylmercury*. Studies comparing ethylmercury and methylmercury suggest that they are processed differently in the human body. Ethylmercury is broken down and excreted much more rapidly than methylmercury. Therefore, ethylmercury (the type of mercury found in the influenza vaccine) is much less likely than methylmercury (the type of mercury in the environment) to accumulate in the body and cause harm.

What Is Thimerosal?

Thimerosal is a mercury-based preservative that has been used for decades in the United States in multidose vials (vials containing more than one dose) of medicines and vaccines.

Why Is Thimerosal Used as a Preservative in Vaccines?

Thimerosal is added to vials of vaccine that contain more than one dose to prevent the growth of bacteria and fungi in the event that they get into the vaccine. This may occur when a syringe needle enters a vial as a vaccine is being prepared for administration. Contamination by germs in a vaccine could cause severe local reactions, serious illness, or death. In some vaccines, preservatives are added during the manufacturing process to prevent microbial growth.

S. Nabi, *Toxic Effects of Mercury*, DOI 10.1007/978-81-322-1922-4_28,
© Springer India 2014

How Does Thimerosal Work in the Body?

Thimerosal does not stay in the body a long time so it does not build up and reach harmful levels. When thimerosal enters the body, it breaks down to ethylmercury and thiosalicylate, which are easily eliminated.

Is Thimerosal Safe?

Thimerosal has a proven track record of being very safe. Data from many studies show no convincing evidence of harm caused by the low doses of thimerosal in vaccines.

What Are the Possible Side Effects of Thimerosal?

The most common side effects are minor reactions like redness and swelling at the injection site. Although rare, some people may be allergic to thimerosal. Research shows that most people who are allergic to thimerosal will not have a reaction when thimerosal is injected under the skin.

Does Thimerosal Cause Autism?

Research *does not* show any link between thimerosal in vaccines and autism, a neurodevelopmental disorder. Although thimerosal was taken out of childhood vaccines in 2001, autism rates have gone up, which is the opposite of what would be expected if thimerosal caused autism.

Do MMR Vaccines Contain Thimerosal?

No. Measles, mumps, and rubella (MMR) vaccines do not and never did contain thimerosal. Varicella (chicken pox), inactivated polio (IPV), and pneumococcal conjugate vaccines have also never contained thimerosal.

Do All Flu Vaccines Contain Thimerosal?

No. Influenza (flu) vaccines are currently available in both thimerosal-containing and thimerosal-free versions. The total amount of flu vaccine without thimerosal as a preservative at times has been limited, but availability will increase as vaccine manufacturing capabilities are expanded. In the meantime, it is important to keep in mind that the benefits of influenza vaccination outweigh the theoretical risk, if any, of exposure to thimerosal.

Thimerosal in Vaccines

Thimerosal is a mercury-containing organic compound (an organomercurial). Since the 1930s, it has been widely used as a preservative in a number of biological and drug products, including many vaccines, to help prevent potentially life-threatening contamination with harmful microbes. Over the past several years, because of an increasing awareness of the theoretical potential for neurotoxicity of even low levels of organomercurials and because of the increased number of thimerosal-containing vaccines that had been added to the infant immunization schedule, concerns about the use of thimerosal in vaccines and other products have been raised. Indeed, because of these concerns, the Food and Drug Administration has worked with, and continues to work with, vaccine manufacturers to reduce or eliminate thimerosal from vaccines.

Thimerosal has been removed from or reduced to trace amounts in all vaccines routinely recommended for children 6 years of age and younger, with the exception of inactivated influenza vaccine (Table 28.1). A preservative-free version of the inactivated influenza vaccine (contains trace amounts of thimerosal) is available in limited supply at this time for use in infants, children, and pregnant women. Some vaccines such as Td,

Table 28.1 Thimerosal content of vaccines routinely recommended for children 6 years of age and younger

Vaccine	Trade name (manufacturer)	Thimerosal status concentration[a] (mercury)	Approval date for thimerosal-free or thimerosal-/preservative-free (trace thimerosal)[b] formulation
DTaP	Infanrix (GlaxoSmithKline Biologicals)	Free	Never contained more than a trace of thimerosal, approval date for thimerosal-free formulation 9/29/2000
	Daptacel (Sanofi Pasteur, Ltd.)	Free	Never contained thimerosal
	Tripedia (Sanofi Pasteur, Inc.)	Trace (≤0.3 μg Hg/0.5 mL dose)	03/07/01
DTaP-HepB-IPV	Pediarix (GlaxoSmithKline Biologicals)	Free	Never contained more than a trace of thimerosal, approval date for thimerosal-free formulation 1/29/2007
DTaP-IPV/Hib	Pentacel (Sanofi Pasteur, Ltd.)	Free	Approved June 20, 2008, never contained thimerosal
DTaP-IPV	KINRIX (GlaxoSmithKline Biologicals)	Free	Approved October 8, 2009, never contained thimerosal
Pneumococcal conjugate	Prevnar (Wyeth Pharmaceuticals Inc.)	Free	Never contained thimerosal
	Prevnar 13 (Wyeth Pharmaceuticals Inc.)	Free	Approved February 24, 2010, never contained thimerosal
Inactivated poliovirus	IPOL (Sanofi Pasteur, SA)	Free	Never contained thimerosal
Varicella (chicken pox)	Varivax (Merck & Co., Inc.)	Free	Never contained thimerosal
Mumps, measles, and rubella	M-M-R-II (Merck & Co., Inc.)	Free	Never contained thimerosal
Mumps, measles, rubella, and Varicella	ProQuad (Merck & Co., Inc.)	Free	Approved September 6, 2005, never contained thimerosal
Hepatitis A	Havrix (GlaxoSmithKline Biologicals)	Free	Never contained thimerosal
	Vaqta (Merck & Co., Inc.)	Free	Never contained thimerosal
Hepatitis B	Recombivax HB (Merck & Co., Inc.)	Free	08/27/99
	Engerix-B (GlaxoSmithKline Biologicals)	Free	03/28/00, approval date for thimerosal-free formulation 1/30/2007

(continued)

Table 28.1 (continued)

Vaccine	Trade name (manufacturer)	Thimerosal status concentration[a] (mercury)	Approval date for thimerosal-free or thimerosal-/preservative-free (trace thimerosal)[b] formulation
Haemophilus influenzae type b conjugate (Hib)	ActHIB (Sanofi Pasteur, SA) OmniHIB (GlaxoSmithKline)	Free	Never contained thimerosal
	PedvaxHIB (Merck & Co., Inc.)	Free	Approval date for thimerosal-free formulation 08/99
	Hiberix (GlaxoSmithKline Biologicals)	Free	Approved August 19, 2009, never contained thimerosal
Hib/hepatitis B combination	Comvax (Merck & Co., Inc.)	Free	Never contained thimerosal
Seasonal trivalent influenza	Fluzone (multidose presentation) (Sanofi Pasteur, Inc.)	0.01 % (12.5 μg/0.25 mL dose, 25 μg/0.5 mL dose)[c]	
	Fluzone (single-dose presentation) (Sanofi Pasteur, Inc.)[d]	Free	12/23/2004
	Fluvirin (multidose presentation) (Novartis Vaccines and Diagnostics Ltd.)	0.01 % (25 μg/0.5 mL dose)	
	Fluvirin (single-dose presentation) (Novartis Vaccines and Diagnostics Ltd.) (preservative free)	Trace (<1 ug Hg/0.5 mL dose)	09/28/01
	Fluarix (single-dose presentation) (GlaxoSmithKline Biologicals)	Free	Approved 10/19/09, never contained thimerosal
	Afluria (multidose presentation) (CSL Limited)	0.01 % (24.5 μg/0.5 mL dose)	
	Afluria (single-dose presentation) (CSL Limited)	Free	Approved 11/10/09, never contained thimerosal
Seasonal influenza, live	FluMist (MedImmune Vaccines, Inc.)	Free	Never contained thimerosal
Rotavirus	RotaTeq (Merck & Co., Inc.)	Free	Approved February 3, 2006, never contained thimerosal
	Rotarix (GlaxoSmithKline Biologicals)	Free	Approved April 3, 2008, never contained thimerosal

[a]Thimerosal is approximately 50 % mercury (Hg) by weight. A 0.01 % solution (1 part per 10,000) of thimerosal contains 50 μg of Hg per 1 mL dose or 25 μg of Hg per 0.5 mL dose

[b]The term "trace" has been taken in this context to mean 1 μg of mercury per dose or less

[c]Children 6 months old to less than 3 years receive a half-dose of vaccine, i.e., 0.25 mL; children 3 years of age and older receive 0.5 mL

[d]A trace thimerosal-containing formulation of Fluzone was approved on 9/14/02 and has been replaced with the formulation without thimerosal

Table 28.2 Preservatives used in US licensed vaccines

Preservative	Vaccine examples (trade name; manufacturer)
Thimerosal	TT (one)
	Influenza multidose presentations (several)
Phenol	Typhoid Vi polysaccharide (Typhim Vi; Sanofi Pasteur, SA)
	Pneumococcal polysaccharide (Pneumovax 23; Merck & Co., Inc.)
Benzethonium chloride (phemerol)	Anthrax (BioThrax; Emergent BioDefense Operations Lansing Inc.)
2-Phenoxyethanol	IPV (IPOL; Sanofi Pasteur, SA)

which is indicated for older children (≥7 years of age) and adults, are also now available in formulations that are free of thimerosal or contain only trace amounts. Vaccines with trace amounts of thimerosal contain 1 μg or less of mercury per dose (Table 28.2).

In the following text, a discussion of preservatives, the use of thimerosal as a preservative, guidelines on exposure to organomercurials (primarily methylmercury), thimerosal toxicity, recent and future FDA actions, and the conclusions of the Institute of Medicine's most recent review of thimerosal in vaccines are presented. This narrative on thimerosal contains references to the literature and links to other sites for readers who wish additional information; for quick reference, a number of frequently asked questions (FAQs) and answers are provided.

Preservatives in Vaccines

To begin, we need to answer two questions – what are preservatives and why are they used in vaccines. For our purposes, preservatives may be defined as compounds that kill or prevent the growth of microorganisms, particularly bacteria and fungi. They are used in vaccines to prevent microbial growth in the event that the vaccine is accidentally contaminated, as might occur with repeated puncture of multidose vials. In some cases, preservatives are added during manufacture to prevent microbial growth; with changes in manufacturing technology, however, the need to add preservatives during the manufacturing process has decreased markedly.

The US Code of Federal Regulations (the CFR) requires, in general, the addition of a preservative to multidose vials of vaccines; indeed, worldwide, preservatives are routinely added to multidose vials of vaccine. Tragic consequences have followed the use of multidose vials that did not contain a preservative and have served as the impetus for this requirement. One particularly telling incident from Australia is described by Sir Graham S. Wilson in his classic book, *The Hazards of Immunization*.

In January 1928, in the early stages of an immunization campaign against diphtheria, Dr. Ewing George Thomson, medical officer of Health of Bundaberg, began the injection of children with toxin–antitoxin mixture. The material was taken from an India-rubber-capped bottle containing 10 mL of TAM. On January 17, 20, 21, and 24, Dr. Thomson injected subcutaneously a total of 21 children without ill effect. On January 27, a further 21 children were injected. Of these children, 11 died on January 28 and 1 on January 29 (Wilson 1967).

This disaster was investigated by a Royal Commission, and the final sentence in the summary of their findings reads as follows:

> The consideration of all possible evidence concerning the deaths at Bundaberg points to the injection of living staphylococci as the cause of the fatalities.

From this experience, the Royal Commission recommended that biological products in which the growth of a pathogenic organism is possible should not be issued in containers for repeated use unless there is a sufficient concentration of antiseptic (preservative) to inhibit bacterial growth.

The US requirement for preservatives in multidose vaccines was incorporated into the CFR in January 1968, although many biological products had contained preservatives, including thimerosal, prior to this date. Specifically, the CFR states:

> Products in multi-dose containers shall contain a preservative, except that a preservative need not be

> added to Yellow Fever Vaccine; Polio-virus Vaccine, Live Oral; viral vaccine labeled for use with the jet injector; dried vaccines when the accompanying diluent contains a preservative; or to an Allergenic Product in 50 percent or more volume (v/v) glycerin **[21 CFR 610.15(a)]**.The CFR also requires that the preservative used shall be sufficiently non-toxic so that the amount present in the recommended dose of the product will not be toxic to the recipient, and in combination used it shall not denature the specific substance in the product to result in a decrease below the minimal acceptable potency within the dating period when stored at the recommended temperature. **[21 CFR 610.15(a)]**

Preservatives cannot completely eliminate the risk of contamination of vaccines. The literature contains several reports of bacterial contamination of vaccines despite the presence of a preservative, emphasizing the need for meticulous attention to technique in withdrawing vaccines from multidose vials (Bernier et al. 1981; Simon et al. 1993). The need for preservatives in multidose vials of vaccines is nonetheless clear. Several preservatives are used in US licensed vaccines, and these are listed in. It is important to note that the FDA does not license a particular preservative; rather, the product containing that preservative is licensed, with safety and efficacy data generally collected in the context of a license application for a particular product.

Thimerosal as a Preservative

Thimerosal, which is approximately 50 % mercury by weight, has been one of the most widely used preservatives in vaccines. It is metabolized or degraded to ethylmercury and thiosalicylate. Ethylmercury is an organomercurial that should be distinguished from methylmercury, a related substance that has been the focus of considerable studies. At concentrations found in vaccines, thimerosal meets the requirements for a preservative as set forth by the *United States Pharmacopeia*; that is, it kills the specified challenge organisms and is able to prevent the growth of the challenge fungi (U.S. Pharmacopeia 2001). Thimerosal in concentrations of 0.001 % (1 part in 100,000) to 0.01 % (1 part in 10,000) has been shown to be effective in clearing a broad spectrum of pathogens. A vaccine containing 0.01 % thimerosal as a preservative contains 50 μg of thimerosal per 0.5 mL dose or approximately 25 μg of mercury per 0.5 mL dose.

Prior to its introduction in the 1930s, data were available in several animal species and humans providing evidence for its safety and effectiveness as a preservative (Powell and Jamieson 1931). Since then, thimerosal has been the subject of several studies and has a long record of safe and effective use preventing bacterial and fungal contamination of vaccines, with no ill effects established other than minor local reactions at the site of injection.

While the use of mercury-containing preservatives has declined in recent years with the development of new products formulated with alternative or no preservatives, thimerosal has been used in some immune globulin preparations, antivenins, skin test antigens, and ophthalmic and nasal products, in addition to certain vaccines. Under the FDA Modernization Act of 1997, the FDA compiled a list of regulated products containing mercury, including those with thimerosal (Federal Register 1999). It is important to note that this list was compiled in 1999; some products listed are no longer manufactured and many products have been reformulated without thimerosal. Updated lists of vaccines and their thimerosal content can be found in Tables 28.1 (routinely recommended pediatric vaccines) and 28.3 (expanded list of vaccines).

Guidelines on Exposure to Organomercurials

Mercury is an element that is dispersed widely around the Earth. Most of the mercury in the water, soil, plants, and animals is found as inorganic mercury salts. Mercury accumulates in the aquatic food chain, primarily in the form of the methylmercury, an organomercurial. Organic forms of mercury are more easily absorbed when

Table 28.3 Thimerosal and expanded list of vaccines – (*updated June 20, 2012*) thimerosal content in currently manufactured US licensed vaccines

Vaccine	Trade name	Manufacturer	Thimerosal concentration[a]	Mercury
Anthrax	BioThrax	Emergent BioDefense Operations Lansing, LLC	0	0
DTaP	Tripedia[b]	Sanofi Pasteur, Inc.	≤0.00012 %	≤0.3 µg/0.5 mL dose
	Infanrix	GlaxoSmithKline Biologicals	0	0
	Daptacel	Sanofi Pasteur, Ltd.	0	0
DTaP-HepB-IPV	Pediarix	GlaxoSmithKline Biologicals	0	0
DT	No trade name	Sanofi Pasteur, Inc.	<0.00012 % (single dose)	<0.3 µg/0.5 mL dose
		Sanofi Pasteur, Ltd.[c]	0.01 %	25 µg/0.5 mL dose
Td	No trade name	MassBiologics	≤0.00012 %	≤0.3 µg mercury/0.5 mL dose
	Decavac	Sanofi Pasteur, Inc.	≤0.00012 %	≤0.3 µg mercury/0.5 mL dose
	No trade name	Sanofi Pasteur, Ltd.	0	0
Tdap	Adacel	Sanofi Pasteur, Ltd.	0	0
	Boostrix	GlaxoSmithKline Biologicals	0	0
TT	No trade name	Sanofi Pasteur, Inc.	0.01 %	25 µg/0.5 mL dose
Hib	ActHIB	Sanofi Pasteur, SA	0	0
	Hiberix	GlaxoSmithKline Biologicals	0	0
	PedvaxHIB liquid	Merck & Co., Inc.	0	0
Hib/HepB	Comvax[d]	Merck & Co., Inc.	0	0
Hepatitis B	Engerix-B	GlaxoSmithKline Biologicals	0	0
	Pediatric/adolescent		0	0
	Adult			
	Recombivax HB	Merck & Co., Inc.	0	0
	Pediatric/adolescent		0	0
	Adult (adolescent)		0	0
	Dialysis			
Hepatitis A	Havrix	GlaxoSmithKline Biologicals	0	0
	Vaqta	Merck & Co., Inc.	0	0
HepA/HepB	Twinrix	GlaxoSmithKline Biologicals	0	0
IPV	IPOL	Sanofi Pasteur, SA	0	0
	Poliovax	Sanofi Pasteur, Ltd.	0	0
Influenza	Afluria	CSL Limited	0 (single dose)	0 (0.5 mL single dose)
			0.01 % (multidose)	24.5 µg (0.5 mL multidose)
	Agriflu	Novartis Vaccines and Diagnostics Ltd.	0	0

(continued)

Table 28.3 (continued)

Vaccine	Trade name	Manufacturer	Thimerosal concentration[a]	Mercury
	Fluzone[e] (multidose presentation)	Sanofi Pasteur, Inc.	0.01 %	25 μg/0.5 mL dose
	Fluvirin (multidose vial)	Novartis Vaccines and Diagnostics Ltd.	0.01 %	25 μg/0.5 mL dose
	Fluzone (single-dose presentations)	Sanofi Pasteur, Inc.	0	0
	Fluvirin (single-dose prefilled syringe)	Novartis Vaccines and Diagnostics Ltd.	0	0
	Fluarix	GlaxoSmithKline Biologicals	0	0
	FluMist and FluMist Quadrivalent	MedImmune Vaccines, Inc.	0	0
	FluLaval	ID Biomedical Corporation of Quebec	0.01 %	25 μg/0.5 mL dose
Japanese encephalitis	IXIARO	Intercell AG	0	0
MMR	MMR-II	Merck & Co., Inc.	0	0
Meningococcal	Menomune A, C, AC and A/C/Y/W-135	Sanofi Pasteur, Inc.	0.01 % (multidose)	25 μg/0.5 dose
			0 (single dose)	0
	Menactra A, C, Y and W-135	Sanofi Pasteur, Inc.	0	0
	Menveo	Novartis Vaccines and Diagnostics Inc.	0	0
Pneumococcal	Prevnar 13 (pneumoconjugate)	Wyeth Pharmaceuticals Inc.	0	0
	Pneumovax 23	Merck & Co., Inc.	0	0
Rabies	IMOVAX	Sanofi Pasteur, SA	0	0
	Rabavert	Novartis Vaccines and Diagnostics	0	0
Smallpox (Vaccinia), live	ACAM2000	Acambis, Inc.	0	0
Typhoid fever	Typhim Vi	Sanofi Pasteur, SA	0	0
	Vivotif	Berna Biotech, Ltd.	0	0
Varicella	Varivax	Merck & Co., Inc.	0	0
Yellow fever	Y-F-Vax	Sanofi Pasteur, Inc.	0	0
Zoster vaccine live	Zostavax	Merck & Co., Inc.	0	0

[a]Thimerosal is approximately 50 % mercury (Hg) by weight. A 0.01 % solution (1 part per 10,000) of thimerosal contains 50 μg of Hg per 1 mL dose or 25 μg of Hg per 0.5 mL dose

[b]Sanofi Pasteur's Tripedia may be used to reconstitute ActHib to form TriHIBit. TriHIBit is indicated for use in children 15–18 months of age

[c]This vaccine is not marketed in the United States

[d]Comvax is not licensed for use less than 6 weeks of age because of decreased response to the Hib component

[e]Children under 3 years of age receive a half-dose of vaccine, i.e., 0.25 mL (12.5 μg mercury/dose)

ingested and are less readily eliminated from the body than are inorganic forms of mercury. Humans are exposed to methylmercury primarily from the consumption of seafood (Mahaffey et al. 1997).

Methylmercury is a neurotoxin. The toxicity of methylmercury was first recognized during the late 1950s and early 1960s when industrial discharge of mercury into the Minamata Bay, Japan, led to the widespread consumption of mercury-contaminated fish (Harada 1995). Epidemics of methylmercury poisoning also occurred in Iraq during the 1970s when seed grain treated with a methylmercury fungicide was accidentally used to make bread (Bakir et al. 1973). During these epidemics, fetuses were found to be more sensitive to the effects of methylmercury than adults. Maternal exposure to high levels of methylmercury resulted in infants exhibiting severe neurological injury including a condition resembling cerebral palsy, while their mothers showed little or no symptoms. Sensory and motor neurological dysfunction and developmental delays were observed among some children who were exposed in utero to lower levels of methylmercury.

More recently, several epidemiological studies have examined the effect of low dose dietary exposure to methylmercury, with inconsistent results. Studies from the Faroe Islands reported that subtle cognitive deficits (e.g., performance on attention, language, and memory tests), detectable by sophisticated neuropsychometric testing, were associated with methylmercury levels previously thought to be safe (Grandjean et al. 1997). Studies in the Seychelles, evaluating more global developmental outcomes, did not reveal any correlation between abnormalities and methylmercury levels (Davidson et al. 1998).

Various agencies have developed guidelines for safe exposure to methylmercury, including the US Environmental Protection Agency (Mahaffey et al. 1997), US Agency for Toxic Substances and Disease Registry (ATSDR 1999), the FDA (Federal Register 1979), and the World Health Organization (WHO 1996). These exposure levels range from 0.1 μg/kg body weight/day (EPA) to 0.47 μg/kg body weight/day (WHO). The range of recommendations is due to varying safety margins, differing emphasis placed on various sources of data, the different missions of the agencies, and the population that the guideline is intended to protect. All guidelines, however, fall within the same order of magnitude. While these guidelines may be used as screening tools in risk assessment to evaluate the "safety" of mercury exposures, they are not meant to be bright lines above which toxicity will occur. However, as exposure levels increase in multiples of these guidelines, there is increasing concern on the part of the public health community that adverse health consequences may occur (Mahaffey 1999).

To address the issue of conflicting methylmercury exposure guidelines, the Congress asked the National Academy of Sciences to study the toxicological effects of methylmercury and provide recommendations on the establishment of a scientifically appropriate methylmercury reference dose. Their report concluded that the EPA's current reference dose, the RFD, for methylmercury, 0.1 μg/kg/day is a scientifically justifiable level for the protection of human health. The FDA is considering this and other data relevant to its exposure guideline for methylmercury.

Thimerosal Toxicity

The various mercury guidelines are based on epidemiological and laboratory studies of methylmercury, whereas thimerosal is a derivative of ethylmercury. Because they are different chemical entities – ethyl- and methylmercury – different toxicological profiles are expected. There is, therefore, an uncertainty that arises in applying the methylmercury-based guidelines to thimerosal. Lacking definitive data on the comparative toxicities of ethyl- versus methylmercury, FDA considered ethyl- and methylmercury as equivalent in its risk evaluation. There are some data and studies bearing directly on thimerosal toxicity and these are summarized in this section.

Allergic responses to thimerosal are described in the clinical literature, with these responses manifesting themselves primarily in the form of delayed-type local hypersensitivity reactions, including redness and swelling at the injection site (Cox and Forsyth 1988; Grabenstein 1996). Such reactions are usually mild and last only a few days. Some authors postulate that the thiosalicylate component is the major determinant of allergic reactions (Goncalo et al. 1996). In a clinical setting, however, it is usually not possible to determine whether local reactions are caused by thimerosal or other vaccine components.

The earliest published report of thimerosal use in humans was published in 1931 (Powell and

Jamieson 1931). In this report, 22 individuals received 1 % solution of thimerosal intravenously for unspecified therapeutic reasons. Subjects received up to 26 mg thimerosal/kg (1 mg equals 1,000 μg) with no reported toxic effects, although two subjects demonstrated phlebitis or sloughing of skin after local infiltration. Of note, this study was not specifically designed to examine toxicity; 7 of 22 subjects were observed for only 1 day, the specific clinical assessments were not described, and no laboratory studies were reported.

Several cases of acute mercury poisoning from thimerosal-containing products were found in the medical literature with total doses of thimerosal ranging from approximately 3 mg/kg to several hundred mg/kg. These reports included the administration of immune globulin (gamma-globulin) (Matheson et al. 1980) and hepatitis B immune globulin (Lowell et al. 1996), chloramphenicol formulated with 1,000 times the proper dose of thimerosal as a preservative (Axton 1972), thimerosal ear irrigation in a child with tympanostomy tubes (Rohyans et al. 1994), thimerosal treatment of omphaloceles in infants (Fagan et al. 1977), and a suicide attempt with thimerosal (Pfab et al. 1996). These studies reported local necrosis, acute hemolysis, disseminated intravascular coagulation, acute renal tubular necrosis, and central nervous system injury including obtundation, coma, and death (IOM).

Several animal studies have evaluated the toxicity of thimerosal. In 1931, Powell and Jamieson reported acute toxicity studies in several animal species. Maximum tolerated doses not associated with death of the animals were 20 mg thimerosal/kg (rabbits) and 45 mg/kg (rats). Blair evaluated the administration of thimerosal intranasally for 190 days and observed no histopathological changes in the brain or kidney (Blair et al. 1975). Magos et al. directly compared the toxicity of ethyl- versus methylmercury in adult male and female rats administered five daily doses of equimolar concentrations of ethyl- or methylmercury by gavage (Magos et al. 1985). Magos concluded that ethylmercury, the mercury derivative found in thimerosal, is less neurotoxic than methylmercury, the mercury derivative for which the various guidelines are based.

One final piece of data regarding thimerosal is worth noting. At the initial National Vaccine Advisory Committee-sponsored meeting on thimerosal in 1999, concerns were expressed that infants may lack the ability to eliminate mercury. More recent NIAID-supported studies at the University of Rochester and National Naval Medical Center in Bethesda, MD, examined levels of mercury in blood and other samples from infants who had received routine immunizations with thimerosal-containing vaccines (Pichichero et al. 2002). Blood levels of mercury did not exceed safety guidelines for methylmercury for all infants in these studies. Further, mercury was cleared from the blood in infants exposed to thimerosal faster than would be predicted for methylmercury; infants excreted significant amounts of mercury in stool after thimerosal exposure, thus removing mercury from their bodies. These results suggest that there are differences in the way that thimerosal and methylmercury are distributed, metabolized, and excreted. Thimerosal appears to be removed from the blood and body more rapidly than methylmercury. NIAID is sponsoring a follow-up study with larger numbers of infants in Buenos Aires where thimerosal-containing vaccines are still administered to children.

Recent and Future FDA Action

FDA has been actively addressing the issue of thimerosal as a preservative in vaccines. Under the FDA Modernization Act (FDAMA) of 1997, the FDA conducted a comprehensive review of the use of thimerosal in childhood vaccines. Conducted in 1999, this review found no evidence of harm from the use of thimerosal as a vaccine preservative, other than local hypersensitivity reactions (Ball et al. 2001).

As part of the FDAMA review, the FDA evaluated the amount of mercury an infant might receive in the form of *ethyl*mercury from vaccines under the US recommended childhood immunization schedule and compared these levels with existing guidelines for exposure to *methyl*mercury, as there are no existing guidelines for ethylmercury, the metabolite of thimerosal. At the time of this review in 1999, the maximum cumulative exposure to mercury from vaccines in the recommended childhood immunization schedule was within acceptable limits for the methylmercury exposure guidelines set by FDA, ATSDR, and WHO. However, depending on the vaccine formulations used and the weight of the infant, some infants could have been exposed to cumulative levels of mercury during the first 6 months of life that exceeded EPA recommended guidelines for safe intake of methylmercury.

As a precautionary measure, the Public Health Service (including the FDA, National Institutes of Health (NIH), Centers for Disease Control and Prevention (CDC) and Health Resources and Services Administration (HRSA) and the American Academy of Pediatrics) issued two joint statements, urging vaccine manufacturers to reduce or eliminate thimerosal in vaccines as soon as possible (CDC 1999). The US Public Health Service agencies have collaborated with various investigators to initiate further studies to better understand any possible health effects from exposure to thimerosal in vaccines.

Available data has been reviewed in several public forums including the Workshop on Thimerosal held in Bethesda in August 1999 and sponsored by the National Vaccine Advisory Committee, two meetings of the Advisory Committee on Immunization Practices of the CDC, held in October 1999 and June 2000, and the Institute of Medicine's Immunization Safety Review Committee in July 2001 and May 2004. Through its Vaccine Safety Datalink, the CDC has examined the incidence of autism as a function of the amount of thimerosal a child received from vaccines. Preliminary results indicated no change in autism rates relative to the amount of thimerosal a child received during the first 6 months of life (from 0 μg to greater than 160 μg). A weak association was found with thimerosal intake and certain neurodevelopmental disorders (such as attention deficit hyperactivity disorder) in one study but was not found in a subsequent study. Additional studies are planned in these areas.

Much progress has been made to date in removing or reducing thimerosal in vaccines. New pediatric formulations of hepatitis B vaccines have been licensed by the FDA, Recombivax HB (Merck, thimerosal free) in August 1999, and Engerix-B (GlaxoSmithKline, thimerosal free) in January 2007. In March 2001, the FDA approved a second DTaP vaccine formulated without thimerosal as a preservative (Aventis Pasteur's Tripedia, trace thimerosal). Aventis Pasteur, Ltd. was also approved to manufacture a thimerosal-free DTaP vaccine, Daptacel, in 2002. In September 2001, Chiron/Evans was approved for manufacturing a preservative-free formulation of their influenza vaccine, Fluvirin, that contained trace thimerosal. In September of 2002, Aventis Pasteur, Inc. was approved to manufacture a preservative-free formulation of their influenza vaccine, Fluzone, that contained trace thimerosal, and in December 2004, a thimerosal-free formulation of Fluzone was approved. Two Td vaccines are also available in preservative-free formulations, Aventis Pasteur, Inc.'s Decavac and Aventis Pasteur, Ltd.'s Td vaccine. Also, Aventis Pasteur, Inc.'s DT vaccine is now available only in a preservative-free formulation. These changes have been accomplished by reformulating products in single-dose vials that do not contain a preservative. At present, all routinely recommended vaccines for US infants are available only as thimerosal-free formulations or contain only trace amounts of thimerosal ($\leq$1 than micrograms mercury per dose), with the exception of inactivated influenza vaccine. Inactivated influenza vaccine for pediatric use is available in a thimerosal preservative-containing formulation and in formulations that contain either no thimerosal or only a trace of thimerosal, but the latter is in more limited supply; see Table 28.1. A more extensive tabulation of vaccines and thimerosal content may be found in Table 28.3.

The Safety Review of Thimerosal-Containing Vaccines and Neurodevelopmental Disorders Conducted by the Institute of Medicine

In 2001, the Institute of Medicine convened a committee (the Immunization Safety Review Committee) to review selected issues related to immunization safety. The IOM has, to date, completed reviews in two areas. The first review by this committee focused on a potential link between autism and the combined mumps, measles, and rubella vaccine. The second review focused on a potential relationship between thimerosal use in vaccines and neurodevelopmental disorders (IOM 2001). This latter issue was brought to the fore primarily as the result of the hypothesis, formulated by S. Bernard and others from Cure Autism Now, that autism is a novel form of mercury poisoning (Bernard et al. 2001); this hypothesis, linking autism to mercury, was based on a comprehensive review of the scientific literature on mercury toxicity.

In its report of October 1, 2001, the IOM's Immunization Safety Review Committee concluded that the evidence was inadequate to either accept or reject a causal relationship between thimerosal exposure from childhood vaccines and the neurodevelopmental disorders of autism, attention deficit hyperactivity disorder (ADHD), and speech or language delay. Additional studies were needed to establish or reject a causal relationship. The Committee did conclude that the hypothesis that exposure to thimerosal-containing vaccines could be associated with neurodevelopmental disorders was biologically plausible.

The Committee believed that the effort to remove thimerosal from vaccines was "a prudent measure in support of the public health goal to reduce mercury exposure of infants and children as much as possible." Furthermore, in this regard, the Committee urged that "full consideration be given to removing thimerosal from any biological product to which infants, children, and pregnant women are exposed."

In 2004, the IOM's Immunization Safety Review Committee issued its final report, examining the hypothesis that vaccines, specifically the MMR vaccines and thimerosal-containing vaccines, are causally associated with autism. In this report, the committee incorporated new epidemiological evidence from the United States, Denmark, Sweden, and the United Kingdom and studies of biological mechanisms related to vaccines and autism since its report in 2001. The committee concluded that this body of evidence favors rejection of a causal relationship between thimerosal-containing vaccines and autism and that hypotheses generated to date concerning a biological mechanism for such causality are theoretical only. Further, the committee stated that the benefits of vaccination are proven and the hypothesis of susceptible populations is presently speculative and that widespread rejection of vaccines would lead to increases in incidences of serious infectious diseases like measles, whooping cough, and Hib bacterial meningitis.

The FDA is continuing its efforts to reduce the exposure of infants, children, and pregnant women to mercury from various sources. Discussions with the manufacturers of influenza virus vaccines (which are now routinely recommended for pregnant women and children 6–23 months of age) regarding their capacity to potentially increase the supply of thimerosal-reduced and thimerosal-free presentations are ongoing. Discussions are also underway with regard to other vaccines. Of note, all hepatitis B vaccines for the United States, including for adults, are now available only as thimerosal-free or trace thimerosal-containing formulations. In addition, all immune globulin preparations including hepatitis B immune globulin and Rho (D) immune globulin preparations are manufactured without thimerosal. For additional information on the issue of thimerosal in vaccines, see Frequently Asked Questions (FAQs).

Establishment Safety Debate: Thimerosal in Vaccines Versus Admittedly Dangerous Methylmercury

The vaccine purveyors often argue that "ethylmercury" compounds (thimerosal) are safe, while admitting that "methylmercury" compounds are harmful. Hence, industry uses the FDA-approved purportedly "safe" ethylmercury compound found in thimerosal.

Having thoroughly reviewed two key published studies on mercury metabolism, one in rats and one in human infants, what Dr. Paul G. King noticed and articulated, among other realities, in his latest posting is that

> During the metabolic process in the human and animal bodies the supposedly "harmless" ethylmercury compound, Thimerosal, is metabolized (converted) into the toxic and "harmful" methylmercury. And then in turn, the harmful methylmercury is metabolized (converted) into the most harmful, long-term-toxic, "inorganic" mercury that is retained in bodily tissue.

In the rat study, which Dr. King cites, the lab rats were raised for the purpose of this study and had no reported mercury levels in their blood before the experiment.

There were three groups in the study:

1. A test group of thimerosal (ethylmercury)-treated rats
2. A test group of methylmercury chloride-treated rats
3. A control group of rats treated with a "water" placebo

At the end of the experiment, as expected, the water-treated control group had no reported levels of mercury in their blood or organs.

The group treated with methylmercury chloride (which vaccine purveyors routinely "sound bite" as the harmful organic mercury as compared to the "safe" thimerosal, ethylmercury), as expected, had both methylmercury and inorganic mercury in their blood and organs (Note: "Inorganic" mercury is the end product of mercury metabolism).

> The methylmercury subject group confirmed that the metabolic pathway for mercury in the human and animal body consists in the reduction/conversion of the harmful methylmercury into a more harmful "inorganic" mercury which is tissue-bound, and long-term-toxic. Hence, both the originating substance (methylmercury) and its conversion/reduction, inorganic mercury was found.

The thimerosal group: Unexpectedly, the rats treated with thimerosal (ethylmercury) were found to have three types of mercury in the blood samples and their organs, ethylmercury (the originating "supposedly harmless" compound), methylmercury (the admittedly harmful compound), and inorganic mercury (the most harmful, tissue-bound end product of mercury metabolism).

> This observation begs an answer to the question: Where did the "methylmercury" come from since this group was only originally and solely treated with Thimerosal (an "ethylmercury" compound)?

Based on the published findings in the three groups of rats, the metabolic pathway for organic mercury involves the conversion of ethylmercury (thimerosal) into "methylmercury" and then the further reduction of "methylmercury" into inorganic mercury.

It may be that some of the "ethylmercury" (from thimerosal) are also directly converted into inorganic mercury. However, there are apparently no studies in either humans or other animals that establish the biochemical conversion of ethylmercury compounds directly into the "inorganic" mercury.

In conclusion, in simple layman's terms, these studies, as brought to light by Dr. King, establish that ethylmercury (thimerosal), a "supposedly harmless" compound of mercury according to the vaccine establishment, is converted in the rat and apparently in the human infant into "methylmercury" which, the establishment admits, is a harmful form of mercury. It is then further reduced to the long-term most harmful, "inorganic mercury" that bioaccumulates in the tissues and organs.

Based on these findings, we can conclude that injecting the thimerosal (ethylmercury), found in flu shots, into pregnant women (exposing the in utero fetus to mercury and the egregious recommendation that children should be vaccinated, annually with thimerosal-preserved inactivated influenza vaccines from 6 months of age until 18 years of age) is a perplexing interwoven government/pharmaceutical health strategy that afflicts, debilitates, and destroys the lives of individuals and their families in the United States and in any other nation that (a) recommends inactivated influenza vaccines for pregnant women and children, (b) permits those flu shots to be thimerosal preserved, and (c) follows the US recommendations for annual flu shots.

P.S. One more interesting observation: At sacrifice, the rats in the group that had been treated with thimerosal–ethylmercury (supposed the harmless compound) had significantly higher levels of the long-term, harmful "inorganic" mercury in their brains than the rats in the methylmercury chloride-treated group.

Perhaps, we should inject the harmful methylmercury directly into our children's arms instead of thimerosal (ethylmercury). It appears that if we use the harmful form of mercury right off the bat, there is less inorganic mercury in our children's brains. But, on the other hand, if the establishment wants more inorganic mercury in our children's brains and organs (which it appears they do), let's stick with thimerosal. The pharmaceutical companies stand to gain when we make our children sicker and sicker by injecting them first in utero and then consistently, year after year, with a known and extremely dangerous neurotoxin. And both the government and the pharmaceutical companies stand to gain when they injury the emerging generation with vaccines, misdiagnose them as mentally ill (autistic, ADD, ADHD, bipolar, OCD, etc.), and then dumb them down with a lifelong supply of psyche drugs which, according to the Mayo Clinic, is resulting in the birth of an epidemic of deformed offspring.

References

ATSDR (1999) Toxicological profile for mercury. Agency for Toxic Substances and Disease Registry. U.S. Department of Health and Human Services. Public Health Service, Atlanta

Axton JMH (1972) Six cases of poisoning after a parenteral organic mercurial compound (merthiolate). Postgrad Med J 48:417–421

Bakir F, Damluji SF, Amin-Zaki L, Murtadha M, Khalidi A, Al-Rawi NY, Tikriti S, Dhahir HI, Clarkson TW, Smith JC, Doherty RA (1973) Methyl mercury poisoning in Iraq. Science 181:230–241

Ball LK, Ball R, Pratt RD (2001) An assessment of thimerosal use in childhood vaccines. Pediatrics 107(5):1147–1154

Bernard S, Enayati A, Redwood L, Roger H, Binstock T (2001) Autism: a novel form of mercury poisoning. Med Hypotheses 56:462–471

Bernier RH, Frank JA, Nolan TF (1981) Abscesses complicating DTP vaccination. Am J Dis Child 135:826–828

Blair AMJN, Clark B, Clarke AJ, Wood P (1975) Tissue concentrations of mercury after chronic dosing of squirrel monkeys with thimerosal. Toxicology 3:171–176

Centers for Disease Control and Prevention (1999) Notice to readers: thimerosal in vaccines: a joint statement of the American Academy of Pediatrics and the Public Health Service. Morb Mort Wkly Rep 48:563–565

Cox NH, Forsyth A (1988) Thimerosal allergy and vaccination reactions. Contact Dermatitis 18:229–233

Davidson PW, Myers GJ, Cox C, Axtell C, Shamlaye C, Sloan-Reeves J, Cernichiari E, Needham L, Choi A, Wang Y, Berlin M, Clarkson TW (1998) Effects of prenatal and postnatal methylmercury exposure from fish consumption on neurodevelopment: outcomes at 66 months of age in the Seychelles child development study. JAMA 280:701–707

Fagan DG, Pritchard JS, Clarkson TW, Greenwood MR (1977) Organ mercury levels in infants with omphaloceles treated with organic mercurial antiseptic. Arch Dis Child 52:962–964

Federal Register, 19 Jan 1979, 44, 3990

Federal Register, 19 Nov 1999, 64:63323–63324

Goncalo M, Figueiredo A, Goncalo S (1996) Hypersensitivity to thimerosal: the sensitivity moiety. Contact Dermatitis 34:201–203

Grabenstein JD (1996) Immunologic necessities: diluents, adjuvants, and excipients. Hosp Pharm 31:1387–1401

Grandjean P, Weihe P, White RF, Debes F, Araki S, Yokoyama K, Murata K, Sorensen N, Dahl R, Jorgensen PL (1997) Cognitive deficit in 7-year-old children with prenatal exposure to methylmercury. Neurotoxicol Teratol 19:417–428

Harada M (1995) Minamata disease: methylmercury poisoning in Japan caused by environmental pollution. Crit Rev Toxicol 25:1–24

IOM (Institute of Medicine) (2001) Thimerosal-containing vaccines and neurodevelopmental disorders. Washington, DC: National Academy Press

Lowell HJ, Burgess S, Shenoy S, Peters M, Howard TK (1996) Mercury poisoning associated with hepatitis B immunoglobulin. Lancet 347:480

Magos L, Brown AW, Sparrow S, Bailey E, Snowden RT, Skipp WR (1985) The comparative toxicology of ethyl- and methylmercury. Arch Toxicol 57: 260–267

Mahaffey KR (1999) Methylmercury: a new look at the risks. Public Health Rep 114:397–413

Mahaffey KR, Rice G, et al (1997) An assessment of exposure to mercury in the United States: mercury study report to congress. Washington, DC: U.S. Environmental Protections Agency. Document EPA-452/R097-006

Matheson DS, Clarkson TW, Gelfand EW (1980) Mercury toxicity (acrodynia) induced by long-term injection of gammaglobulin. J Pediatr 97:153–155

Pfab R, Muckter H, Roider G, Zilker T (1996) Clinical course of severe poisoning with thiomersal. Clin Toxicol 34:453–460

Pichichero ME et al (2002) Lancet 360:1737–1741

Powell HM, Jamieson WA (1931) Merthiolate as a germicide. Am J Hyg 13:296–310

Rohyans J, Walson PD, Wood GA, MacDonald WA (1994) Mercury toxicity following merthiolate ear irrigations. J Pediatr 104:311–313

Simon PA, Chen RT, Elliot JA, Schwartz B (1993) Outbreak of pyogenic abscesses after diphtheria and tetanus toxoids and pertussis vaccine. Pediatr Infect Dis J 12:368–371

U.S. Pharmacopeia 24 (2001) Rockville: U.S. Pharmacopeial Convention

WHO (1996) Food and Agriculture Organization, International Atomic Energy Agency Expert Group. Trace elements in human nutrition and health. WHO, Geneva

Wilson GS (1967) The hazards of immunization. The Athlone Press, New York, pp 75–84

Thimerosal and Occurrence of Autism

29

Autism: Where the "randomness of life" collides and clashes with an individual's need for the sameness~

(Eileen Miller)

Thimerosal is the preservative of choice for vaccine manufacturers. First introduced by Eli Lilly and Company in the late 1920s and early 1930s, the company began selling it as a preservative in vaccines in the 1940s. Thimerosal contains 49.6 % mercury by weight and is metabolized or degraded into ethylmercury and thiosalicylate. Mercury, or more precisely, ethylmercury, is the principal agent that kills contaminants. Unfortunately, mercury also kills much more than that.

The Department of Defense classifies mercury as a hazardous material that could cause death if swallowed, inhaled, or absorbed through the skin. Studies indicate that mercury tends to accumulate in the brains of primates and other animals after they are injected with vaccines. Mercury poisoning has been linked to cardiovascular disease, autism, seizures, mental retardation, hyperactivity, dyslexia, and many other nervous system conditions. That's why the FDA rigorously limits exposure to mercury in foods and drugs. Some common sources of mercury include dental amalgam fillings, various vaccines, and certain fish contaminated by polluted ocean waters.

The toxicity of mercury has never been in question. The real question is precisely how much mercury-laced thimerosal is toxic, and what are the possible consequences for our children at low doses?

Eli Lilly and Co. supposedly answered this question for us back in 1930. Concluding thimerosal to be of "a very low order of toxicity . . . for man," the company hired its own doctors to perform thimerosal experiments in Indianapolis City Hospital on meningitis patients during a severe outbreak in 1929. This 60-year-old evidence was still quoted on the company's brochures as recently as 1990. Andrew Waters, who is involved in a lawsuit against Eli Lilly, claims that most critical studies on the toxicity of thimerosal were suppressed by the company until now.

Banned Around the World, But Not in the United States

That might explain why thimerosal was eliminated in many countries 20 years ago. In 1977, a Russian study found that adults exposed to ethylmercury, the form of mercury in thimerosal, suffered brain damage years later. Studies on thimerosal poisoning also describe tubular necrosis and nervous system injury, including obtundation, coma, and death. As a result of these findings,

S. Nabi, *Toxic Effects of Mercury*, DOI 10.1007/978-81-322-1922-4_29,
© Springer India 2014

Russia banned thimerosal from children's vaccines in 1980. Denmark, Austria, Japan, Great Britain, and all the Scandinavian countries have also banned the preservative.

Eli Lilly stuck to its "scientific" facts, but the truth began slipping between the cracks in 1999. After the number of immunizations rose to 12–15 per child, the public finally became privy to the possible dangers of thimerosal. One 1999 study revealed that some infants, due to a genetic or developmental factor, lack the ability to eliminate mercury. Trace amounts of mercury in these infants, when accumulated over several vaccines, could pose a severe health risk. Some vaccines, such as vaccines for hepatitis B, contained as much as 12.5 mcg of mercury per dose.

That's more than 100 times the EPA'S upper limit standard when administered to infants

Hepatitis B vaccines aren't the only immunizations under suspicion. According to Burton Goldberg in *Alternative Medicine*, scientists are finding stronger and stronger links between thimerosal and neurological damage. One report by Dr. Vijendra Singh of the Department of Pharmacology at the University of Michigan found a higher incidence of measles, mumps, and rubella vaccine (MMR) antibodies in autistic children.

The National Vaccine Information Center in Vienna, Virginia, has noted a strong association between the MMR vaccine and autistic features. Reporting similar findings, the Encephalitis Support Group in England claims that children who became autistic after the MMR vaccine started showing autistic symptoms as early as 30 days after vaccination. The diphtheria, pertussis, and tetanus vaccine (DPT) given at 2, 4, and 6 months has triggered autistic symptoms, as well.

When the FDA finally formally released this information in 1999, the news came too little, too late for some parents. The damage had already been done.

Links Between Autism and Thimerosal

Autism affects 500,000–1.5 million Americans and has grown at an annual rate of 10–17 % since the late 1980s. California found a 273 % increase in autism between 1987 and 1998. Maryland reported a 513 % increase in autism between 1993 and 1998, and several dozen other states reported similar findings. Some scientists say the estimated number of cases of autism has increased 15-fold – 1,500 % – since 1991, when the number of childhood vaccinations doubled. Whereas one in every 2,500 children was diagnosed with autism before 1991, one in 166 children now have the disease.

This increase in reported autism cases eerily parallels the increase in the number and frequency of thimerosal-containing vaccinations administered to infants. As of today, children are given as many as 21 immunizations in the first 15 months of life. After a number of scientists and concerned activists noticed the correlation, an investigation was launched to get to the heart of the matter.

Statistical Evidence Links Thimerosal with Nervous System Disorders

In June 2000, federal officials and industry representatives were assembled by the Centers for Disease Control and Prevention to discuss the disturbing evidence. According to Tom Verstraeten, an epidemiologist who had analyzed the data on the CDC's database, thimerosal appeared to be responsible for a dramatic increase in autism and other neurological disorders. Verstraeten told those at the meeting that a number of earlier studies indicate a link between thimerosal and speech delays, attention deficit hyperactivity disorder, hyperactivity, and autism.

Verstraeten offered no possible cause for this correlation but held that the statistical evidence linking vaccines and neurological disorders was strong. Dr. Bill Weil, a consultant for the American Academy of Pediatrics, and Dr. Richard Johnston, an immunologist and pediatrician from the University of Colorado, presented similar concerns to the group. However, given no causal relationship, the CDC and industry representatives were quick to discredit the evidence.

Consequently, the CDC paid the Institute of Medicine (IOM) to conduct another study on thimerosal. According to Robert F. Kennedy Jr., this study was fixed in order to "whitewash" previous findings. In its 2001 report, the IOM's Immunization Safety Review Committee did conclude that the link between thimerosal and neurodevelopmental disorders was biologically plausible, though the evidence neither proved nor negated it. The Committee stated that phasing out thimerosal from vaccines was "a prudent measure in support of the public health goal to reduce mercury exposure of infants and children as much as possible." However, these findings offered no imperative. The data presented at the 2000 meeting was withheld from publication, and the link between thimerosal and autism remained "inconclusive."

But what does "inconclusive" mean? Well, that depends on who you talk to. According to the FDA, these "inconclusive" findings negate the risk of a causal relationship between thimerosal and autism. Even Tom Verstraeten, one of the presenters of epidemiological evidence at the CDC meeting, seemingly changed his tune a bit. In 2000, Verstraeten vigorously campaigned against thimerosal based upon his "inconclusive" correlation, but after he was hired by GlaxoSmithKline, the doctor changed his position. The same evidence from 2000, in Verstraeten's eyes, became "neutral" in 2003. After criticism for this apparent flip-flop, Verstraeten wrote a letter to the editor of *Pediatrics* in 2004 backing the CDC's actions and his own research methods.

Merck Continues Selling Vaccines with Thimerosal

Without an imperative to eradicate thimerosal immediately, vaccine manufacturers like Merck & Co. seemingly took their time in reducing thimerosal levels in vaccines. After a large public outcry in 1999, Merck & Co. began decreasing or eliminating the amount of thimerosal in its vaccines. In September 1999, Merck announced that its new line of vaccines were preservative-free but still continued to distribute the remainder of thimerosal-preserved vaccines until 2001. Only after a congressional inquiry in 2002 did they stop distributing their stockpile. Rep. Dave Weldon, R-Fla., called Merck's actions "misleading."

While officials at the Centers for Disease Control claim evidence is lacking to support the possible risks of thimerosal, Dr. Mark Geier, a Maryland geneticist and vaccinologist, along with his son and research partner David Geier, says the CDC has chosen to ignore the science. According to Dr. Geier, more than 5,000 articles have been published that question the safety of thimerosal in vaccines.

The Geiers analyzed the data and determined that the more thimerosal a child receives, the greater his or her chances are of being autistic. The CDC says the Geiers misused information from a CDC database that was not intended to help prove theories. Given no real causal mechanism linking thimerosal and autism, the game seems to have become one of slanting the data to suit the needs of government and industrial interests. Even Verstraeten has admitted that these "inconclusive" findings certainly don't rule out the possibility of finding a link in the future.

Grassroots Action Against Vaccine Manufacturers

Given the dearth of health organizations owning up to the dangers of thimerosal, many parents followed their gut instincts and took legal action against vaccine manufacturers. More than 4,200 families have filed lawsuits claiming thimerosal caused injuries to their children. These lawsuits often have two goals: First, to seek reparations for the loss of consortium (basically meaning that an autistic child creates emotional and psychological burdens on their family life), and, second, to ensure that these companies exercise more concern for public health and less concern for their own bottom line. The lawsuits are slow in producing results. The first constraint on these lawsuits is the National Childhood Vaccine Act of 1986. This act stipulates that victims cannot seek redress in the courts without first filing a claim for recovery in the federal Vaccine Court. The statute of limitations for this is within 3 years of "the first symptom or manifestation of onset or of the significant aggravation of a [vaccine-related] injury."

In the cases of many thimerosal victims, the link between autism and vaccines didn't appear until 6 years after the first vaccine was administered. While this statute has stopped some claims against vaccine manufacturers, including such big firms as Aventis, GlaxoSmithKline, Merck, and Johnson & Johnson, many judges are now allowing suits against Eli Lilly, the maker of thimerosal, to stand. While the Vaccine Act shields vaccine manufacturers, one judge reasons that the legislation does not protect the production of thimerosal because it is a "component."

The burden of proof in court is also extremely problematic for most of these suits. Given the supposed lack of scientific data, lawyers are hard-pressed to prove the link between thimerosal and autism. In what seems like an underhanded move, the CDC sold its data to a private company, ensuring that lawyers could not access it under the Freedom of Information Act.

In the past 5 years, the Congress has also aided vaccine manufacturers, supposedly for "security" reasons. In 2002, *a mysterious piggyback on the 2002 Homeland Security bill freed drug companies of liability in lawsuits regarding thimerosal.* Called the "Eli Lilly Protection Act" by outraged parents and activists, the then House Majority Leader Dick Armey told CBS News he snuck the amendment in to keep vaccine makers from going out of business. Armey claimed it was a matter of national security. "We need their vaccines if the country is attacked with germ weapons."

Ironically, foreign biological terrorism hasn't been a big problem for American citizens, but those whose lives (and the lives of their children) have been affected if not ruined by the harmful effects of thimerosal would undoubtedly say these potentially harmful vaccines are indeed a problem. Armey's piggyback bill was repealed in 2003, but that didn't stop lawmakers from continuing to protect the vaccine industry.

Senate Majority Leader Bill Frist is no stranger to the thimerosal debate, having received USD 873,000 in contributions from the pharmaceutical industry and USD 10,000 from Eli Lilly. Frist's position allowed him to attempt to help the industry from the inside, according to Kennedy. Kennedy reports that on five occasions, Frist tried to seal the government's vaccine-related documents and shield Eli Lilly from subpoenas. Frist also introduced a provision in the 2005 Senate Bill S-3 called the "Protecting United States in the War on Terror Act" that would effectively insulate the pharmaceutical industry from liability for thimerosal poisoning. Pharmaceutical manufacturers, including Merck, GlaxoSmithKline, Aventis, Wyeth, and Eli Lilly, can basically get off scot-free for their actions, even as more and more evidence suggests that top company officials were aware of the possible dangers and did nothing.

A secret memo leaked to the Los Angeles Times reportedly implicates one vaccine manufacturer, Merck & Co., for knowing that thimerosal could pose serious threats to infants. Allegedly, Dr. Maurice Hilleman, one of Merck's top scientists, warned the president of Merck of a possible threat as early as 1991. Dr. Hilleman told executives that 6-month-old children receiving regular immunizations frequently received mercury doses 87 times higher than guidelines for the maximum consumption of mercury. Given today's more prudent mercury standards, those thimerosal doses would be 400 times that of safe levels. Dr. Hilleman recommended in the memo that thimerosal be discontinued.

Not only do government and industry officials seem to be trying to downplay the possible harms of thimerosal, the media is also denying the issue coverage. Just recently, ABC flip-flopped on whether it will air interviews with Robert Kennedy Jr., a leading critic of thimerosal. ABC has been accused of suppressing the interviews because of its ties to the pharmaceutical industry.

The Thimerosal Debate Continues

Along with the enormous amount of controversy surrounding this issue, the 5-year-old plea for "more research" may have finally produced some results. Burton Goldberg notes that a defect in the myelinization process (insulation of nerve fibers) could explain mercury's propensity to cause autism and neurological damage. This may also account for the frequent development of epilepsy in older autistic children.

Scientists are also working on biological links that support the strong correlations. Researchers at the Northeastern University, working with scientists from the University of Nebraska, Tufts University, and Johns Hopkins University, may have recently found the mechanism by which thimerosal interferes with brain activity. If these researchers are right, vaccine manufacturers could do little to keep the damaging effects of thimerosal hidden.

Pharmacy professor Richard Deth and colleagues found that exposure to thimerosal potently interrupts growth factor signaling, causing adverse effects on the transfer of carbon atoms. These carbon atoms play a significant role in regulating normal DNA function and gene expression and are critical to proper neurological development. Additionally, the scientists recently obtained more insight into the mechanism by which thimerosal interferes with folate-dependent methylation. The mechanism inhibits the biosynthesis of the active form of vitamin B12 (methylcobalamin), a vitamin now being administered to autistic children.

The Experts Speak on Mercury, Vaccines, and Thimerosal

Now, all childhood vaccines have at least one mercury-free version, and I urge parents to ask for those versions if they choose to vaccinate their children. Injecting mercury into children, especially infants whose immune systems are still underdeveloped (hepatitis B shots are typically given at birth, before the immune system has developed), can be an assault to the immune system (*What Your Doctor May Not Tell You About Autoimmune Disorders* by Stephen B Edelson MD, p. 65).

In 1999, studies began to surface showing that multidose vial vaccines, such as the MMR and hepatitis B vaccines, contained enough thimerosal to expose vaccinated children to 62.5 ug of mercury per visit to the pediatrician. This is 100 times the dose considered safe by the Federal Environmental Protection Guidelines for infants! Worse yet, some infants will receive doses even higher; because thimerosal tends to settle in the vial. If it is not shaken up before being drawn, the first dose will contain low concentrations of mercury and the last dose will contain enormously high concentrations. If your baby is the unlucky one that gets the last dose, serious brain injury can result (*Health and Nutrition Secrets* by Russell L Blaylock MD, p. 166).

Thousands of families say they can demonstrate with videotapes and photos that their children were normal prior to being vaccinated, reacted badly to the vaccines, and became autistic shortly thereafter. The number of vaccines given before age 2 has risen from 3 in 1940, when autism occurred in perhaps one case per 10,000 births, to 22 different vaccines given before the age of 2 in the year 2000 (*Building Wellness with DMG* by Roger V Kendall PhD, p. 104).

We know that certain forms of mercury, such as methylmercury and phenylmercury, are highly lipid soluble, which makes the brain especially susceptible to mercury accumulation. These forms of mercury are found in vaccines as the preservative thimerosal. Once in the brain, it tends to attach itself to protein structures, especially to the cell membrane, where it can disrupt membrane functions. By binding to the cell membrane, mercury changes the membrane's fluid-like quality, making it stiffer and causing the cell to age faster. The brain is unique in that neurons depend on special microscopic tube-like structures within the cell, appropriately called neurotubules, for their function. These neurotubules are manufactured by the cell from a substance called tubulin. We know that mercury interacts with tubulin causing it to unravel. Studies in rats have shown that doses of mercury corresponding to those seen in humans can cause a 75 % increase in tubulin inhibition (*Health and Nutrition Secrets* by Russell L. Blaylock MD, p. 53).

In the case of the susceptible newborn infant and toddler, multiple exposures to mercury-containing and multiple antigen vaccines are highly suspect in the causation of multiple organ injury. The GI tract, the liver, the pancreas, the kidneys, the immune system, and the brain are major sites of mercury absorption. Researchers have clearly shown a chronic inflammatory bowel disease due to vaccine strain measles in a subset of children with autism (Thompson et al. 1988) (*Disease Prevention and Treatment* by Life Extension Foundation, p. 153).

Studies of autistic children have frequently shown very high levels of mercury, with no other source but vaccines found for the exposure. These levels are equal to those seen in adults during toxic industrial exposures. Several autism clinics have found dramatic improvements in the behavior and social interactions in children from whom the mercury was chelated. Results depended on how soon the mercury was removed following exposure, but permanent damage can be caused if the metal is not chelated soon enough. Still, even in cases of severe damage, because of the infant brain's tremendous reparative ability, improvements are possible. The problem of autism involves numerous body systems including the gastrointestinal, immune, and nervous systems; as a result, we see numerous infections and magnified effects of malnutrition. Intrepid workers in the shadows, that is, outside the medical establishment, have worked many miracles with these children using a multidisciplinary scientific approach completely ignored by the orthodoxy. Some children have even experienced a return to complete physiological normalcy (*Health and Nutrition Secrets* by Russell L. Blaylock MD, p. 166).

Mercury and autism mercury toxicity is a suspected cause of a steep rise – a tenfold increase between 1984 and 1994 – in diagnosed cases of autism in children around the world, according to some scientists. Specifically, the culprit is thimerosal, a mercury-based compound used as a preservative in vaccines commonly administered to babies and infants. Thimerosal-free vaccines are available. If you have a child who will be receiving vaccinations, ask for and make sure thimerosal-free vaccines are used. Kelp, with its essential minerals (especially calcium and magnesium), helps remove unwanted metal deposits (*Prescription for Dietary Wellness* by Phyllis A. Balch, p. 198).

The pertussis vaccine (DPT) may cause 45,000 cases of autism per year in United States, affecting 15 cases out of 10,000 vaccinations; also caused by the measles–mumps–rubella vaccine (MMR) that causes mental impairment, gastrointestinal damage, and increased mortality in 6–12 months from impaired immunity; nine out of ten cases were not breast-fed; eating dairy products caused parasites in the autistic (take Vermex; contact Dr. Nelson in Mexico for control of parasites in children with autism). There were over 500,000 victims of autism residing in the United States in 1994. The pertussis vaccination is not used in Sweden, which has virtually 0 cases of autism, as does Holland. This mental illness afflicts environmentally and socially nonreactive persons, of withdrawn personality, with inability to speak, violent tantrums, insomnia, and actions such as bolting across a road with no regard for the dire consequences, may be caused by infant antibiotic use in ear infections with subsequent yeast overgrowth, by cumulative genetic brain damage, vitamin deficiencies, or milk and additive allergies. Immune disorders in autism include white blood cell neutrophil myeloperoxidase enzyme deficiency for insufficient hypochlorite ions to kill yeast, genetic type from chromosome 17 mutation or biotinidase deficiency, or acquired type from lead poisoning, folic acid or B12 deficiency, infection, or leukemias (*Anti-aging Manual* by Joseph B Marion, p. 450).

Multiple vaccinations, especially in newborns, are another major source of childhood mercury exposure because of the mercury-containing thimerosal preservative. Over 22 vaccinations are now recommended for children before the age of 2 (*Health and Nutrition Secrets* by Russell L Blaylock MD, p. 64).

In addition, there is some anecdotal evidence that autism may be tied to diet. One theory is that, in very rare cases, a child's immune system could be weakened by the measles–mumps–rubella vaccination (MMR), which is usually administered before a child turns 2. As a result of this weakening, the theory goes, the child's digestive system is unable to break down certain food proteins, leading to abnormal brain development. Proponents of this theory believe that putting the child on a diet that eliminates certain foods, such as wheat and dairy products, could in certain cases reverse the course of the disease. This theory remains speculative, however, and research needs to be done to determine its validity. In fact, a 2001 report issued by an Institute of Medicine committee examining studies about the health effects of the MMR vaccine in young children suggests that there is no proven link between the vaccine and autism. The committee recommends that there be no change in immunization practices that require children to be immunized during early childhood (*The Immune Advantage* by Ellen Mazo and Keith Berndtson MD, p. 292).

Rather than calling for an all-out immediate ban on thimerosal-containing vaccines, they suggested that parents continue to have their children vaccinated with mercury-contaminated vaccines until new stocks of uncontaminated vaccine could be made available. Here are two doctors' unions that had to be beat over the head with an overwhelming amount of data that mercury-contaminated vaccines were harming children far worse than the actual diseases against which the vaccine was intended to protect them, only to have them suggest that parents continue to harm their children just to satisfy their vaccination obsession. Are you surprised to discover that recent investigations have found that several doctor members of vaccine boards were either receiving grants from vaccine manufacturers or held stock in the companies? They were willing to sacrifice the health of millions of children just to fill their pockets with cash. These people should be looking through bars, not serving on boards (*Health and Nutrition Secrets* by Russell L Blaylock MD, p. 167).

Vaccines may afflict 45,000 cases of autism per year in United States, which afflicts 15 victims in every 10,000 births: There are now 500,000 of these victims in the United States. In Sweden, not using the pertussis vaccine, there is virtually no autism (and likewise in Holland) (*Anti-aging Manual* by Joseph B Marion, p. 600).

Many symptoms of autism are similar to those of mercury poisoning. Immune dysfunction, visual disturbances, and motor dysfunction are seen in both. Treating autistic children for removal of mercury and other heavy metals has shown significant improvement in their autistic symptoms. Most autistic individuals have poor liver detoxification, low antioxidant levels, and low levels of glutathione. Vaccines are effective, but the production and use of vaccines should proceed more cautiously. Currently manufactured vaccines still contain harmful substances like mercury. The link between vaccines and autism is far stronger than the medical community is willing to admit, and more research in this area should be an urgent priority (*Building Wellness with DMG* by Roger V Kendall PhD, p. 105).

Studies indicate that autism may be the result of adverse reactions to childhood vaccinations. Dr. Alan Cohen, an environmental physician from Connecticut, notes that high levels of autism and attention deficit hyperactivity disorder (ADHD) did not occur until the mandatory use of childhood vaccinations and suggests that there may be a connection between certain vaccines and the onset of these conditions (*Complete Encyclopedia of Natural Healing* by Gary Null PhD, p. 46).

Almost from the inception of vaccination programs, manufacturers added a mercury preservative called thimerosal to vaccines. The practice continued until recently and was stopped only because of the outcry from thousands of concerned parents and numerous experts in the field. The American Academy of Pediatrics and the American Academy of Family Practice did not warn parents or pediatricians that the mercury was dangerous until they were forced to. That mercury was toxic to cells had been known for over 60 years, but manufacturers apparently were more worried about lawsuits (*Health and Nutrition Secrets* by Russell L Blaylock MD, p. 165).

In fact, a 2001 report issued by an Institute of Medicine committee examining studies about the health effects of the MMR vaccine in young children suggests that there is no proven link between the vaccine and autism. The committee recommends that there be no change in immunization practices that require children to be immunized during early childhood. Another disorder affecting the brain, Alzheimer's disease, may also have an immune connection. Alzheimer's is a degenerative disease that slowly attacks nerve cells in the brain. It eventually results in the loss of all memory and mental functioning. Scientists are currently investigating the role that the immune system plays in producing an overabundance of the amino acid glutamate, a powerful nerve cell killer. Another immune connection that researchers are investigating is the idea that Alzheimer's might be triggered, in part, by a virus (*The Immune Advantage* by Ellen Mazo and Keith Berndtson MD, p. 292).

In the past 10 years, the number of autistic children has risen between 200 % and 500 % in every state in the United States. This sharp increase in autism followed the introduction of MMR vaccine in 1975. Representative Dan Burton's healthy grandson was given injections for nine diseases in 1 day. These injections were followed by autism (*A Physicians Guide to Natural Health Products that Work* by James Howenstine MD, p. 267).

> Probably 20 % of American children, one in five, suffers from a "development disability,"" according to Harris Coulter, Ph.D., Founder and Director of the Center for Empirical Medicine, in Washington, D.C. "This is a stupefying figure and we have inflicted it on ourselves.' Development disabilities' are nearly always generated by encephalitis. And the primary cause of encephalitis in the U.S. and other industrialized countries is the childhood vaccination program. To be specific, a large proportion of the millions of U.S. children and adults suffering from autism, seizures, mental retardation, hyperactivity, dyslexia, and other branches of the hydra-headed entity called 'development disabilities' owe their disorders to one of the vaccines against childhood diseases. (*Alternative Medicine* by Burton Goldberg, p. 1101)

Martin noted that the increased incidence of chronic fatigue syndrome, attention deficit hyperactivity disorder, autism, and other behavior-linked illnesses "may be an inadvertent consequence of stealth virus vaccine contaminants" (*AIDS and Ebola* by Leonard Horowitz, p. 493).

Just for perspective if we go back to 1971 up to 1980, we see that California consistently added 100–200 new cases a year; but in the year 2002, California added 3,577 new cases. Since 1980, the documented start of California's autism epidemic, the number of new cases has steadily increased. If we break down those statistics, it means that from 1994 to 1995, California only added on average 2 new autistic children a day into its system. In 2001, it was a rate of 8 new autistic children added a day; in 2002, it jumped up to 10 children a day. Mercury-containing vaccines are still in use today, including the most recently recommended addition to the childhood immunization schedule, two shots of flu vaccine for infants, bringing the total number of vaccines up to 41 in California that a child will receive before the age of 2. It will take a few years to start seeing the effect of the phasing out of the mercury-containing preservative thimerosal from childhood vaccines on this autism epidemic. Many symptoms of autism are similar to those of mercury poisoning. Immune dysfunction, visual disturbances, and motor dysfunction are seen in both. Treating autistic children for removal of mercury and other heavy metals has shown significant improvement in their autistic symptoms. Most autistic individuals have poor liver detoxification, low antioxidant levels, and low levels of glutathione (*Building Wellness with DMG* by Roger V Kendall PhD, p. 105).

Since the 1990s, there has been a tenfold or 1,000 % increase in autism, an increase which has been linked by some researchers to the organic mercury preservative commonly found in baby vaccines. A greatly increased incidence of juvenile diabetes has been correlated to specific vaccination sequences and to the number of vaccines given. In some Australian Aboriginal communities, every second child died shortly after vaccination (*The Natural Way to Heal* by Walter Last, p. 309).

The best current estimates are that autism occurs in 40–67 children per 10,000 live births. This means that the prevalence of autism has increased 1,000 % in the last decade. According to the figures released in January 2003 by the

California Department of Developmental Services, California experienced an astounding 31 % increase in the number of new children (*Building Wellness with DMG* by Roger V Kendall PhD, p. 104).

Reference

Thompson CM, Markesbery WR, Ehmann WD, Mao YX, Vance DE (1988) Regional brain trace-element studies in Alzheimer's disease. Neurotoxicology 9:1–7

30 Methylmercury and Lupus

Lupus: From the bitterness of disease man learns the sweetness of health.

(Catalan proverb)

Systemic Lupus Erythematosus

Recently, twin sisters Jennie Gunhammar, a photographer, and, her sister, Jessie, a lecturer at the London School of Economics, both died within a year of each other of the autoimmune disease, systemic lupus erythematosus (SLE or lupus) aged just 35 (Jennie and Jessie pictured right). The story hit the UK headlines because Jennie's long-term boyfriend, Andrew Clinch, has set up a charity in her name and held an art auction to raise money for a specialist unit at London's St. Thomas' Hospital.

The Incidence of Lupus

The tragedy is that these young, talented, and beautiful women are far from alone because systemic lupus erythematosus is not a rare disease at all.

The numbers of those affected in the United States had been estimated to be around 500,000, but a recent telephone survey commissioned by the Lupus Foundation of United States suggested that up to two million Americans could be affected by the illness. Using this revised figure would mean that approximately 400,000 UK citizens have SLE – ten times the current estimate.

Although lupus can be diagnosed at any age, like other autoimmune diseases, it is primarily a disease affecting young women of childbearing age, and estrogen is thought to play a large role in the illness. The female-to-male ratio in children is 3:1, in adults it can be as high as 15:1, and then it declines in postmenopausal women to approximately 8:1, averaging out at nine times as many women as men affected by SLE.

The illness is also more common in people of Asian and Hispanic descent than among Caucasians and up to four times more common among people of Afro-Caribbean descent. The highest rate of incidence in the United States was found in the predominantly black population surrounding Birmingham, Alabama, where as many as one in every 200 women were found to have SLE.

The incidence of all autoimmune diseases – including lupus – is also increasing dramatically with estimates suggesting between a three- and eightfold increase in recent decades although whether this is due to better diagnosis or to an increasing frequency of the disease is not known.

The other alarming fact is that autoimmune illnesses – once extremely rare – are now being diagnosed in childhood. The rates of childhood type 1 diabetes and autoimmune thyroid disorders such as Graves' disease, for example, are escalating at an alarming rate.

Autoimmune disorders are diseases associated with the developed world, and people

S. Nabi, *Toxic Effects of Mercury*, DOI 10.1007/978-81-322-1922-4_30,
© Springer India 2014

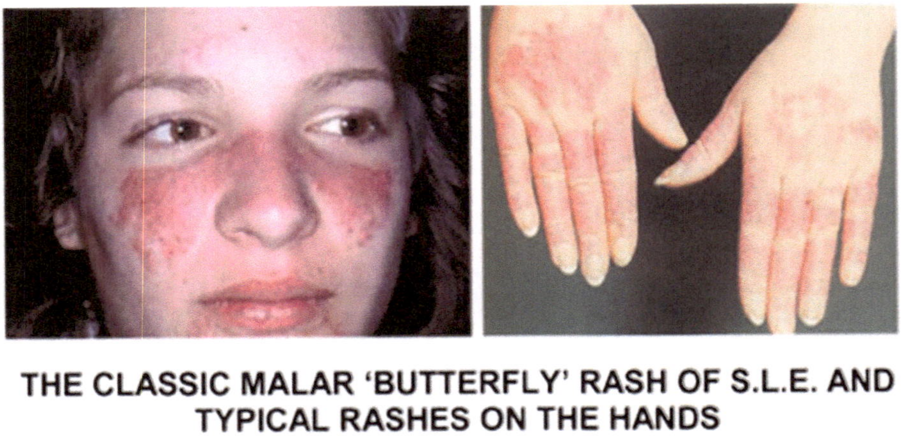

Fig. 30.1 Symptoms of lupus

migrating from low-incidence to high-incidence countries acquire immune disorders with a high incidence at the first generation. No socioeconomic link has been found. Lupus and other autoimmune diseases are also more common in urban than rural areas.

The allopathic world takes this to be evidence supporting the "hygiene hypothesis" in essence, the idea that the immune system will turn upon the body when not kept occupied fighting infectious organisms.

The incidence of all autoimmune conditions has increased to the extent that they are now one of the leading causes of death among younger women, although nearly 80 % now live for 20 or more years after diagnosis.

Famous SLE sufferers include Michael Jackson (who also had vitiligo); the singers Seal, Elaine Paige, Toni Braxton, and Lady Gaga (who has tested borderline positive for SLE but is currently asymptomatic); and Ferdinand Marcos, the former president of the Philippines who died of the complications of lupus.

What Is Lupus?

Lupus is a systemic autoimmune disease where the body attacks its own tissues. This causes inflammation and tissue damage and can affect any part of the body including the heart and blood vessels, the joints, skin, lungs, liver, kidneys, and nervous system. The disease typically has periods of remission interspersed with flare-ups known as flares which can be life-threatening.

As with so many other disorders, the cause of SLE is a mystery to allopathic medicine and is said to be multifactorial with a number of predisposing factors. The illness tends to run in families, but no causal gene has been identified. The onset of SLE may be associated with infections in general although no specific pathogen has been consistently linked to the disease. Onset may also be associated with taking pharmaceutical drugs or other factors such as collagen breast implants or exposure to ultraviolet light.

The symptoms of SLE vary widely, may imitate a host of other illnesses, and can come and go unpredictably, meaning that accurate diagnosis can be elusive and many may have suffered for years – if not decades – before a diagnosis of SLE is finally made. The symptoms of SLE include:

- Fever
- Malaise and fatigue.
- Arthritic joint pains and swelling especially in the joints of the hand and wrist.
- Muscle pain.
- Rashes on the cheeks and elsewhere which may be made worse by exposure to ultraviolet light as shown below (Fig. 30.1).

- A variety of pulmonary and cardiac disorders.
- Ulcerations of the mouth, nose, throat, and vagina.
- Blood disorders including anemia and low white blood cell and platelet counts.
- Kidney disorders including passing protein and/or blood in the urine which can lead to renal failure.
- Immune dysfunction leading to serious infections.
- Anxiety and depression. Depression affects up to 60 % of women with SLE.
- Alopecia.
- Headaches.
- Various neurological symptoms including weakness, loss of sensation, pins-and-needles sensations, burning pain, and seizures.
- Temporary cognitive dysfunction.
- Mood disorders and psychoses.
- Insomnia.
- Reduced fertility and an increased rate of spontaneous abortion and fetal death in female SLE sufferers.

Allopathic Medicine: The Diagnosis and Treatment of SLE

Cells throughout the body each have a limited life span after which they are dismantled by white blood cells known as macrophages and the components recycled to make new cells. This programmed cell death is known as *apoptosis* and is dysfunctional in people with SLE. In those affected by lupus, the macrophages in the lymphatic system are smaller, fewer in number, die sooner, and rarely contain material from apoptotic cells, and free nuclear debris is often found outside the macrophages.

The triggers for SLE such as viral infections and various drugs all cause damage to the cell which may expose the nucleus and cause the body to produce antinuclear antibodies (ANA), in addition to the generation of circulating immune complexes and activation of the complement system.

Laboratory tests for lupus may include testing positive for antinuclear antibodies (ANA) or anti-extractable nuclear antigen (anti-ENA), although some physicians may also test for other biomarkers such as complement levels, electrolytes, renal function, liver enzymes, and a complete blood count.

Allopathic treatment attempts to control symptoms using immunosuppressants and corticosteroids and to relieve pain using analgesics.

Natural Medicine: The Case for Metal Toxicity as a Cause of Lupus

Toxic metals in general and mercury in particular undermine the immune system, cause massive oxidative damage inducing apoptosis, and produce a host of other destructive biochemical effects throughout the body. Toxic metals can be derived from a number of different sources such as food packaged in aluminum foil, lead and copper piping used for water supplies, and nickel in hydrogenated fats.

In 1974, the World Health Organization acknowledged that heavy metal toxicity was a *major cause of world disease*, and yet this fact still seems to have gone largely unacknowledged by allopathic medicine. Many people have significant exposure to a variety of toxic metals through their restorative dental work and in particular to mercury from dental amalgam fillings.

Dental amalgam never truly sets but becomes a very stiff paste. The galvanic currents that are created by the various metals used both within the amalgam filling itself (silver, copper, tin, etc.) and those used in other restorations such as crowns (gold, palladium, nickel, etc.) create electrical activity within the moisture of the mouth and cause the liquid mercury within the amalgam fillings to evaporate. Please refer to the Smoking Tooth video (http://www.thenaturalrecoveryplan.com/243_Smoking-Tooth-Video/multimedia/details.html) for evidence.

The World Health Organization also determined in 1991 that the majority of mercury exposure (65–90 %) for most people comes from their dental amalgam fillings, although the mercury preservatives used in vaccinations (thimerosal) and eating the larger predatory fish are also key sources.

The Causative Role of Mercury in Immune Dysfunction in Lupus

A high percentage of patients with autoimmune disorders have significant immune reactions to mercury (72–94 % depending upon form), palladium, gold, and nickel using MELISA blood tests. In particular, for those with chronic conditions, fatigue has been shown to be *primarily associated with hypersensitivity to inorganic and organic mercury, nickel, and gold.*

In addition, studies in animals confirm that exposure to both inorganic mercury and organic methylmercury has been shown to induce autoimmune disorders.

The mechanism is that when metal particles enter the body, they bind with proteins. In some people, this protein–metal complex is identified by the immune system as being foreign, and the white blood cells, or lymphocytes, perceiving an attack go into defense mode.

Brain structures such as the hypothalamus, pituitary, and adrenal glands (the HPA axis) are also upregulated, and this constant and chronic stress on all bodily systems can cause exhaustion and ultimately death. This stress *will last as long as the inflammatory process is fuelled by the presence of toxic metals.*

The metalloprotein compounds that mercury forms also alter gene expression which affects cellular respiration, metabolism, and enzymatic processes and have been shown to have a relation to autoimmune reactions in significant numbers of people.

Mercury is also highly toxic to the immune system damaging and inhibiting T and B lymphocytes and neutrophils and inducing tumor necrosis factor-alpha (TNF-α) and the formation of antinuclear antibodies (ANA). TNF-α regulates the immune system and can induce apoptosis and inflammation and inhibits the formation of cancerous tumors and viral replication.

When taken together, the effects of mercury poisoning make the affected individual much more vulnerable to viral, bacterial, fungal, and parasitic infections which are often regarded as the trigger for autoimmune diseases.

Mercury and the Neurological Symptoms of Lupus

Toxic metals cause oxidative damage to the fats present in the brain and nerves and promote the release of inflammatory cytokines. This inflammation disrupts brain neurotransmitters resulting in reduced levels of serotonin, dopamine, and noradrenaline (norepinephrine) affecting mood and sleep.

Mercury in particular is also highly attracted to, and highly destructive of, nervous tissue, and this explains the neurological symptoms such as paralysis and tingling and also the severe psychological symptoms including depression and psychosis. Please refer to the Mercury and Neuron Degeneration video (http://www.thenaturalrecoveryplan.com/244_Mercury-and-Neuron-Degeneration/multimedia/details.html) for evidence.

In addition, the antibodies produced to attack the metalloprotein compounds can result in some of the more serious neurological symptoms of lupus as nerve tissue is targeted.

Heredity and Lupus

The reason why the autoimmune diseases appear to run in families has to do with a specific genetic blood factor (type apoE4) which causes the individual to bioaccumulate mercury, resulting in susceptibility to chronic autoimmune conditions such as lupus as early as age 40. Those with type apoE3 have a somewhat compromised ability to excrete mercury, and those with apoE2 readily excrete mercury and are less susceptible and therefore unlikely to contract an autoimmune disease.

Exposure to mercury will also vary with the individual, and estrogen exacerbates the symptoms of mercury toxicity making women more vulnerable than their male relatives.

The Symptoms of Mercury Poisoning and Lupus Are the Same

Many of the symptoms of autoimmune diseases are the same as those of mercury poisoning including fatigue, insomnia, headaches, rashes, respiratory and cardiovascular disorders, immune dysfunction, oral ulceration, muscle and joint pain, cognitive and mood problems, and many different neurological symptoms.

The body burden of toxic metals can take years to reach some kind of threshold and estrogen appears to be synergistic with mercury which may account for the increased incidence of lupus among women of childbearing age.

Other factors such as the widespread introduction of high-copper (non-gamma II) amalgam in the 1970s may partly account for the rapidly rising numbers of autoimmune disease sufferers. This is because high-copper amalgam emits *at least 50 times* the mercury of the previous formulation in addition to high amounts of toxic copper.

The relentless introduction of more and more vaccinations which often contain the mercury preservative, thimerosal, in addition to other toxic metals such as aluminum which are included as adjuvants (irritants intended to provoke the immune system to respond) is also possibly a key factor.

In the United States, the Centers for Disease Control now recommends 24 vaccines and boosters by the age of just 2, in addition to dozens more vaccinations throughout childhood, for adults undertaking foreign travel, for various occupations (hepatitis B for those exposed to body fluids), and annual flu vaccinations over the age of 50. For some people this literally translates to the best part of a hundred vaccines in a lifetime – all required to keep you "healthy!".

Have we, in short, exchanged protection against some infectious diseases (arguable) for a growing number of people afflicted with lifelong, chronic, disabling, and potentially fatal conditions?

This would also, at least in part, account for the growing number of young children being diagnosed with autoimmune diseases as their immune systems are assaulted by more and more vaccines at an ever younger age.

Lupus Symptoms Improve When Amalgam Fillings Are Removed

Patients with autoimmune conditions are often found to have a lot of high negative current dental amalgam fillings which work to drive metal ions into the tissues of the mouth. Symptoms in sufferers usually improve significantly after dental amalgam filling removal, and levels of autoimmune antibodies also decline. The common finding of mouth ulceration in lupus sufferers may be a direct indicator of the source of the problem.

The anti-amalgam dentist Dr. Hal Huggins has successfully treated over a thousand patients with chronic autoimmune conditions like lupus. He claims that approximately 85 % experienced a significant improvement in their health following a careful program of amalgam filling removal and mercury detoxification to reduce their body burden of mercury.

Mercury Impairs Detoxification

Mercury also reduces the available levels of glutathione, the liver enzyme required in order to detoxify mercury, and many other environmental and endogenous toxins. Toxic metals such as mercury also form a variety of inorganic compounds which can inhibit many cellular enzyme processes, coenzymes, hormones, and blood cells.

A consistent finding in patients with SLE is that they tend to have greatly elevated blood plasma cysteine to sulfate ratios and therefore have insufficient sulfates available to carry out necessary bodily processes. Mercury has been shown to diminish and block sulfur oxidation and thus reduce glutathione levels.

As the body becomes more toxic, the processes of energy production are impaired, the liver detoxification system is burdened, and fatigue and/or muscle pain can develop from toxic stress as the body becomes unable to detoxify harmful waste products or environmental toxins.

Epidemiological Evidence Supports Mercury Toxicity as a Cause of Lupus

The fact that autoimmune diseases such as lupus are almost exclusively diseases of the developed world suggests that the cause is environmental. The incidence of SLE in western Africans is very low compared to black Americans with West African ancestry, which strongly suggests the factor is related to our way of living. Within one generation of immigration, the incidence of autoimmune diseases increases dramatically.

Many suspect the vaccination program – which is mandatory in most states of United States – of being responsible for these findings possibly in association with toxicity from the metals used in dentistry. Note that tooth decay rates also increase dramatically within one generation as the diet of the developed world is adopted necessitating more dental treatment.

The fact that the incidence of SLE is higher in urban areas also suggests an environmental factor. It is possible that rural children receive less medical care (vaccinations) and dental care (amalgam fillings) than their urban cousins and also that they are exposed to less in the way of environmental toxins.

The Natural Medicine View of SLE

The rashes on the hands and face associated with lupus are found at the ends of energy meridians and are the body's attempt at detoxifying the mercury. Suppression of this response using corticosteroids is a triple whammy as far as the body is concerned. This is because not only are its attempts to detoxify being thwarted but adrenal function in specific and endocrine function in general is also steadily being undermined along with the immune response.

Natural medicine regards SLE as being caused by toxicity, and effective treatment has to systematically address the underlying cause. Please refer to Chronic Fatigue, ME and Fibromyalgia: The Natural Recovery Plan (http://www.thenaturalrecoveryplan.com/the_book.php) for a program of amalgam filling replacement and mercury detoxification as it is important to be fully informed.

Studies have shown that total avoidance of dairy products helps alleviate symptoms in many autoimmune disease sufferers, and exclusion of dietary excitotoxins such as monosodium glutamate (MSG) and aspartame is also strongly advised.

Lupus is a very difficult disease to diagnose. Because lupus rarely presents itself the same way in any two people, it is very challenging for those in the medical profession to understand, diagnose, and properly treat the disease. Often, a diagnosis can be long coming, which can be extremely frustrating for both the patient and physician alike. The fact that the symptoms lupus presents may have sudden onset or progress slowly and could be temporary or permanent makes diagnosing lupus confusing and concerning. There are, however, certain signs and symptoms that may cause you to ask, "Could I have lupus?".

Lupus often affects many different systems in the body, and therefore, if you do have lupus, the symptoms you may experience will depend heavily on which part of the body is being affected by the disease. Following is a thorough yet abbreviated list:

- *Brain and Nervous System*: Persistent and unusual headaches, memory loss, or confusion.
- *Lungs*: Lupus can damage the lungs through pleurisy and pneumonitis (inflammation), or pulmonary emboli, resulting in shortness of breath and pain in the chest from deep breathing.
- *Renal System*: About half of systemic lupus erythematosus (SLE) patients will develop some form of kidney inflammation, called lupus nephritis. This inflammation can lead to kidney failure, but like most lupus symptoms, the effect on the kidneys is quite variable and hard to predict. Increased protein (showing as blood) in the urine, swelling of the feet and legs, and high blood pressure can be indicators that the kidneys may be affected.

A Quick Synopsis of Common Lupus Symptoms

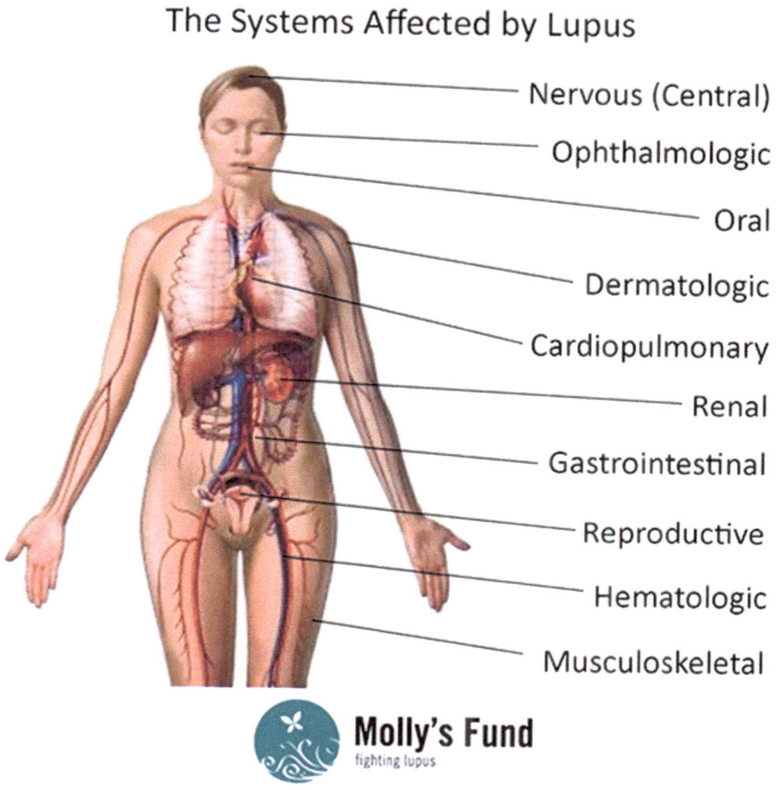

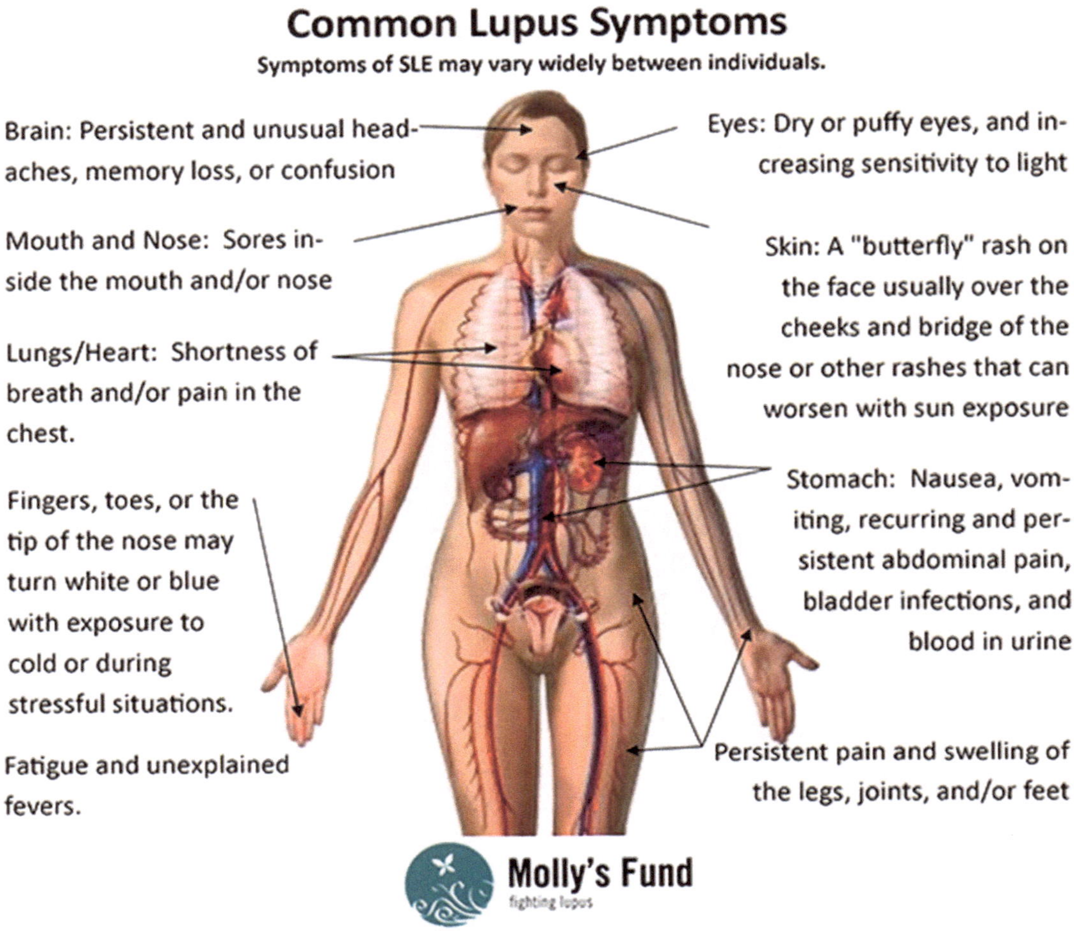

- *Eyes*: Damage to nerves and blood vessels in the eye might be evident, leading to dry or puffy eyes and increasing sensitivity to light.
- *Mouth*: Sores inside the mouth are a common symptom of lupus.
- *Skin*: Lupus is known for producing a distinctive "butterfly" rash on the face, usually over the cheeks and bridge of the nose. These rashes can be exacerbated by sun exposure (photosensitivity). Hives or sores may be present, which worsen with sun exposure, and sudden and unexplained hair loss could also signify lupus.
- *Fingers, Toes, and Tip of the Nose*: Fingers may turn white or blue with exposure to cold or during stressful situations, caused by the constriction of small blood vessels in those areas. Called Raynaud's phenomenon, this condition is closely associated with lupus.
- *Stomach and Digestion*: Lupus can cause or exacerbate ulcerative colitis, pancreatitis, and liver conditions, resulting in nausea, vomiting, and recurring and persistent abdominal pain, bladder infections, and blood in urine.
- *Legs, Joints, and Feet*: Persistent joint pain and swelling is a common lupus symptom; legs and feet may also swell.
- *Fatigue and Unexplained Fevers*

31 Methylmercury and Amyotrophic Lateral Sclerosis

A disease that will challenge your courage, test your faith, your love and your will to fight for your life and your right to stay in control of your comfort and care.

(Debbie Tope)

ALS is a systemic motor neuron disease that affects the corticospinal and corticobulbar tracts, ventral horn motor neurons, and motor cranial nerve nuclei. Approximately 10 % of ALS cases are of the familial type that has been linked to a mutation of the copper/zinc superoxide dismutase gene (Cu-Zn SOD). The majority of ALS cases are of the sporadic type. Based on studies of groups of monozygous twins, animal studies, and ALS patient case studies, the majority of ALS cases does not appear to be genetic but rather have primarily environmental-related causes often affecting genetically susceptible individual. Mutation of the FUS gene or TPD-43 gene has been shown to be one of the major factors in familial and some sporadic ALS.

ALS is not a unique disease with a single cause or factor but instead is a result of damage to motor neurons and the support system that they depend on by a variety of factors. Spinal and bulbar-onset subtypes of the disease appear to be biochemically different and have differences in mechanisms of causality. Some of the mechanisms of neural damage found in ALS include increased free radical generation/oxidative damage, impaired electron transport, disrupted calcium channel function, reactive astrogliosis and dysfunctional transporters for L-glutamate, neurotoxicity, oxidative damage to mitochondrial DNA inhibition of the mitochondrial respiratory chain, autoimmunity, and generalized disruption of metabolism of neuroexcitotoxic amino acids like glutamate, aspartate, and NAAG. The mechanisms by which exposure to mercury and other neurotoxic substances causes all of this will be documented.

The main factors determining whether chronic conditions are induced by metals appear to be exposure and genetic susceptibility, which determines individual's immune sensitivity and ability to excrete and detoxify metals. Very low levels of exposure have been found to seriously affect relatively large groups of individuals who are immune sensitive to toxic metals or have an inability to detoxify metals due to such as deficient sulfoxidation or metallothionein function or other inhibited enzymatic processes related to detoxification or excretion of metals. Those with the genetic allele apoE4 protein in the blood have been found to detox metals poorly and to be much more susceptible to chronic neurological conditions than those with types apoE2 or E3. There are also other similar factors.

Some of the toxic exposures which have been found to be a factor in ALS-like symptoms other than mercury include lead, pyrethrins, agricultural chemicals, Lyme disease, monosodium glutamate (MSG,580), failed root canalled teeth, post-poliomyelitis, pesticides/formaldehyde, and smoking. All have been demonstrated to cause

S. Nabi, *Toxic Effects of Mercury*, DOI 10.1007/978-81-322-1922-4_31,
© Springer India 2014

some of the mechanisms of damage listed above seen in ALS, and since such exposures are common as is exposure to mercury, such exposures appear to synergistically cause the types of damage seen in ALS. A study of approximately 1,000 men and women who died of ALS found that male programmers and laboratory technicians and female machine assemblers may be at increased risk of death from ALS.

This chapter will demonstrate that mercury is the most common of toxic substances which are documented to accumulate through chronic exposure in the neurons affected by ALS and which have been documented to cause all of the conditions and symptoms seen in ALS. It will also be noted that chronic infections such as mycoplasma, echo-7 enterovirus, and Candida albicans also usually affect those with chronic immune deficiencies such as ALS patients and need to be dealt with in treatment. Some studies have also found persons with chronic exposure to electromagnetic fields (EMF) to have higher levels of mercury exposure and excretion and higher likelihood of getting chronic conditions like ALS.

Documentation of High Common Exposures and Accumulation of Mercury in Motor Neurons

Amalgam dental fillings are the largest source of mercury in most people with daily exposures documented to commonly be above government health guidelines. This is due to continuous vaporization of mercury from amalgam in the mouth, along with galvanic currents from mixed metals in the mouth that deposit the mercury in the gums and oral cavity. Mercury has been found in autopsy studies to accumulate in the brain of those with chronic exposures, and levels are directly proportional to the number of amalgam filling surfaces. Due to the high daily mercury exposure and excretion into home and business sewers of those with amalgam, dental amalgam is also the largest source of the high levels of mercury found in all sewers and sewer sludge, and, thus, according to government studies, a significant source of mercury in rivers, lakes, bays, fish, and crops. People also get significant exposure from vaccinations, fish, and dental office vapor.

When amalgam was placed into teeth of monkeys and rats, within 1 year mercury was found to have accumulated in the brain, trigeminal ganglia, spinal ganglia, kidneys, liver, lungs, hormone glands, and lymph glands. People also commonly get exposures to mercury and other toxic metals such as lead, arsenic, and nickel and aluminum from food, water, and other sources. All of these are highly neurotoxic and are documented to cause neurological damage which can result in chronic neurological conditions over time.

Mercury has been found to accumulate preferentially in the primary motor function-related areas involved in ALS – such as the brain stem, cerebellum, rhombencephalon, dorsal root ganglia, and anterior horn motor neurons, which enervate the skeletal muscles.

Mercury, with exposure either to vapor or organic mercury, tends to accumulate in the glial cells in a similar pattern, and the pattern of deposition is the same as that seen from morphological changes. Though mercury vapor and organic mercury readily cross the blood–brain barrier, mercury has been found to be taken up into neurons of the brain and CNS without having to cross the blood–brain barrier, since mercury has been found to be taken up and transported along nerve axons as well through calcium and sodium channels and along the olfactory path. Exposure to inorganic mercury has significant effects on blood parameters and liver function. Studies have found that in a dose-dependent manner, mercury exposure causes reductions in oxygen consumption and availability, perfusion flow, biliary secretion, hepatic ATP concentration, and cytochrome P450 liver content, while increasing blood hemolysis products and tissue calcium content and inducing heme oxygenase, porphyria, and platelet aggregation through interfering with the sodium pump.

Effects of Exposure to Mercury and Toxic Metals

A direct mechanism involving mercury's inhibition of cellular enzymatic processes by binding with the hydroxyl radical (SH) in amino acids appears to be a major part of the connection to allergic/immune reactive conditions such as eczema, psoriasis, rheumatoid arthritis, lupus, scleroderma, allergies, autism, and schizophrenia, as well as to autoimmune conditions such as ALS, Alzheimer's (AD), chronic fatigue (CFS), fibromyalgia (FM), etc. For example, mercury has been found to strongly inhibit the activity of dipeptidyl peptidase (DPP IV) which is required in the digestion of the milk protein casein as well as of xanthine oxidase. Additional cellular level enzymatic effects of mercury's binding with proteins include blockage of sulfur oxidation processes, enzymatic processes involving vitamins B6 and B12, effects on the cytochrome C energy processes, along with mercury's adverse effects on cellular mineral levels of calcium, magnesium, copper, zinc, and lithium. And along with these blockages of cellular enzymatic processes, mercury has been found to cause additional neurological and immune system effects in many by causing immune/autoimmune reactions. Recent studies give a comprehensive review of studies finding a connection between ALS, toxic metals, and autoimmunity. Studies have found the presence of antibodies in ALS patients that interact with motor neurons, inhibiting the sprouting of axons. Immune complexes have also been found in the spinal cords of ALS patients. T cells, activated microglia, and IgG within the spinal cord may be a primary event that leads to lesions and tissue destruction.

Oxidative stress and reactive oxygen species (ROS) have been implicated as major factors in neurological disorders including ALS, motor neuron disease (MND), CFS, FM, Parkinson's (PD), multiple sclerosis (MS), and Alzheimer's (AD). Mercury forms conjugates with thiol compounds such as glutathione and cysteine and causes depletion of glutathione, which is necessary to mitigate reactive damage. One study found that insertion of amalgam fillings or nickel dental materials causes a suppression of the number of T lymphocytes and impairs the T4/T8 ratio. Low T4/T8 ratio has been found to be a factor in autoimmune conditions. Mercury-induced lipid peroxidation has been found to be a major factor in mercury's neurotoxicity, leading to decreased levels of glutathione peroxidation and superoxide dismutase (SOD). Only a few micrograms of mercury severely disturb cellular function and inhibit nerve growth. Metalloprotein (MT) has a major role in regulation of cellular copper and zinc metabolism, metals transport and detoxification, free radical scavenging, and protection against inflammation. Mercury inhibits sulfur ligands in MT and in the case of intestinal cell membranes inactivates MT that normally binds cuprous ions, thus allowing buildup of copper to toxic levels in many and malfunction of the Zn-Cu SOD function. Mercury also causes displacement of zinc in MT and SOD, which has been shown to be a factor in neurotoxicity and neuronal diseases. Exposure to mercury results in changes in metalloprotein compounds that have genetic effects, having both structural and catalytic effects on gene expression. Some of the processes affected by such MT control of genes include cellular respiration, metabolism, enzymatic processes, metal-specific homeostasis, and adrenal stress response systems. Significant physiological changes occur when metal ion concentrations exceed threshold levels. Such MT formation also appears to have a relation to autoimmune reactions in significant numbers of people. Of a population of over 3,000 tested by the immune lymphocyte reactivity test, 22 % was tested positive for inorganic mercury and 8 % for methylmercury, but much higher percentages tested positive among autoimmune condition patients. In the MELISA laboratory, 12 out of 13 ALS patients tested showed positive immune reactivity lymphocyte responses to metals in vitro, indicating metal reactivity a likely major factor in their condition. A recent study assessed the possible causes of high ALS rates in Guam and similar areas and the recent decline in this condition. One of the studies' conclusion was that a likely major factor for the high ALS rates in Guam and similar areas in the past was chronic dietary deficiency since reduced Ca, Mg, and Zn induced excessive absorption of divalent metal cations such as mercury which accelerates oxidant-mediated neuronal degenerations in a genetically susceptible population. The Veterans Administration concluded that higher levels of veterans of Gulf War I than normal contracted ALS. These veterans were subjected to large exposures of toxic metals in vaccines and other toxic exposures, and there is evidence that aluminum hydroxide in vaccines can cause symptoms seen in ALS.

Programmed cell death (apoptosis) is documented to be a major factor in degenerative neurological conditions like ALS, Alzheimer's, MS, Parkinson's, etc. Some of the factors documented to be involved in apoptosis of neurons and immune cells include inducement of the inflammatory cytokine tumor necrosis factor-alpha (TNF-α), reactive oxygen species, and oxidative stress; reduced glutathione levels; liver enzyme effects; inhibition of protein kinase C and cytochrome P450; nitric oxide and peroxynitrite toxicity, excitotoxicity, and lipid peroxidation; excess free cysteine levels; excess glutamate toxicity; excess dopamine toxicity; amyloid-beta generation; and increased calcium influx toxicity and DNA fragmentation and mitochondrial membrane dysfunction.

Chronic neurological conditions such as ALS appear to be primarily caused by chronic or acute brain inflammation. The brain is very sensitive to inflammation. Disturbances in metabolic networks, e.g., immuno-inflammatory processes, insulin–glucose homeostasis, adipokine synthesis and secretion, intracellular signaling cascades, and mitochondrial respiration, have been shown to be major factors in chronic neurological conditions. Inflammatory chemicals such as mercury, aluminum, and other toxic metals as well as other excitotoxins including MSG and aspartame cause high levels of free radicals, lipid peroxidation, inflammatory cytokines, and oxidative stress in the brain and cardiovascular systems.

In amyotrophic lateral sclerosis (ALS), non-neuronal cells play key roles in disease etiology and loss of motor neurons via noncell-autonomous mechanisms. Reactive astrogliosis and dysfunctional transporters for L-glutamate are common hallmarks of ALS pathology. Oxidative and excitotoxic insults exert differential effects on spinal motor neurons and astrocytic glutamate transporters in the progression of ALS. Excitotoxicity in ALS affects both motor neurons and astrocytes, favoring their local interactive degeneration. Mercury and other toxic metals inhibit astrocyte function in the brain and CNS, causing increased glutamate and calcium-related neurotoxicity. Mercury and increased glutamate in the plasma activate free radical-forming processes like xanthine oxidase which produce oxygen radicals and oxidative neurological damage. Nitric oxide-related toxicity caused by peroxynitrite formed by the reaction of NO with superoxide anions, which results in nitration of tyrosine residues in neurofilaments and manganese superoxide dismutase (SOD), has been found to cause inhibition of the mitochondrial respiratory chain, inhibition of the glutamate transporter, and glutamate-induced neurotoxicity involved in ALS. A recent study has linked some cases of sporadic ALS with the failure to edit key residues in ionotropic glutamate receptors, resulting in excessive influx of calcium ions into motor neurons which in turn triggers cell death. The study suggests that edited AMPA glutamate (GluR2) receptor subunits serve as gatekeepers for motor neuron survival.

These inflammatory processes damage cell structures including DNA, mitochondria, and cell membranes. They also activate microglia cells in the brain, which control brain inflammation and immunity. Once activated, the microglia secrete large amounts of neurotoxic substances such as glutamate, an excitotoxin, which adds to inflammation and stimulates the area of the brain associated with anxiety. Inflammation also disrupts brain neurotransmitters resulting in reduced levels of serotonin, dopamine, and norepinephrine. Some of the main causes of such disturbances that have been documented include vaccines, mercury, aluminum, other toxic metals, MSG, aspartame, etc. High levels of aluminum exposure along with low levels of other minerals such as calcium and magnesium have been documented to cause neurological degeneration and appear to be the cause of high ALS and Parkinson's in the past in Guam. There is evidence that aluminum hydroxide in vaccines can cause symptoms such as those seen in ALS. Aluminum has been found to be a factor in some Alzheimer's and Parkinson's cases.

Programmed cell death (apoptosis) is documented to be a major factor in degenerative neurological conditions like ALS, Alzheimer's,

MS, Parkinson's, etc. Some of the factors documented to be involved in apoptosis of neurons and immune cells include mitochondrial membrane dysfunction. Mitochondrial DNA mutations or dysfunction is fairly common, found in at least 1 in every 200 people, and toxicity effects affect this population more than those with less susceptibility to mitochondrial dysfunction. Mercury depletion of GSH and damage to cellular mitochondria and the increased lipid peroxidation in protein and DNA oxidation in the brain appear to be a major factor in conditions such as ALS, Parkinson's disease, autism, etc.

Reduced levels of magnesium and zinc are related to metabolic syndrome, insulin resistance, and brain inflammation and are protective against these conditions. Mercury and cadmium inhibiting magnesium and zinc levels as well as inhibiting glucose transfer are other mechanisms by which mercury and toxic metals are factors in metabolic syndrome and insulin resistance/diabetes.

TNF-α (tumor necrosis factor-alpha) is a cytokine that controls a wide range of immune cell response in mammals, including cell death (apoptosis). This process is involved in inflammatory and degenerative neurological conditions like ALS, MS, Parkinson's, rheumatoid arthritis, etc. Cell signaling mechanisms like sphingolipids are part of the control mechanism for the TNF-α apoptosis mechanism. Glutathione is an amino acid that is a normal cellular mechanism for controlling apoptosis. When glutathione is depleted in the brain, reactive oxidative species increased, and CNS and cell signaling mechanisms are disrupted by toxic exposures such as mercury, neuronal cell apoptosis results, and neurological damage. Mercury has been shown to induce TNF-α, deplete glutathione, and increase glutamate, dopamine, and calcium-related toxicity, causing inflammatory effects and cellular apoptosis in neuronal and immune cells. Mercury's biochemical damage at the cellular level includes DNA damage; inhibition of DNA and RNA synthesis; alteration of protein structure; alteration of the transport and signaling functions of calcium; inhibition of glucose transport and of enzyme function and transport of other essential nutrients; induction of free radical formation; depletion of cellular glutathione (necessary for detoxification processes); inhibition of glutathione peroxidase enzyme; inhibition of glutamate uptake; inducement of peroxynitrite; lipid peroxidation damage, causing abnormal migration of neurons in the cerebral cortex and immune system damage; inhibition of functional methylation; and inducement of inflammatory cytokines and autoimmunity.

Exposure to mercury vapor and methylmercury is well documented to commonly cause conditions involving tremor, with populations exposed to mercury experiencing tremor levels on average proportional to exposure level. However, bacteria, yeasts, and vitamin B12 methylate inorganic mercury to methylmercury in the mouth and intestines, and mercury inhibits functional methylation in the body, a necessary process.

Mercury exposure causes high levels of oxidative stress/reactive oxygen species (ROS), which has been found to be a major factor in apoptosis and neurological disease including dopamine or glutamate-related apoptosis.

Mercury and quinones form conjugates with thiol compounds such as glutathione and cysteine and cause depletion of glutathione, which is necessary to mitigate reactive damage. Such conjugates are found to be highest in the brain substantia nigra with similar conjugates formed with L-Dopa and dopamine in Parkinson's disease. Mercury depletion of GSH and damage to cellular mitochondria and the increased lipid peroxidation in protein and DNA oxidation in the brain appear to be a major factor in Parkinson's disease and a factor in other neurological conditions.

Mercury blocks the immune function of magnesium and zinc, whose deficiencies are known to cause significant neurological effects. The low Zn levels result in deficient Cu-Zn superoxide dismutase (Cu-ZnSOD), which in turn leads to increased levels of superoxide due to toxic metal exposure. This is in addition to mercury's effect on metallothionein and copper homeostasis as previously discussed. Copper is an essential trace metal which plays a fundamental role in the biochemistry of the nervous system. Several chronic neurological conditions involving copper metabolic disorders are well documented like Wilson's disease and Menkes disease. Mutations in the copper/zinc enzyme superoxide dismutase (SOD) have been shown to be a major factor in the motor neuron degeneration in conditions like familial ALS. Exposures to toxic metals such as mercury and cadmium have been found to cause such effects, and similar effects on Cu-Zn SOD have been found to be a factor in other conditions such as autism, Alzheimer's, Parkinson's, and nonfamilial ALS. This condition can result in zinc-deficient SOD and oxidative damage involving nitric oxide, peroxynitrite, and lipid peroxidation, which have been found to affect glutamate-mediated excitability and apoptosis of nerve cells and effects on mitochondria. These effects can be reduced by zinc supplementation, as well as supplementation with antioxidants and nitric oxide-suppressing agents and peroxynitrite scavengers such as vitamin C, vitamin E, lipoic acid, coenzyme Q10, carnosine, Ginkgo biloba, N-acetyl-cysteine, melatonin, etc. In a study involving over one million participants, a 23 % reduction in the risk of the disease was found among those who used vitamin E supplements for 2–4 years and a 36 % reduction occurred among those who used the supplements for 5 years or more compared to those who did not supplement with the vitamin. For those whose vitamin E from diet was among the top 25 % of participants, a 21 % lower adjusted risk of ALS was noted. This effect increased with greater dietary vitamin E intake among women, with those in the top 25 % having a 43 % lower risk than that experienced by those whose intake was lowest. Vitamin E has attracted significant attention from ALS researchers as a result of its antioxidant properties. Vitamin E protects cell membranes against a process known as lipid peroxidation. Lipid peroxidation is the breakdown of the cell membrane and appears to play a role in degenerative diseases such as ALS. Another study in humans indicated that vitamin E can help prevent ALS because of its antioxidant properties.

Ceruloplasmin in plasma can be similarly affected by copper metabolism dysfunction, like SOD function, and is often a factor in neurodegeneration.

Motor neuron dysfunction and loss in amyotrophic lateral sclerosis (ALS) have been attributed to several different mechanisms, including increased intracellular calcium, glutamate dysregulation and excitotoxicity, oxidative stress and free radical damage, nitric oxide-related toxicity caused by peroxynitrite, mitochondrial damage/dysfunction, neurofilament aggregation and dysfunction of transport mechanisms, and autoimmunity. These alterations and effects are not mutually exclusive but rather are synergistic, and increased calcium and altered calcium homeostasis appears to be a common denominator. Mercury forms strong bonds with the –SH groups of proteins causing alteration of the transport of calcium and causes mitochondrial release of calcium. This results in a rapid and sustained elevation in intracellular levels of calcium. Calcium plays a major role in the extreme neurotoxicity of mercury and methylmercury. Both inhibit cellular calcium ATPase and calcium uptake by brain microsomes at very low levels of exposure. Protein kinase C (PKC) regulates intracellular and extracellular signals across neuronal membranes, and both forms of mercury inhibit PKC at micromolar levels, as well as phorbol ester binding. They also block or inhibit calcium L-channel currents in the brain in an irreversible and concentration-dependent manner. Mercury vapor or inorganic mercury exposure affects the posterior cingulate cortex and causes major neurological effects with sufficient exposure. Metallic mercury is much more potent than methylmercury in such actions, with 50 % inhibition in animal studies at 13 ppb. Mercury is seen to be a factor in all of these known mechanisms of neural degeneration seen in ALS and other motor neuron conditions.

Spatial and temporal changes in intracellular calcium concentrations are critical for controlling gene expression and neurotransmitter release in neurons. Mercury alters calcium homeostasis and calcium levels in the brain and affects gene expression and neurotransmitter release through its effects on calcium. Mercury inhibits sodium and potassium (N, K) ATPase in a dose-dependent manner and inhibits dopamine and norepinephrine uptake by synaptosomes and nerve impulse transfer. Mercury also interrupts the cytochrome oxidase system, blocking the ATP energy function, lowering immune growth factor IGF-I levels, and impairing astrocyte function. Astrocytes are common cells in the CNS involved in the feeding and detox of nerve cells. Increases in inflammatory cytokines such as caused by toxic metals trigger increased free radical activity and damage to astrocyte and astrocyte function. IGF-I protects against brain and neuronal pathologies like ALS, MS, and fibromyalgia by protecting the astrocytes from this destructive process.

Na(+),K(+)-ATPase is a transmembrane protein that transports sodium and potassium ions across cell membranes during an activity cycle that uses the energy released by ATP hydrolysis. Mercury is documented to inhibit Na(+),K(+)-ATPase function at very low levels of exposure. Studies have found that in ALS cases, there was a reduction in serum magnesium and RBC membrane Na(+)-K+ATPase activity and an elevation in plasma serum digoxin. The activity of all serum free radical scavenging enzymes, concentration of glutathione, alpha-tocopherol, iron binding capacity, and ceruloplasmin decreased significantly in ALS, while the concentration of serum lipid peroxidation products and nitric oxide increased. The inhibition of Na+-K+ATPase can contribute to increase in intracellular calcium and decrease in magnesium, which can result in (1) defective neurotransmitter transport mechanism, (2) neuronal degeneration and apoptosis, (3) mitochondrial dysfunction, (4) defective Golgi body function, (5) and protein processing dysfunction. It is documented in this paper that mercury is a cause of most of these conditions seen in ALS.

Mercury exposure also degrades the immune system resulting in more susceptibility to viral, bacterial, or parasitic effects along with Candida albicans which are often present in those with chronic conditions and require treatment. Four such commonly found in ALS patients are mycoplasma, echo-7 enterovirus, Candida albicans, and parasites. One clinic found that over 85 % of patients with ALS tested have mycoplasma infection, often *M. pneumoniae*, but in Gulf War veterans mostly a man-made variety used in bioterrorism agents – *M. fermentans*. Mercury from amalgam interferes with production of cytokines that activate macrophage and neutrophils, disabling early control of viruses or other pathogens and leading to enhanced infection. While the others are also being commonly found, mycoplasma has been found in 85 % of ALS patients by clinics treating such conditions. Mycoplasma appears to be a cofactor with mercury in the majority of cases and shifts the immune T cell balance toward inflammatory cytokines. Treatment of these chronic infections is required and documented to cause improvement in such patients.

Mercury lymphocyte reactivity and effects on amino acids such as glutamate in the CNS induce CFS-type symptoms including profound tiredness, musculoskeletal pain, sleep disturbances, gastrointestinal and neurological problems along with other CFS symptoms, and fibromyalgia. Mercury has been found to be a common cause of fibromyalgia, which, based on a Swedish survey, occurs in about 12 % of women over 35 and 5.5 % of men. ALS patients have been found to have a generalized deficiency in metabolism of the neuroexcitotoxic amino acids like glutamate, aspartate, NAAG, etc.

The brain has elaborate protective mechanisms for regulating neurotransmitters such as glutamate, which is the most abundant of all neurotransmitters. When these protective regulatory mechanisms are damaged or affected, chronic neurological conditions such as ALS can result. Glutamate is the most abundant amino acid in the body and, in the CNS, acts as excitatory neurotransmitter, which also causes inflow of calcium. Astrocytes, a type of cell in the brain and CNS with the task of keeping clean the area around nerve cells, have a function of neutralizing excess glutamate by transforming it to glutamic acid. If astrocytes are not able to rapidly neutralize excess glutamate, then a buildup of glutamate and calcium occurs, causing swelling and neurotoxic effects. Mercury and other toxic metals inhibit astrocyte function in the brain and CNS causing increased glutamate and calcium-related neurotoxicity which is responsible for much of the fibromyalgia symptoms and a factor in neural degeneration in MS and ALS. This is also a factor in conditions such as CFS, Parkinson's, and ALS. Animal studies have confirmed that increased levels of glutamate (or aspartate, another amino acid excitatory neurotransmitter) cause increased sensitivity to pain and higher body temperature – both found in CFS/fibromyalgia. Mercury and increased glutamate activate free radicals forming processes like xanthine oxidase which produce oxygen radicals and oxidative neurological damage. Nitric oxide-related toxicity caused by peroxynitrite formed by the reaction of NO with superoxide anions, which results in nitration of tyrosine residues in neurofilaments and manganese superoxide dismutase (SOD), has been found to cause inhibition of the mitochondrial respiratory chain, inhibition of the glutamate transporter, and glutamate-induced neurotoxicity involved in ALS.

In addition to the documentation showing the mechanisms by which mercury causes the conditions and symptoms seen in ALS and other neurodegenerative diseases, many studies of patients with major neurological or degenerative diseases have found direct evidence mercury and amalgam fillings play a major role in development of conditions such as ALS. Such supplements including N-acetyl-cysteine (NAC), vitamins E and C, zinc, and creatinine have been found to offer significant protection against cell apoptosis and neurodegeneration in neurological conditions such as ALS.

Medical studies and doctors treating chronic conditions like fibromyalgia have found that supplements which cause a decrease in glutamate or protect against its effects have a positive effect on fibromyalgia and other chronic neurological conditions. Some that have been found to be effective include CoQ10, Ginkgo biloba and pycnogenol, NAC, vitamin B6, methylcobalamin (B12), L-carnitine, choline, ginseng, vitamins C and E, nicotine, and omega 3 fatty acids (fish and flaxseed oil). A study demonstrated protective effects of methylcobalamin, a vitamin B12 analogue, against glutamate-induced neurotoxicity and similarly for iron in those who are iron deficient.

In a study of the brains of persons dying of ALS, spherical and crescent-shaped intraneuronal inclusions (SCI) were distributed in association with each other among the parahippocampal gyrus, dentate gyrus of the hippocampus, and amygdala, but not any non-motor-associated brain regions. The occurrence of SCI in both the second and third layers of the parahippocampal gyrus and amygdala was significantly correlated to the presence of dementia in ALS cases. Mercury has been found to accumulate in these areas of the brain and to cause adverse behavioral effects in animal studies and humans.

Another neurological effect of mercury that occurs at very low levels is inhibition of nerve growth factors, for which deficiencies result in nerve degeneration. Only a few micrograms of mercury severely disturb cellular function and inhibit nerve growth. Prenatal or neonatal exposures have been found to have lifelong effects on nerve function and susceptibility to toxic effects. Prenatal mercury vapor exposure which results in levels of only four parts per billion in newborn rat brains was found to cause decreases in nerve growth factor and other effects. This is a level that is common in the population with several amalgam fillings or other exposures. There is also evidence that fetal or infant exposure causes delayed neurotoxicity evidenced in serious effect at middle age. Insulin-like growth factor I (IGF-I) is positively correlated with growth hormone levels and has been found to be the best easily measured marker for levels of growth hormone, but males have been found more responsive to this factor than women. IGF-I controls the survival of spinal motor neurons affected in ALS during development as well as later in life. IGF-I and insulin levels have been found to be reduced in ALS patients with evidence this is a factor in ALS. Several clinical trials have found IGF-I treatment is effective at reducing the damage and slowing the progression of ALS and Alzheimer's with no medically important adverse effects. It has also been found that in chronically ill patients, the levels of pituitary and thyroid hormones that control many bodily processes are low and that supplementing both thyrotropin-releasing hormone and growth control hormone is more effective at increasing all of these hormone levels in the patient.

Extremely toxic anaerobic bacteria from root canals or cavitations formed at incompletely healed tooth extraction sites have also been found to be common factors in fibromyalgia and other chronic neurological conditions such as Parkinson's and ALS, with condensing osteitis which must be removed with a surgical burr along with 1 mm of bone around it. Cavitations have been found in 80 % of sites from wisdom tooth extractions tested and 50 % of molar extraction sites tested. The incidence is likely somewhat less in the general population. Medical studies and doctors treating fibromyalgia have found that supplements which cause a decrease in glutamate or protect against its effects have a positive effect on fibromyalgia and other chronic neurological conditions like ALS. Some that have been found to be effective include vitamin B6, methylcobalamin (B12), L-carnitine, choline, ginseng, Ginkgo biloba, vitamins C and E, CoQ10, nicotine, and omega 3 fatty acids (fish and flaxseed oil).

Clinical tests of patients with ALS, MND, Parkinson's, Alzheimer's, lupus (SLE), and rheumatoid arthritis have found that the patients generally have elevated plasma cysteine to sulfate ratios, with the average being 500 % higher than controls, and in general being poor sulfur oxidizers. This means that these patients have blocked enzymatic processes for converting the basic cellular fuel cysteine to sulfates and glutathione and thus insufficient sulfates available to carry out necessary bodily processes. Mercury has been shown to diminish and block sulfur oxidation and thus reducing glutathione levels which is the part of this process involved in detoxifying and excretion of toxics like mercury. Glutathione is produced through the sulfur oxidation side of this process. Low levels of available glutathione have been shown to increase mercury retention and increase toxic effects, while high levels of free cysteine have been demonstrated to make toxicity due to inorganic mercury more severe. The deficiency in conjugation and detoxification of sulfur-based toxins in the liver results in toxic metabolites and progressive nerve damage over time. Mercury has also been found to play a part in inducing intolerance and neuronal problems through blockage of the P-450 enzymatic process. Patients with some of these conditions have found that bathing in Epsom salts (magnesium sulfate) offers temporary relief for some of their symptoms by providing sulfates that avoid the blocked metabolic pathway. A test that some doctors treating conditions like ALS usually prescribed to measure the cysteine to sulfate ratio and other information useful in diagnosis and treatment is the Great Smokies Diagnostic Labs comprehensive liver detox test. The test results come with some recommendations for treatment. A hair test for toxic metals is also usually ordered to determine toxic exposures that might be involved. A more definitive test such as MELISA for immune reactivity to toxics is available by sending blood to a European lab. Other labs also have other useful tests such as immune reactivity biocompatibility tests, ELISA, or organic acid panels or amino acid panels. Treatment using IV glutathione, vitamin C, and minerals has been found to be very effective in the stabilizing and amelioration of some of these chronic neurological conditions by neurologist such as Perlmutter in Florida.

In one subtype of ALS, damaged, blocked, or faulty enzymatic superoxide dismutase (SOD) processes appear to be a major factor in cell apoptosis involved in the condition. Mercury is known to damage or inhibit SOD activity.

Prevention and Treatment of ALS

Tick-borne encephalitis, such as Lyme disease, has been found to cause ALS symptoms in a significant portion of untreated acute cases. Lyme disease is widespread in the United States. Large numbers of patients diagnosed with ALS and other neurological conditions have been found to have treatable tick-borne encephalitis, and many have recovered after treatment. Anyone diagnosed with degenerative neurological symptoms should be investigated with the possibility of Lyme disease or post-polio encephalitis. Poliomyelitis also has a chronic state that resembles ALS.

Since elevated plasma cysteine has been reported in some ALS patients, sulfite and cysteine toxicity may be involved in other cases of ALS. Patients with ALS with nonmutant SOD should be tested for sulfite toxicity, cysteine, glutamate, and GSH levels, and whether they have low levels of GSH metabolism enzymes. During the time when strict dietary and supplement measures normalized a patient's whole blood GSH, blood cysteine, and urine sulfite, the patient did not experience additional physical decline.

Total dental revision (TDR) which includes replacing amalgam fillings, extracting root canalled teeth, and treating cavitations has been found to offer significant health improvements to many with ALS and other autoimmune conditions. Root canals and cavitations have been found to harbor anaerobic bacteria which give off toxins of extreme toxicity which block enzymatic processes at the cellular level causing degenerative processes according to the medical labs that do the tests, similar to mercury's effects but in some cases even more toxic . IGF-1 treatments have also been found to alleviate some of the symptoms of ALS. Medical studies and doctors treating fibromyalgia have found that supplements which cause a decrease in glutamate or protect against its effects have a positive effect on fibromyalgia. Some that have been found to be effective in treating metals-related autoimmune conditions include vit B6, coenzyme Q10, methylcobalamin (B12), SAMe, L-carnitine, choline, ginseng, Ginkgo biloba, vitamins C and E, nicotine, and omega 3 fatty acids (fish and flaxseed oil).

One dentist with severe symptoms similar to ALS improved after treatment for mercury poisoning, and others treated for mercury poisoning or using TDR have also recovered or significantly improved. The Edelson Clinic in Atlanta which treats ALS patients reports similar experience, and the Perlmutter Clinic has also had some success with treatment of ALS and other degenerative neurological conditions.

While there are many studies documenting effectiveness of chemical chelators like DMSA and DMPS at reducing metals levels and alleviating adverse effects for most conditions, and many thousands of clinical case results, there is also some evidence from animal studies that these chelators can result in higher levels of mercury in the motor neurons in the short term which might be a problem for ALS patients. Thus, other detox options might be preferable for ALS patients until enough clinical evidence is available treating ALS patients with them with mercury toxicity. Another chelator used for clogged arteries, EDTA, forms toxic compounds with mercury and can damage brain function. The use of EDTA may need to be restricted in those with high Hg levels. N-acetylcysteine (NAC) has been found to be effective at increasing cellular glutathione levels and chelating mercury. Experienced doctors have also found additional zinc to be useful when chelating mercury as well as counteracting mercury's oxidative damage. Zinc induces metallothionein which protects against oxidative damage and increases protective enzyme activities and glutathione which tend to inhibit lipid peroxidation and suppress mercury toxicity. Also, lipoic acid, LA, has been found to dramatically increase excretion of inorganic mercury (over 12-fold) but to cause decreased excretion of organic mercury and copper. Lipoic acid has a protective effect regarding lead or inorganic mercury toxicity through its antioxidant properties but should not be used with high copper until copper levels are reduced. LA and NAC (N-acetyl-cysteine) also increase glutathione levels and protect against superoxide radical/peroxynitrite damage, so thus have an additional neuroprotective effect. Zinc is a mercury and copper antagonist and can be used to lower copper levels and protect against mercury damage. Lipoic acid has been found to have protective effects against cerebral ischemic reperfusion, excitotoxic amino acid (glutamate) brain injury, mitochondrial dysfunction, and diabetic neuropathy.

Antioxidants such as carnosine; coenzyme Q10; vitamins B, C, D, and E; Ginkgo biloba; superoxide dismutase (SOD); N-acetyl-cysteine (NAC); alpha-lipoic acid; and pycnogenol have also been found protective against degenerative neurological conditions. Other supplements found to be protective against neuronal degenerative conditions include acetyl-L-carnitine, EFAs

(DHA/EPA), DHEA, CoQ10, magnesium, vitamins B1 and B5, Hydergine, and octacosanol. Such supplements only offer limited protection and reductions in progression of ALS without other measures that deal with underlying mechanisms of causality. In a study involving over one million participants, a 23 % reduction in the risk of the disease was found among those who used vitamin E supplements for 2–4 years and a 36 % reduction occurred among those who used the supplements for 5 years or more compared to those who did not supplement with the vitamin.

Other supplements that appear useful in conditions involving neurotoxicity or muscle function degeneration include creatine and lithium. In the motor cortex of the ALS group, the N-acetylaspartate (NAA)/creatine (Cr(t)) metabolite ratio was lower than in our control group, indicating NAA loss. Upon creatine supplementation, we observed that creatine supplementation causes an increase in the diminished NAA levels in ALS motor cortex as well as an increase of choline levels in both ALS and control motor cortices. This indicates an improvement in function of the pathological ALS skeletal muscles related to changes of the mitochondrial respiratory chain which appears to affect motor neuron survival. In another study by the NAS, lithium carbonate at 150 mg twice daily significantly reduced the degeneration of ALS patients. A recent study demonstrated that combined treatment with lithium and valproic acid elicits synergistic neuroprotective effects against glutamate excitotoxicity in cultured brain neurons. Combined lithium and valproate treatment delays disease onset, reduces neurological deficits, and prolongs survival in an amyotrophic lateral sclerosis mouse model. Methylcobalamin and SAMe have also been found to provide some protection against neurotoxicity.

Two experimental treatments for ALS that have shown some effectiveness at reducing disease progression are recombinant human insulin-like growth factor and Orap (Pimozide).

Mercury Pollution Prevention

32

Mercury pollution from power plants is a national problem that requires a national response.

(Tom Allen quotes)

Mercury Pollution Prevention

Mercury is toxic to humans. The most common route of mercury exposure in humans is eating fish contaminated by methylmercury, according to the federal Centers for Disease Control. People can be exposed to mercury through its historical use in equipment and other products. Preventing exposure to mercury requires the shared efforts of the government, businesses, and individuals.

What DEQ Is Doing to Prevent Mercury Pollution

Hazardous Waste Regulation: Every business in Idaho is required to track the volume of wastes generated, determine whether or not each is hazardous, and ensure that all wastes are properly disposed off according to federal, state, and local requirements. As the state agency delegated responsibility for administering the federal Resource Conservation and Recovery Act in Idaho, DEQ works closely with businesses to assure that hazardous wastes, including mercury, are transported, stored, and disposed off safely.

BMPs for Mercury-Containing Equipment at Public Drinking Water Systems: As the state agency delegated responsibility for administering the federal Safe Drinking Water Act in Idaho, DEQ works closely with local public health districts and public drinking water systems to assure that the water we drink is free of contaminants including mercury. In 2003–2004, DEQ conducted an evaluation of equipment used by public drinking water systems in Idaho to identify mercury-containing equipment, including electric switches, sensors, gauges, and meters, with a potential to contaminate drinking water. A technical guide was completed and is available to provide system operators with best management practices to prevent mercury-containing equipment from contaminating drinking water.

National Vehicle Mercury Switch Replacement Program: DEQ participates in the National Vehicle Mercury Switch Replacement Program. Idaho autorecyclers and salvage yards have been invited to voluntarily participate in this program to remove mercury switches from vehicles before the vehicles are crushed and recycled. The End of Life Vehicle Solutions Corporation implements this program.

School Labs Chemical Management and Cleanup Project: In a study conducted by DEQ in 2003–2004, it was revealed that a number of schools in Idaho have amassed large stockpiles of hazardous products on school grounds and lack knowledge of secure management and safe disposal procedures for hazardous chemicals and waste. Mercury was among the hazardous

S. Nabi, *Toxic Effects of Mercury*, DOI 10.1007/978-81-322-1922-4_32,
© Springer India 2014

materials discovered at some school labs. DEQ's Chemical Roundup Program was developed to assist schools in understanding and implementing best practices for managing and disposing off their hazardous chemicals and wastes, including mercury.

Public Education and Outreach: DEQ is working to increase public awareness of risks associated with mercury exposure and ways to reduce releases of mercury into the environment.

What You Can Do to Prevent Mercury Pollution

Here are some things you can do to help reduce the threat mercury poses to our environment and health.

Buy Mercury-Free

- Choose products that do not contain mercury. Look for digital thermometers and electronic thermostats.
- Ask your dentist to employ mercury-free composites for any dental work and inquire if your insurance company will pay for the mercury-free materials (if not, speak with your employer or insurance company).
- Ask your local pharmacy or hardware store to consider phasing out the sale of mercury-containing products.

Discard Products Safely

- Separate mercury-containing products, such as electronic equipment with monitors (including televisions), fluorescent lighting, thermometers, thermostats, old paint (pre-1991), and batteries (pre-1995), from regular garbage.
- Do not remove mercury switches from products, such as thermostats; it is safer to keep or recycle the product when it is intact.
- Take any mercury-containing products that you have collected to your local hazardous waste collection facility. Be careful that mercury thermometers are well protected from breakage. Call your local municipality for information on where to bring mercury-containing waste or link to DEQ's online Recycling Directory to find a collection site near you.
- Mercury thermostats can be recycled for free at locations throughout Idaho. Search by zip code on the Thermostat Recycling Corporation's website for the recycling site nearest you.

Conserve Energy

- Turn the heat back and use air conditioning only when necessary. An electronic programmable thermostat can help do this for you. Burning less coal and oil (which naturally contain mercury) for electricity will emit less mercury into the environment.
- Purchase only energy-efficient products, such as compact fluorescent lights (instead of incandescent bulbs). Even though fluorescents contain a small amount of mercury, they will help reduce energy use and related pollution. Ensure that spent fluorescent bulbs are recycled through a household hazardous waste facility.
- Ensure your home is properly sealed and well insulated.

Avoid Exposure to Mercury at Home, School, and Work

- Never play or let your children play with liquid mercury.
- Never use an ordinary vacuum cleaner to clean up mercury. The vacuum cleaner will release mercury vapor into the air and increase exposure. The vacuum cleaner will also be contaminated and have to be thrown away.
- Never use a broom to clean up mercury. It will break the mercury into smaller droplets and spread them.
- Never pour mercury down a drain. It may cause plumbing problems and cause pollution of the septic tank or sewage treatment plant.

- Never wash mercury-contaminated items in a washing machine. Mercury may contaminate the machine and/or pollute sewage.
- Never walk around if your shoes might be contaminated with mercury. Contaminated clothing can also spread mercury around.

Introduction

Mercury is a toxic metal that occurs naturally in the environment. The silvery white inorganic form is most familiar because of its widespread use in fever thermometers. The properties of inorganic mercury that make it useful in medicine (response to temperature and pressure) also permit vaporization and resulting toxic effects. Organic forms of mercury circulate in the environment and may change from one form to another in the process. Although all forms of mercury are toxic to humans, organic forms generally are much more toxic than inorganic forms. The organic forms are of particular concern when they enter the food chain, since these are primarily neurotoxins that can damage the brain, nervous system, and other organs. The inorganic forms primarily affect the kidneys. The developing brain of a fetus or child is especially vulnerable to organic mercury exposure.

Human exposure to mercury can occur through inhalation, ingestion, or skin contact. The most common routes of exposure in the healthcare setting include inhalation of inorganic mercury vapor after a spill or accidental skin contact with mercury. Accidental spills of liquid mercury can increase the levels of mercury in the air or wastewater of a healthcare facility. Small droplets of spilled mercury may lodge in cracks, mix with dust, or go down drains. Mercury may adhere to fabrics, shoe soles, watches, or jewelry, on which it can be transported to other locations. A small spill of mercury in a carpeted patient room can become a major cleanup challenge. The mercury used in healthcare can have a broad impact on the surrounding community if spills or waste enter the groundwater and work their way into the food chain. For all these reasons, hospitals are working to reduce or even eliminate the use of mercury.

Common Uses of Mercury in Healthcare Settings

Mercury-containing products can be found almost anywhere in the healthcare setting. They range from medical instruments and clinical laboratory chemicals to electrical equipment and cleaning solutions. Pharmaceutical manufacturers have used thimerosal, a mercury product, as a common preservative. Following are examples of medical and nonmedical uses for mercury in the healthcare setting.

Medical Uses

- Thermometers
- Sphygmomanometers
- Esophageal dilators (also called bougie tubes)
- Cantor tubes and Miller Abbott tubes (used to clear intestinal obstructions)
- Feeding tubes
- Dental amalgam
- Laboratory chemicals (fixatives, stains, reagents, preservatives)
- Medical batteries

Nonmedical Uses

- Cleaning solutions with caustic soda or chlorine contaminated with mercury during the production process
- Batteries
- Fluorescent lamps and high-intensity lamps
- Thermostats (nonelectronic)
- Pressure gauges
- Some electrical switches used for lights and appliances

Memorandum of Understanding

Hospitals are collaborating to reduce and ultimately eliminate mercury. In June 1998, the Environmental Protection Agency (EPA) and the American Hospital Association (AHA) signed a Memorandum of Understanding (MOU) which has as one of its stated goals the virtual elimination of mercury waste by the year 2005.

Primary Voluntary Goals

- Virtual elimination of mercury-containing waste from hospitals' waste stream by 2005

- The reduction of the overall volume of waste (both regulated and nonregulated) by 33 % by 2005 and 50 % by 2010
- The identification of hazardous substances for pollution prevention and waste reduction opportunities

Taking the Mercury-Free Pledge

Taking the mercury-free pledge is an excellent way for facilities to begin their mercury pollution prevention efforts. Such a pledge needs senior management support and provides an opportunity for healthcare facilities to gain recognition for their pollution prevention efforts.

Several groups have passed resolutions and adopted environmental management policies that call for mercury elimination. In March 2000, the California Medical Association adopted a resolution that promotes the eventual elimination of mercury and urges medical product suppliers to develop mercury-free replacement products. Catholic Healthcare West has adopted mercury-free language in its purchasing policies.

Solutions and Strategies for Pollution Prevention

The EPA's approach to pollution prevention is outlined in its educational program (Mercury in Medical Facilities) as a series of priority steps, the first being prevention. The primary strategies include source reduction and recycling or waste minimization. Waste that cannot be recycled must then be treated, and the final and last resort is disposal and release.

Mercury Source Reduction

The healthcare industry is making progress in eliminating mercury pollution. Fewer healthcare institutions have operating incinerators. Large medical waste disposal companies are relying less on incineration and more on steam sterilization autoclaves and other alternatives. Hospitals and medical waste disposal companies are working to ensure appropriate source separation so that mercury is pulled from the waste stream before it is sent for incineration. Such source separation programs may require significant initial educational and implementation efforts, but they are almost guaranteed to be successful because the mercury load in the waste stream should be essentially eliminated. The few items involving essential uses of mercury would be relatively easy to track and control.

Mercury Substitution

There is minimal risk of mercury exposure during the normal use and proper handling of products; however, problems may occur if the mercury in a product is exposed to air or if a product is not properly discarded.

For most mercury-containing products in the hospital, the preferred management practice is to replace the item with a mercury-free product. However, it may not be possible to replace all of the hospital's mercury products at once, and in a few cases, there may be no reliable, cost-effective substitute. In such cases, the best management practice is to have effective procedures in place for handling and recycling or disposal of the mercury-containing products. Recycling is recommended. Disposal should be the last resort – it is expensive and increases the potential for mercury to be dispersed into the environment.

Strategies for Managing Mercury in Healthcare Facilities

- Using alternatives for products that contain mercury
- Recycling mercury-containing products when they can no longer be used
- Correct handling and disposal of mercury, mercury-containing equipment, and laboratory chemicals
- Establishing protocols for proper cleanup of spills involving mercury
- Enforcing compliance with institutional policies

Mercury Spill Prevention

Follow proper procedures when cleaning or refilling instruments that contain mercury:

- Clean or refill instruments over a tray to contain any spills. Never handle mercury over a sink. Reserve the room for mercury use only. Restrict traffic in the area.
- Clean and calibrate all mercury-containing equipment according to the manufacturer's recommended handling procedures and the procedures recommended by your hospital's safety officer.
- Train all workers who use mercury devices about the properties and hazards of mercury, safe handling procedures, and specific policies related to mercury recycling and disposal.

Minimizing the impact of a spill is part of spill prevention. Mercury devices should be used only in rooms without carpeting or other floor coverings not smooth and easily cleaned. Mercury devices should not be used in units with beds that have high structures or projections that can break wall-mounted sphygmomanometers or in areas where patients cannot be moved.

Mercury Spill Response

- Be prepared for a spill in any area of the hospital where mercury-containing devices are used.
- Have a mercury vacuum cleaner or mercury spill kit readily available to consolidate spilled mercury and limit the amount of mercury released into the air. The mercury vacuum cleaner is designed to clean up liquid mercury spills: An activated carbon filter in the vacuum absorbs and contains the mercury vapors. Never use a regular vacuum cleaner to clean up mercury, as it will vaporize the mercury and blow it into the air.
- In the event of a large spill, patients must be removed from the room during cleanup. The room must remain vacant until the air is free of mercury vapor.
- Staff members must have special training in order to clean up mercury spills.

Healthcare facilities should have mercury spill policies addressing:

- Availability of a competent staff person, trained in mercury spill cleanup, on all shifts.
- Availability of protective equipment and clothing for cleanup staff
- Incorporation of OSHA requirements
- The circumstances in which patient(s), visitors, and staff should be evacuated from the area before cleanup
- Guidelines for determining when the spill has been adequately cleaned and the area can safely be occupied
- The type of equipment to be used for the size and type of spill and type of flooring (linoleum, carpet, cement)
- Availability of manufacturer's instructions for all equipment to be used for cleanup procedures
- Strategies for waste disposal
- Method for documentation of the incident report describing the spill, the cleanup method used unusual circumstances, employee exposure, and follow-up monitoring

Additional policies should be developed for:

- Handling mercury spills during a medical procedure
- Servicing mercury-containing medical equipment
- Education and training of staff who may come in contact with mercury-containing products

What Can I Do to Prevent Mercury Pollution?

Residents

- Mercury is being phased out of many retail products sold in Minnesota. As a consumer, educate yourself; do not purchase mercury-containing items if a substitute is available. When an item containing mercury is purchased, dispose it off properly when it becomes waste. Here are some ways that you can reduce mercury pollution:
  - Reduce your use of mercury-containing items.

- Choose non-mercury alternatives.
- Dispose off any item that contains mercury properly. In the WLSSD area, this means brining the item to the Household Hazardous Waste Facility. Disposal is free for residents. Businesses are charged a small fee through the Clean Shop program.
- Use less electricity: The largest source of mercury emission in Minnesota is the burning of coal for power generation.

Businesses and Industries

- Mercury may be an unintended contaminant in some industrial acids or other materials. Look for mercury-free alternatives in both raw materials and measurement devices.
- Dispose off mercury-containing items through a licensed hazardous waste disposal contractor or use the WLSSD Clean Shop services if qualified.
- Blueprint for Mercury Elimination for Wastewater Treatment Plants

Dentists

- Dental amalgam contains mercury. Dental offices were once a large source of mercury to wastewater treatment plants. Offices in the WLSSD service area have installed separators to prevent amalgam waste from entering the sewer system. This has dramatically reduced the amount of mercury entering the treatment plant at WLSSD, reducing emission to water sources in our area.